ADVANCES IN IMMUNOLOGICAL TREATMENT OF CANCER

Cellular Mechanisms and Regulation

ADVANCES IN IMMUNOLOGICAL TREATMENT OF CANCER

Cellular Mechanisms and Regulation

Rishabha Malviya, PhD
Arun Kumar Singh, MPharm
Deepika Yadav, MPharm

AAP | APPLE ACADEMIC PRESS

First edition published 2025

Apple Academic Press Inc.
1265 Goldenrod Circle, NE,
Palm Bay, FL 32905 USA

760 Laurentian Drive, Unit 19,
Burlington, ON L7N 0A4, CANADA

CRC Press
2385 NW Executive Center Drive,
Suite 320, Boca Raton FL 33431

4 Park Square, Milton Park,
Abingdon, Oxon, OX14 4RN UK

Library and Archives Canada Cataloguing in Publication

Title: Advances in immunological treatment of cancer : cellular mechanisms and regulation / Rishabha Malviya, PhD, Arun Kumar Singh, MPharm, Deepika Yadav, MPharm.
Names: Malviya, Rishabha, author. | Singh, Arun Kumar, MPharm, author. | Yadav, Deepika, 1997- author.
Description: First edition. | Includes bibliographical references and index.
Identifiers: Canadiana (print) 2024042154X | Canadiana (ebook) 20240421558 | ISBN 9781774917053 (hardcover) | ISBN 9781774918609 (softcover) | ISBN 9781003533955 (ebook)
Subjects: LCSH: Cancer—Immunotherapy.
Classification: LCC RC271.I45 M35 2025 | DDC 616.99/4061—dc23

Library of Congress Cataloging-in-Publication Data

CIP data on file with US Library of Congress

ISBN: 978-1-77491-705-3 (hbk)
ISBN: 978-1-77491-860-9 (pbk)
ISBN: 978-1-00353-395-5 (ebk)

About the Authors

Rishabha Malviya, PhD
Associate Professor, Department of Pharmacy, School of Medical and Allied Sciences, Galgotias University, India

Rishabha Malviya, PhD, has 12 years of research experience and is presently working as Associate Professor in the Department of Pharmacy, School of Medical and Allied Sciences, Galgotias University, India. His areas of interest include formulation optimization, nanoformulation, targeted drug delivery, localized drug delivery, and characterization of natural polymers as pharmaceutical excipients. He has authored more than 150 research and review papers for national and international journals of repute. He has authored and edited over 40 books and over 30 book chapters published by Scrivener Publishing, Wiley, CRC Press/Taylor and Francis, Springer, River Publisher, Apple Academic Press, IOP, and others. His cumulative impact factor is 190. He has over 50 patents to his name. Dr. Malviya has received an Outstanding Reviewer Award from Elsevier. His name has been included in the Stanford University's World's Top 2% Scientists, 2020 by Elsevier BV and Stanford University. He is reviewer, editor, and editorial board member of more than 50 national and international journals. He was an invited author for the magazine *Atlas of Science* as well as a pharma magazine dealing with the B2B industry, *Ingredients South Asia.* He completed his BPharm at Uttar Pradesh Technical University and MPharm (Pharmaceutics) at Gautam Buddha Technical University, Lucknow, Uttar Pradesh, India. His PhD (Pharmacy) work was in novel formulation development techniques.

Arun Kumar Singh, MPharm
Department of Pharmacy, School of Medical & Allied Sciences, Galgotias University, Greater Noida, Uttar Pradesh, India

Arun Kumar Singh, MPharm, is affiliated with the Department of Pharmacy, School of Medical & Allied Sciences, Galgotias University, Greater Noida, Uttar Pradesh, India. His areas of interest include nano-formulation, blockchain, IoT, machine learning, cancer, artificial intelligence, and big data. He has authored four books with Scrivener Publishing/ Wiley, IOP publishing and Apple Academic Press/ Taylor and Francis Groups. He has published five chapters with renowned international publishers. He has also published more than eight SCI papers with total 43 cumulative impact factor. His strength is his research skill, innovative thinking, leadership qualities, decision-making ability, and positive thinking.

Deepika Yadav, MPharm
Department of Pharmacy, School of Medical & Allied Sciences, Galgotias University, Greater Noida, Uttar Pradesh, India

Deepika Yadav earned her BPharm from K R Mangalam University, Gurugram, Haryana and MPharm from Galgotias University, Greater Noida, India. She has authored two books with Apple Academic Press/ Taylor and Francis Group and Scrivener Publishing/Wiley as well as book chapters with other reputed international publishers. She is also working on two books with other renowned international publishers.

Contents

Foreword

It is with great pleasure that I pen this foreword for the book authored by Dr. Rishabha Malviya and his colleagues. This is a fantastic book covering a subject that has important implications for public health. Simply reading the book's title should suffice: *Advances in Immunological Treatment of Cancer: Cellular Mechanisms and Regulation.*

In this book, the authors explore the cutting-edge research behind cancer's driving mechanism of immune suppression, as well as some cutting-edge therapeutic approaches to combating it.

There has been a lot of research into the mechanisms of immune suppression and how they relate to cancer and cancer treatment. This book offers novel approaches to cancer therapy with a special emphasis on the emerging new principles of immune suppression that drive the disease. Further, the book explains the fundamental principles of immune suppression that are driving research in this area. It also shows that conventional chemotherapeutic drugs and molecularly targeted therapeutics can both have significant antitumor effects in animal models.

Comprehensively covering immune escape as a critical trait in cancer and how specific combinations of immunotherapy and chemotherapy can be used to attack this trait, this book is a must-read for basic, translational, and clinical researchers looking to improve the treatment of advanced disease.

I believe that this book will be useful to its readers because it provides a comprehensive analysis of immunology's potential role in the fight against cancer, the immune system's role in regulating tumor growth, and the impact of the immune system on tumor immunogenicity.

—Prof. (Dr.) Kamla Pathak
Dean, Faculty of Pharmacy
Uttar Pradesh University of Medical Sciences,
Saifai, Etawah, Uttar Pradesh, India

Preface

The novel principles of immune suppression that drive cancer are explored in this book, along with some novel therapeutic strategies for counteracting these mechanisms. New studies have identified the mechanisms that contribute to tumor immune suppression, adding to the growing body of evidence that immune escape is a critical feature of cancer. It also demonstrates that conventional chemotherapeutic drugs are as effective as molecular targeted therapeutics in eliminating tumors in animal models. The book focuses on the emerging new principles of immune suppression that drive cancer and provides radical new ideas for improving therapy by targeting these principles.

The purpose of the book, *Advances in Immunological Treatment of Cancer: Cellular Mechanisms and Regulation,* is to provide a clinically relevant overview of cancer immunology and immunotherapy. In recent years, researchers have paid more attention to the role that immune suppression plays in cancer and cancer treatment. Immune suppression's role in cancer and its treatment has received more attention in the recent years. In this book, we delve into the novel principles of immune suppression that drive cancer, as well as some novel therapeutic strategies for counteracting these mechanisms. In light of the growing evidence that immune escape is a crucial feature of cancer, recent studies have defined the underlying mechanisms that contribute to tumor immune suppression. Also, it proves that standard chemotherapeutic drugs are as effective as molecular targeted therapeutics in killing tumors in animal models. Those in the fields of basic, translational, and clinical cancer research should make it a point to read this book, as it provides a thorough overview of immune escape as a critical trait in cancer, and how targeted combinations of immunotherapy and chemotherapy to combat this trait may significantly improve the treatment of advanced disease.

The first chapter provides an overview of immunoregulation via allogeneic stem cell transplantation; the second discusses the various resistance mechanisms in cancer immunology; the third offers new hope for survival via immunoediting; the fourth details the development of antigen and phenotypic plasticity molecules for cancer immunotherapy; the fifth discusses toll-like receptors; and the sixth details self-to-tumor immune interactions.

Methods for reducing tumor growth and altering tumor immunogenicity are discussed in Chapter 7, and the many ways in which conventional chemotherapy can boost the immune system are detailed in Chapter 8. The function of innate and adaptive immunity in immune vigilance is explored in Chapter 9.

This book is a must-read for researchers in basic, translational, and clinical cancer fields, as it provides a comprehensive overview of immune escape as a critical trait in cancer and how targeted combinations of immunotherapy and chemotherapy to combat this trait may significantly improve the treatment of advanced disease.

CHAPTER 1

Introduction of Immunoregulation through Allogeneic Stem Cell Transfusion

ABSTRACT

Since allogeneic hematopoietic stem cell transplantation (alloHSCT) may produce a strong graft-versus-leukemia (GVL) immune response, which may reject cancer cells in vivo and restore normal donor immunity, it has the potential to cure many types of hematologic malignancies. The mechanisms of GVL have been elucidated thanks to studies in animals and people, showing how alloHSCT produces clinically substantial tumor immunity. These studies demonstrate the critical roles played by NK cells, B cells, and T cells in setting off the GVL response. Possible antigenic triggers for the GVL response include tumor-associated antigens and minor histocompatibility antigens. Allogeneic HSCT has been shown to have therapeutic advantages, but understanding how GVL operates would allow us to develop safer and more effective drugs to enhance these effects. In order to improve the efficacy of cancer immunity in the autologous scenario, it may be necessary to adjust specific aspects of the approach that help in the elimination of tumor cells after allogeneic HSCT.

1.1 INTRODUCTION

Remission rates are quite high after allogeneic hematopoietic stem cell transplantation for a variety of hematologic malignancies[1–5] (HSCT). High-dose myeloablative chemotherapy and/or whole-body irradiation are recommended before allogeneic HSCT to eliminate any remaining tumor cells.[6,7] Donor hematopoietic stem cells, when infused, migrate to the bone marrow

Advances in Immunological Treatment of Cancer: Cellular Mechanism and Regulation.
Rishabha Malviya, Arun Kumar Singh, and Deepika Yadav (Authors)

and engraft there, potentially repairing the immune and hematopoietic systems. Since the restoration of donor immune cells is important for the elimination of recipient cancer cells (graft versus leukemia), inadequate or excessive conditioning will make HSCT ineffective.[8] The targets and effectors for GVL have been better understood because of the large amounts of clinical experience and laboratory research over the last two decades. This chapter will analyze how allogeneic HSCT produces antitumor immunity in animals.[9] When healthy, donor hematopoietic cells from many lineages are engrafted, the body's immune system becomes more aggressive against the tumor. Cancer cells may also express host-specific targets and minor histocompatibility antigens,[10] in addition to tumor-associated antigens (mHA). There is still a lack of knowledge about the antigenic targets and cellular subsets causing anticancer immunity in the allogeneic context.[11] Donor immune cells' capacity to react to recipient tumor cells after transplant may be impaired by many clinical variables [including graft-versus-host disease (GVHD) and concomitant immunosuppressive medication]. Researchers now have a better knowledge of how clinically effective tumor immunity is produced following HSCT thanks to animal studies and clinical data analysis that have helped define the mechanisms of GVL.[12] Allogeneic transplantation is therapeutically effective; nevertheless, it is not without hazards and toxicity, which might be mitigated with treatments based on these principles.[13] Altering the methods used to eradicate tumor cells after allogeneic HSCT has the potential to boost cancer immunity in the autologous scenario.[14,15]

1.2 HEMATOPOIETIC REGENERATION AFTER ALLOGENEIC HEMATOPOIETIC STEM CELL TRANSPLANTATION

Malignant patients undergo allogeneic HSCT in the hopes of treating their conditions by the permanent elimination of recipient cancer cells. To do this, it is required to transplant the immune system of a healthy donor into the patient.[16] The donor must recognize tumor-associated antigens, generate an effective rejection response, and maintain a sufficient memory response to avoid disease recurrence. Important for immunologic control of the recipient's tumor cells is how quickly and well the donor's immune system is restored following transplant.[17,18]

Donor immune reconstitution following transplant may be affected by several variables, some of which are shown in Figure 1.1. How well a patient responds to a transplant depends on several factors, including the severity and toxicity of the transplant preparative regimen, the duration and dosage of immunosuppressive medications used to prevent and treat graft-versus-host disease (GVHD), and the extent to which the immunosuppressive medications affect the recipient.[19] Children may recover their T-cell immunity more rapidly than adults after organ transplants[20,21] because their thymus glands are more functioning in childhood. Immune deficiency and post-transplant infection are more common in those over the age of 65 who have had substantial chemotherapy and have less thymic tissue.[22,23] Bone marrow microenvironmental abnormalities are a hallmark of myelodysplasia (MDS). While engraftment of donor stem cells is often sluggish in MDS patients, it is successful in the long run.[24] This is because, following successful engraftment with normal hematopoietic stem cells, a normal marrow microenvironment and normal hematopoietic function may be reinstated. Patients who have had high-dose myeloablative therapy before HSCT almost always rely on donor stem cells for early engraftment and long-term maintenance of hematopoiesis.[25] Less rigorous conditioning therapies increase the likelihood of patients having transplants developing mixed hematopoietic chimerism. Although a patient's immune system may become a chimaera following a transplant, making the patient immune to recipient cells but unable to generate effective antitumor immunity, milder conditioning procedures are connected with less damage.[26,27]

Immune reconstitution may be influenced by several circumstances, one of which is allogeneic stem cell transplantation.[28] Human leukocyte antigen (HLA) compatibility, the quantity and quality of hematopoietic stem cells and other graft cells, and the outcomes of in vitro modification of the stem cell graft before the infusion are all crucial.[29,30] These factors may significantly affect the efficiency of the antitumor response elicited by the donor cells and, hence, the success of immune reconstitution in a specific patient.[31]

HLA-identical siblings, HLA-matched unrelated donors, partially HLA-mismatched related or unrelated donors, and HLA-mismatched umbilical cord blood are all acceptable sources of hematopoietic stem cells.[32,33] In such cases, the degree of HLA-matching impacts both the likelihood of developing GVHD after a transplant and the efficacy of the subsequent T-cell immune reconstitution.[34] The capacity of a product to promote immunological engraftment and post-transplantation recovery is proportional to the number of hematopoietic stem cells present in the product.[35] A patient's entire hematological and immunological capabilities may be preserved forever by the engraftment of high-purity CD34+ cells or grafts containing a limited number of hematopoietic stem cells (such as cord blood products).[36,37] However, posttransplant infections are common for these patients because of the time it takes for their immune systems to fully recover.[38]

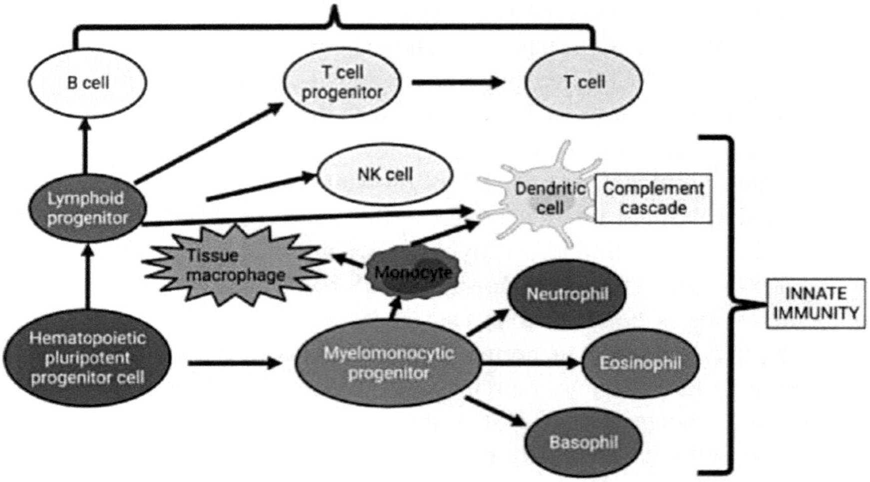

FIGURE 1.1 Schematic representation of mechanism of action in immunity.

However, immune reconstitution after myeloablative therapy requires the production of new immune cells from undifferentiated hematopoietic progenitor cells, rather than the mature T, B, and natural killer (NK) cells provided by the stem cell product, which aid in maintaining early post-transplant immunological activity.[39] Cellular reconstitution is the chain of actions that takes place after a myeloablative stem cell transplant (shown schematically in Fig. 1.2). In the majority of cases, a patient's NK cells will return to health before any of the other lymphoid subsets.[40] In the first post-transplant month, NK cells outnumber all other lymphoid cells combined.[41] This is particularly the case when recipients get immature T-cell-free bone marrow. CD4+ cells often lag behind CD8+ cells during T-cell reconstitution. Therefore, the CD4+/CD8+ ratio is unusually low for a relatively extended length of time after HSCT.[42]

Three to four months following HSCT, CD8+ T-cell levels return to normal, but it takes 6–12 months for CD4+ T-cell counts to normalize. The number of B cells in the body's peripheral blood normally returns to normal about 1 year after a transplant.[43] Unlike late T cells, which are naive and are formed by the differentiation of donor-derived hematopoietic progenitor cells, early T cells are mature donor cells that proliferate inside the stem cell graft.[44]

Despite age-related decreases in thymic activity, the T-cell receptor (TCR) repertoire of a healthy individual may still be rather varied. Foreign antigens are better able to evade the immune system and lay the ground for an efficient

T-cell response to external pathogens when the T-cell repertoire is limited.[45] In a similar vein, if one's TCR repertoire were too uniform, one's capacity to mount tailored T-cell responses against tumor-associated antigens would be diminished. Quantifying the frequency with which different TCR Vb genes are employed in a person's peripheral T cells is one approach to gaining a feel for the diversity of their TCR repertoire.[46,47] Also, spectrum typing may be used to evaluate the heterogeneity of CDR3 regions produced by T-cell populations. Many groups, including our own, have employed these techniques to study T-cell repertoire recovery after myeloablative treatment and HSCT.[48] These results led researchers to conclude that early post-transplant TCR repertoire diversity was low. If all donor cells engraft well, the patient's TCR repertoire begins to improve and, in many cases, returns to reasonably normal levels by the 1-year post-HSCT mark.[49,50] Superior thymic function in youngsters likely accounts for their accelerated development of this trait. It seems that post-transplant problems including graft-versus-host disease and mixed hematopoietic chimerism, as opposed to full donor hematopoiesis, are linked to a lag in the restoration of the patient's normal TCR repertoire.[51] When patients with recurrent illness after allogeneic HSCT restore a large number of TCRs, they are more likely to mount an efficient immune response against tumor cells, leading to the remission of disease.[52]

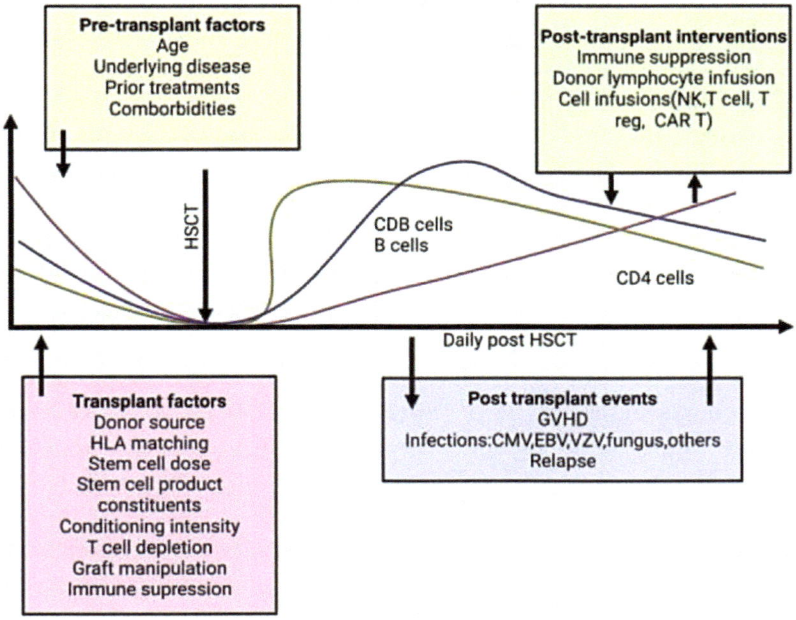

FIGURE 1.2 Allogeneic HSCT: a timeline of immune reconstitution.

1.3 ALLOGENEIC HSCT TRIGGERS GVL RESPONSES THROUGH DONOR LYMPHOCYTE INFUSIONS

Clinical studies show that GVL was mediated by unstimulated donor T lymphocytes present in the allogeneic stem cell product,[53] and hence DLI is often employed as a therapeutic option for patients with recurrent hematologic malignancy following allogeneic HSCT. Three patients with recurrent CML after allogeneic HSCT showed excellent clinical responses to DLI (1990).[54] Studies in humans suggest that DLI alone, without the use of chemotherapy or radiation, may result in disease remission, and donor-derived immune effector cells have been shown to play a critical role in the establishment of powerful tumor immunity. The efficiency of DLI in producing GVL has been shown in later investigations,[55] lending credence to this claim. DLI's effects on individuals with myeloid leukemias were shown to be long-lasting and reproducible, according to a study by Kolb et al (1995).[56] A study involving 25 North American bone marrow transplant programs found that donor leukocyte infusions (DLI) resulted in complete remissions in 60% of chronic myelogenous leukemia patients, with higher response rates in cytogenetic and chronic-phase relapse patients.[57] Results from clinical research have been reliable. To begin with, research has shown that DLI treatment is effective for around 80% of those with CML. Central nervous system illnesses may have a minor effect on people with DLI. Approximately 50% of the participants in the researchers' study had a response to treatment for multiple myeloma, but only 40% experienced a response for CLL.[58,59]

While there is a high response rate among chronic leukemia patients, the rate is much lower among those with acute leukemia (10-15 percent).[60] Those with fewer health problems and a slower rate of acute illness onset seem to benefit the most from DLI. Patients with cytogenetic evidence of recurrent CML without overt hematologic recurrence had response rates of >90%. In contrast to the high response rate seen with recurrent stable-phase CML, only around 5–10% of individuals with advanced CML (blast crisis/ accelerated phase) react to DLI.[61,62] Secondly, it often takes 2–4 months after a single infusion for the therapeutic benefits of DLI to become apparent.[63,64] When the number of naive T cells is low, this lengthy delay may reflect the time needed to produce an effective response. Alternately, the delay may represent the time necessary for the malignant clone to recover from the impact of having its component cells lysed.[65] Furthermore, DLI responses may persist for a long period, at least for specific diseases.

Long-term follow-up of 66 patients who had full remission after DLI therapy was described by the researchers.[66] Fifty-nine people were diagnosed with chronic myeloid leukemia, whereas the remaining twenty-seven did not. Patients with CML had an 83, 76, and 73% survival rate after 1, 2, and 3 years, respectively.[67,68] Since these responses tend to be more long-lasting when molecular remission is achieved, this demonstrates that the effectiveness of DLI is linked to the immunologic elimination of the malignant stem cell clone.

Due to its positive effects in some patients with relapsed illness, DLI has been the subject of several investigations into its potential as a prophylactic medicine to stop the return of the malignant disease. To lower the number of T lymphocytes in the stem cell graft, Ferra et al. used DLI after transplanting pure CD34 peripheral blood stem cells.[69–70] By administering donor T-cell infusions at predetermined intervals after T-cell depletion BMT, Barrett et al. (1998) were able to reduce the severity of acute GVHD without compromising the GVL benefit. Nonmyeloablative preparative regimens heavily rely on GVL effects to minimize residual malignancy,[71] and hence implementing such techniques is crucial.

Taken together, these studies provide several strategies for using DLI in clinical practice to boost tumor immunity and avoid recurrence after allogeneic HSCT.[72] Currently, there are several clinical studies underway to compare various methods.[73] The most prevalent side effect of DLI is graft-versus-host disease (GVHD), and several strategies have been devised to reduce its severity without reducing the advantages of GVL. Dana-Farber Cancer Institute's researchers[74] looked at the feasibility of depleting DLI products of CD8+ cells and then replenishing them with a certain number of CD4+ donor cells.

It has been shown that CD8-depleted DLI infusions reduce the risk of GVHD and maintain steady GVL activity.[75] In a prospective study, participants who had undergone myeloablative conditioning and T-cell reduction before a stem cell transplant were randomly assigned to receive either unmodified or CD8-depleted DLI.[76] When compared to patients who did not get CD8-depleted DLI, those who did have a much lower risk of GVHD and no increased risk of recurrence. Multiple clinical and immunologic outcomes in vivo have been associated with the injection of CD4+ donor T cells. In addition to converting partial donor hematopoiesis to full donor hematopoiesis and restoring complete donor hematopoiesis, hematopoietic stem cell transplantation has been shown to improve T-cell neogenesis and T-cell receptor (TCR) repertoire size.[77,78] Patients who perform well after receiving CD4+ DLI have been discovered to have CD8+ donor T-cell clones that are reactive to cancer. These and previous studies demonstrate that CD4+ donor T cells may mediate GVL without causing severe GVHD, which might be used to target malignant recipient cells in vivo.[79] The precise process by which donor CD4+ T cells stimulate CD8+ effector T cells in vivo is not yet fully known.[80]

1.4 IMPORTANT FUNCTIONS OF DONOR T CELLS IN MEDIATING GVL

Several clinical and laboratory research have started outlining the immunologic pathways that contribute to GVL[81] after it was shown that GVL is crucial in removing malignant recipient cells after transplant. DLI products cover the whole spectrum of immune cells, from natural killer (NK) cells to B cells and dendritic cells.[82] However, T cells are often credited as being the primary reason for DLI's effectiveness. Truitt and Atasoylu (1991) showed that CD4+ and CD8+ T-cell populations both contribute to GVL activity in vivo using a mouse model. Both CD4+ T cells and CD8+ T cells are required for efficient immunological response, as shown by the fact that depletion of either type resulted in a diminished GVL response.[83,84] According to studies,[85,86] both CD4+ and CD8+ T-cell subsets in allogeneic HSCT patients include leukemia-reactive T cells. The specific involvement of distinct T-cell populations as mediators of GVL is mainly understood, and only a small number of T-cell clones have been discovered in patients following allogeneic HSCT.[87]

One of the two approaches shown schematically in Figure 1.3 is often used for the extraction and characterization of T-cell clones that generate GVL/GVHD.[88] You can see this concept in action in Sequence A. Donor T cells are stimulated with the target cells of the final recipient, and then cell lines or clones are selected for their ability to specifically identify recipient cells without eliciting a response in normal donor cells.[89,90] The initial population of responding T cells may come from an uninfected donor or the patient's peripheral blood (after engraftment).[91] To determine whether a T-cell clone will respond to normal or malignant cells in the recipient, it must be exposed to both the tumor cells and a variety of different normal hematological and nonhematopoietic cell types.[92] Identification of the appropriate MHC-restricting alleles and determination of whether class I or class II major histocompatibility complex (MHC) molecules present the target antigen may be required when tissue restriction patterns are observed.[93] It is unusual for further research to be done to confirm the peptide epitope recognized by the clone[108] and disclose the genetic variation driving identification by donor T-cell clones. Peptide-MHC tetramers and peptide ELISPOT testing may be utilized to identify the specific T-cell clone in peripheral blood and tissue samples after the identification of the peptide epitopes recognized by these T-cell clones.[94] After molecular sequencing of the modified TCR genes in selected clones, new PCR-based techniques for identifying and quantifying particular clones in serial samples received from the patient after transplant may be devised.[95] Researchers have found several interesting T-cell clones using this method,[96] and many studies have shown the target selectivity of these cells. The first step is to decide on a group of proteins or peptides to analyze. These targets are used as stimulating antigens during the isolation and expansion of T-cell clones in vitro.[97,98] Once antigen-specific T cells have been cloned, researchers may try to identify the MHC limiting allele and determine whether the target peptide is expressed at high enough levels in the recipient tissues for the clone to recognize them. Cloned T cells may be tested for specificity in both the human recipient and nonhuman primates.[99] Both methods have shown encouraging results that provide a new understanding of the role of T cells in graft-versus-host disease (GVHD) and graft-versus-host immunity (GVH) after allogeneic HSCT. What strategies have limits is also obvious.[100,101] The first approach is advantageous for identifying mHA that reflects distinct genetic variations in the donor and the recipient because it includes selecting clones that are reactive with recipient cells but not donor cells.[102] By using this approach, it is difficult to detect autoreactive clones, which recognize antigens expressed in both the donor and recipient cells.[103] The second strategy, which requires the identification of a target protein in advance, is more suited to characterizing in vivo responses to known tumor-associated antigens or mHA after transplant. There were no novel mHAs or tumor-associated antigens found in our analyses.[104–106] These comprehensive approaches have allowed us to recover leukemia-reactive clones from responding patients and to uncover several immunogenic peptide epitopes recognized by donor T cells following allogeneic HSCT.[107,108] After allogeneic HSCT, individuals with GVHD have been shown to contain clones of cytotoxic T lymphocytes (CTLs) with an anti-mHA bias.[109] These CTLs are so antigen-specific that they may lyse newly generated leukemic cells and suppress the growth of leukemic progenitor cells in vitro.[110]

It has been demonstrated that highly specific T-cell clones may eliminate transplanted leukemia cells in NOD/SCID animals,[111,112] and human CTL clones specific for mHA may preferentially target leukemia stem cells. It has been shown that DLI can be utilized to identify leukemia-reactive CTL in leukemia patients. Increased numbers of cytolytic CD8+ donor T lymphocytes that selectively target recipient hematopoietic cells were seen in patients with recurrent chronic myeloid leukemia.[113] Researchers reported comparable findings in a DLI-responsive myeloma patient. Two studies[114] found that patients whose CD8 activity had been impaired by treatment with CD8-depleted DLI benefited from infusions of CD4 donor T cells. One kind of GVL effector discovered post-DLI is the mHA-specific CTL clone HA1. Patients with relapsed leukemia following allogeneic HSCT have had mHA-specific CTL clones isolated, expanded in vitro, and infused into them. Some of these patients have shown remarkable clinical progress.[115,116] These and other studies have shown the capacity of CD4+ and CD8+ donor T cells to mediate GVL and decrease the number of lingering tumor cells in allogeneic HSCT patients.[117–120]

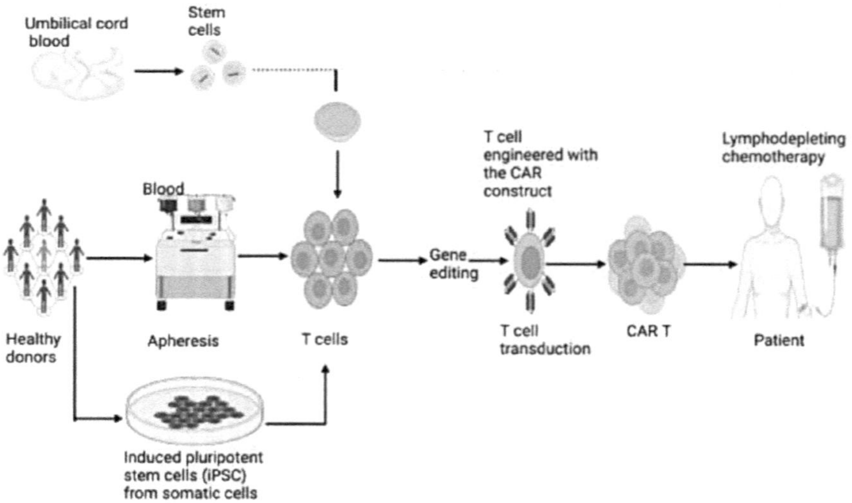

FIGURE 1.3 Other methods for identifying donor T cells after allogeneic HSCT.

1.5 ALLOGENEIC HSCT: DONOR T-CELL TARGET ANTIGENS

Following allogeneic HSCT, there has been much study of the antigens targeted by donor T cells. Since donor T cells are the primary mediators of GVL and GVHD, understanding these immunologic processes and developing techniques to untangle their effects in vivo will likely require identifying the particular peptide epitopes recognized by T cells.[121] Allogeneic HSCT offers many benefits, but it also has a significant risk of graft-versus-host disease (GVHD). As the author learns more about the antigen differences between GVL and GVHD,[122–125] the author may develop novel techniques that should improve the HSCT results for patients. By delving into the mechanisms of GVL, the author can determine whether the same ideas can be employed to elicit autologous tumor immunity or if they are unique to the allogeneic HSCT setting.

When there is a lack of HLA similarity between the donor and the recipient, GVHD is more likely to develop after transplantation because donor T cells will be targeting recipient MHC antigens.[126] Both the donor and the recipient in an allogeneic stem cell transplant were formerly required to be unrelated to one another on both sides of their families.[127–130] However, most transplants performed today are between persons with a high degree of genetic similarity or HLA. Donor T cells that trigger GVHD often target mHA in this way.[131]

Several groups of genes on chromosome 6 work together to produce proteins called major histocompatibility antigens (MHAs). However, minor histocompatibility antigens (mHAs) arise from random changes across the human genome.[132,133] Self-MHC molecules produce self-peptides, and high-affinity receptors that react to these peptides are removed by negative selection during normal thymic development.[134] The thymus synthesizes and processes a vast range of self-proteins, with the resulting peptides being presented on major histocompatibility complex (MHC) class I and class II molecules.[135,136] This is a proven reality, even though our present knowledge can only scratch the surface. During the thymic phase of negative selection, growing T cells are nearly eliminated if they respond to any of these self-peptides. Those CD4+ and CD8+ T cells that make it through thymic selection retain the ability to recognize both "nonself" human peptides and foreign peptides thanks to their extensive TCR repertoire.[137] Due to the high number of different genes present in an outbred population, each member of the group expresses a slightly different set of polymorphic proteins in their cells.[138] Proteasomes cut up polymorphic proteins into fragments called peptides, which are presented by distinct elements of the major histocompatibility complex.[139,140] There is considerable variation in the self-peptide repertoire shown on the surface of MHC molecules, even among individuals who share an HLA. In the same way, identical and syngeneic twins are affected since they have the same HLA type.[141] Due to the dynamic nature of protein expression throughout development and function, each cell type may produce its unique collection of self-peptides. A high number of mature T cells from a donor that can recognize the minor histocompatibility antigens (mHA) of the recipient are transferred in an allogeneic HSCT.[142,143] Instead, donor T cells created from hematopoietic stem cells after transplantation undergo negative selection to remove any cell clones reactive with recipient mHA. If the donor's T cells have returned to health, the recipient of a stem cell transplant who has undergone T-cell depletion is not at risk for developing GVHD. Age-related thymus atrophy[144,145] occurs in both children and adults.

1.6 MEDIATORS OF GRAFT-VERSUS-HOST DISEASE BY DONOR NATURAL KILLER CELLS

Human NK cells have killer inhibitory receptors (KIR) for major histocompatibility complex class I, as well as receptors selective for nonclassical HLA molecules like HLA-E and NK cell ligands, such as CD94/NKG2.[146] Since the KIR, CD94/NKG2, and HLA genes are all located in different parts of the genome, there is no common genetic foundation for NK cell receptor or ligand expression.[147] Allelic variations within KIR genes, rather than structural changes to NK receptors, may account for the observed variation in NK receptor expression.[148] A minimum of one killer immunoglobulin-like receptor (KIR) on each mature NK cell in the body's reserve recognizes its own HLA class I protein, leaving these cells incapable of attacking other self-tissue. One or more NK cells may recognize a cell as self even though it no longer expresses any HLA class I alleles. Loss of HLA is most often caused by viral infections and malignant alterations of different sorts.[149] Each NK cell receptor has a cytoplasmic tail or adaptor molecules that facilitate intracellular communication to either activate or inhibit NK cell effector function upon receptor activation. Although KIRs play a critical role in negatively controlling the NK cell response to a target, the number of activating and inhibitory receptors ultimately determines the NK cell response.[150] When enough inhibitory NK receptor ligands are created, normal "self" cells are shielded from NK cell lysis. This line of defence is routinely used by targets to evade lysis by NK cells. Successful allogeneic HSCT relies on the polarity of lysis, which is affected by NK activity.[151] If the recipient's cells do not have the homologous KIR ligand, GVL and GVHD may result from donor NK cell lysis of recipient target cells. But that would be conditional on the kind of tissue that the NK cells are aiming to repair.

If the recipient's NK effector cell lyses the donor cell, graft rejection may ensue. Half of all transplants from unrelated donors are thought to have NK cell alloreactivity owing to a mismatch in HLA alleles.[152]

Donor NK cells may have a role in mediating graft-versus-host disease (GVL) in certain instances. NK cells are the first of the lymphoid lineage cells to reconstitute following allogeneic transplantation, and they are associated with a better relapse-free prognosis. The donor NK cells were tested further and found to be free of any T-cell antigen.[153] These findings show that clones with antileukemia activity following HSCT have NK cell function and phenotype. The researchers examined the NK cell activity of chronic myeloid leukemia (CML) patients who had had allogeneic HSCT in the early post-transplant period. Patients with CML who displayed NK lytic activity against host-derived targets three weeks after transplant had a higher likelihood of survival.

1.7 CONCLUSION

Allogeneic HSCT is used to treat hematologic malignancies in around 10,000 individuals annually throughout the globe. Every year, a growing number of patients can receive allogeneic HSCT because of ongoing clinical research that reduces treatment toxicity and increases the availability of unrelated stem cell products from adult volunteer donors and cryopreserved umbilical cord blood banks. Allogeneic HSCT is quickly replacing standard therapies for a wide variety of hematologic malignancies due to its therapeutic effectiveness in eradicating tumor cells in vivo. It has long been known that the donor's immune system plays a crucial role in the immunologic elimination of malignant cells and the subsequent maintenance of their suppression throughout time. Although still associated with significant toxicity and disease control is not achieved in all patients, allogeneic HSCT is a good predictor of the level of effective tumor immunity and the therapeutic benefits that may be obtained with this strategy.

The principal target antigens of this response and the immunologic pathways responsible for them have come a long way, but there is still more work to be done. Clinically effective immune responses are polyclonal, as revealed by the identification of GVL targets of T and B cell responses, suggesting that they target a wide variety of alloantigens and tumor-associated antigens. Variations in cancer cell antigen production, genetic polymorphisms, and the abundance of major histocompatibility complex (MHC) components all provide support to the idea that many immunological targets are patient-specific.

There is also substantial evidence that tumor rejection requires an integrated effort including cellular, humoral, and innate immunity. Each of these procedures has been associated with the rejection of tumors and the prevention of their return. In conclusion, patients preparing for allogeneic HSCT have a better chance of achieving clinical responses when less intensive chemotherapy and radiation treatment are used. This is because there is less residual disease and cancer cells are developing more slowly. These findings emphasize the necessity for further research into the development of innovative drugs with the potential to increase GVL in an allogeneic setting. However, emerging therapies that target antigens extensively expressed in recipient tissues may increase the risk of GVHD. Targeting tumor-associated antigens or mHA in a specific area has great therapeutic potential for ameliorating GVHD symptoms. Successful use of DLI to prevent relapse after transplant has been shown, even in patients who had experienced relapses in the past. Although donor cells have found useful applications, new and superior approaches are always being developed and tried.

Although GVL is illustrative of the therapeutic potential of tumor immunity, the intensive chemotherapy, alloimmunity, and immunological deficiency required for optimal outcomes are very risky. Those hoping for a reduction in the medication's negative effects will be pleased to hear that progress is being made.

Scientists hope to develop innovative and successful techniques for boosting the clearance of malignant cells after allogeneic HSCT as they learn more about the immune systems that contribute to GVL. Improving patient outcomes is the main goal of these studies, which is why researchers are looking for ways to increase GVL without simultaneously increasing GVHD. The development of autologous techniques to improve effective cancer immunity in vivo may benefit from a better understanding of the pathways required for tumor rejection after allogeneic HSCT.

KEYWORDS

- **allogenic stem cell transfusion**
- **myeloblative**
- **hematopoietic regeneration**
- **GVL response**
- **immunogenicity**

REFERENCES

1. Wu, C. J.; Ritz, J. Induction of Tumor Immunity Following Allogeneic Stem Cell Transplantation. *Adv. Immunol.* **2006,** *90*, 133–173.
2. Bonnet, D.; Warren, E. H.; Greenberg, P. D.; Dick, J. E.; Riddell, S. R. CD8 (+), Minor Histocompatibility Antigen-Specific Cytotoxic T Lymphocyte Clones Eliminate Human Acute Myeloid Leukemia Stem Cells. *Proc. Natl. Acad. Sci. USA* **1999,** *96*, 8639–8644.
3. Molldrem, J. J.; Lee, P. P.; Wang, C. et al. Evidence That Specific T Lymphocytes May Participate in the Elimination of Chronic Myelogenous Leukemia. *Nat Med.* **2000,** *6*, 1018–1023.
4. Wu, C. J.; Yang, X. F.; McLaughlin, S. et al. Detection of a Potent Humoral Response Associated with Immune-Induced Remission of Chronic Myelogenous Leukemia. *J. Clin. Invest.* **2000,** *106*, 705–714.
5. Miklos, D. B.; Kim, H. T.; Miller, K. H. et al. Antibody Responses to H-Y Minor Histocompatibility Antigens Correlate with Chronic Graft-Versus-Host Disease and Disease Remission. *Blood* **2005,** *105*, 2973–2978.
6. Akatsuka, Y.; Morishima, Y.; Kuzushima, K.; Kodera, Y.; Takahashi, T. Minor Histocompatibility Antigens as Targets for Immunotherapy Using Allogeneic Immune Reactions. *Cancer Sci.* **2007,** *98*, 1139–1146.
7. Murata, M.; Warren, E. H.; Riddell, S. R. A Human Minor Histocompatibility Antigen Resulting from Differential Expression Due to a Gene Deletion. *J. Exp. Med.* **2003,** *197*, 1279–1289.
8. Tykodi, S. S.; Fujii, N.; Vigneron, N. et al. C19orf48 Encodes a Polymorphic Tumor-Associated Antigen Recognized by CD8+ Cytotoxic T Cells from Renal Cell Carcinoma Patients After Allogeneic Hematopoietic Cell Transplantation. *Clin. Cancer Res.* **2008.**
9. Takahashi, Y.; Harashima, N.; Kajigaya, S. et al. Regression of Human Kidney Cancer Following Allogeneic Stem Cell Transplantation Is Associated with Recognition of an HERV-E Antigen by T Cells. *J. Clin. Invest.* **2008,** *118*, 1099–1109.
10. Tykodi, S. S.; Warren, E. H.; Thompson, J. A. et al. Allogeneic Hematopoietic Cell Transplantation for Metastatic Renal Cell Carcinoma After Nonmyeloablative Conditioning: Toxicity, Clinical Response, and Immunological Response to Minor Histocompatibility Antigens. *Clin. Cancer Res.* **2004,** *10*, 7799–7811.
11. Demirer, T.; Barkholt, L.; Blaise, D. et al. Transplantation of Allogeneic Hematopoietic Stem Cells: An Emerging Treatment Modality for Solid Tumors. *Nat. Clin. Pract. Oncol.* **2008,** *5*, 256–267.
12. Rabinovich, G. A.; Gabrilovich, D.; Sotomayor, E. M. Immunosuppressive Strategies That Are Mediated by Tumor Cells. *Annu. Rev. Immunol.* **2007,** *25*, 267–296.
13. Di Ianni, M.; Falzetti, F.; Carotti, A.; Terenzi, A.; Castellino, F.; Bonifacio, E. et al. Tregs Prevent GVHD and Promote Immune Reconstitution in HLA-Haploidentical Transplantation. *Blood* **2011,** *117* (14), 3921–3928. DOI: 10.1182/blood-2010-10-311894
14. Edinger, M.; Hoffmann, P.; Ermann, J.; Drago, K.; Fathman, C. G.; Strober, S. et al. CD4+ CD25+ Regulatory T Cells Preserve Graft-Versus-Tumor Activity While Inhibiting Graft-Versus-Host Disease After Bone Marrow Transplantation. *Nat. Med.* **2003,** *9* (9), 1144–1150. DOI: 10.1038/nm915
15. Brunstein, C. G.; Miller, J. S.; Cao, Q.; McKenna, D. H.; Hippen, K. L.; Curtsinger, J. et al. Infusion of Ex Vivo Expanded T Regulatory Cells in Adults Transplanted with

Umbilical Cord Blood: Safety Profile and Detection Kinetics. *Blood* **2011,** *117* (3), 1061–1070. DOI: 10.1182/blood-2010-07-293795

16. Hakim, F. T.; Memon, S. A.; Cepeda, R.; Jones, E. C.; Chow, C. K.; Kasten-Sportes, C. et al. Age-Dependent Incidence, Time Course, and Consequences of Thymic Renewal in Adults. *J. Clin. Invest.* **2005,** *115* (4), 930–939. DOI: 10.1172/JCI200522492
17. Bolotin, E.; Smogorzewska, M.; Smith, S.; Widmer, M.; Weinberg, K. Enhancement of Thymopoiesis After Bone Marrow Transplant by In Vivo Interleukin-7. *Blood* **1996,** *88* (5), 1887–1894.
18. Guimond, M.; Fry, T. J.; Mackall, C. L. Cytokine Signals in T-Cell Homeostasis. *J. Immunother.* **2005,** *28* (4), 289–294. DOI: 10.1097/01.cji.0000165356.03924.e7
19. van Heijst, J. W. J.; Ceberio, I.; Lipuma, L. B.; Samilo, D. W.; Wasilewski, G. D.; Gonzales, A. M. R. et al. Quantitative Assessment of T-Cell Repertoire Recovery After Hematopoietic Stem Cell Transplantation. *Nat. Med.* **2013,** *19* (3), 372–377. DOI: 10.1038/nm.3100
20. Yew, P. Y.; Alachkar, H.; Yamaguchi, R.; Kiyotani, K.; Fang, H.; Yap, K. L. et al. Quantitative Characterization of T-Cell Repertoire in Allogeneic Hematopoietic Stem Cell Transplant Recipients. *Bone Marrow Transpl.* **2015,** *50* (9), 1227–1234. DOI: 10.1038/bmt.2015.133
21. Storek, J.; Joseph, A.; Dawson, M. A.; Douek, D. C.; Storer, B.; Maloney, D. G. Factors Influencing T-Lymphopoiesis After Allogeneic Hematopoietic Cell Transplantation. *Transplantation* **2002,** *73* (7), 1154–1158. DOI: 10.1097/00007890-200204150-00026
22. Storek, J.; Geddes, M.; Khan, F.; Huard, B.; Helg, C.; Chalandon, Y. et al. Reconstitution of the Immune System After Hematopoietic Stem Cell Transplantation in Humans. *Semin. Immunopathol.* **2008,** *30* (4), 425–437. DOI: 10.1007/s00281-008-0132-5
23. Marie-Cardine, A.; Divay, F.; Dutot, I.; Green, A.; Perdrix, A.; Boyer, O. et al. Transitional B Cells in Humans: Characterization and Insight from B Lymphocyte Reconstitution After Hematopoietic Stem Cell Transplantation. *Clin. Immunol.* **2008,** *127* (1), 14–25. DOI: 10.1016/j.clim.2007.11.013
24. D'Orsogna, L. J.; Wright, M. P.; Krueger, R. G.; McKinnon, E. J.; Buffery, S. I.; Witt, C. S. et al. Allogeneic Hematopoietic Stem Cell Transplantation Recipients Have Defects of Both Switched and IGM Memory B Cells. *Biol. Blood Marrow Transpl.* **2009,** *15* (7), 795–803. DOI: 10.1016/j.bbmt.2008.11.024
25. Park, B. G.; Park, C. J.; Jang, S.; Chi, H. S.; Kim, D. Y.; Lee, J. H. et al. Reconstitution of Lymphocyte Subpopulations After Hematopoietic Stem Cell Transplantation: Comparison of Hematologic Malignancies and Donor Types in Event-Free Patients. *Leuk. Res.* **2015,** *39* (12), 1334–1341. DOI: 10.1016/j.leukres.2015.09.010
26. Williams, K. M.; Gress, R. E. Immune Reconstitution and Implications for Immunotherapy Following Haematopoietic Stem Cell Transplantation. *Best Pract. Res. Clin. Haematol.* **2008,** *21* (3), 579–596. DOI: 10.1016/j.beha.2008.06.003
27. Suryani, S.; Fulcher, D. A.; Santner-Nanan, B.; Nanan, R.; Wong, M.; Shaw, P. J. et al. Differential Expression of CD21 Identifies Developmentally and Functionally Distinct Subsets of Human Transitional B Cells. *Blood* **2010,** *115* (3), 519–529. DOI: 10.1182/blood-2009-07-234799
28. Mackall, C.; Fry, T.; Gress, R.; Peggs, K.; Storek, J.; Toubert, A. Background to Hematopoietic Cell Transplantation, Including Post Transplant Immune Recovery. *Bone Marrow Transpl.* **2009,** *44* (8), 457–462. DOI: 10.1038/bmt.2009.255

29. Corre, E.; Carmagnat, M.; Busson, M.; de Latour, R. P.; Robin, M.; Ribaud, P. et al. Long-Term Immune Deficiency After Allogeneic Stem Cell Transplantation: B-Cell Deficiency Is Associated with Late Infections. *Haematologica* **2010,** *95* (6), 1025–1029. DOI: 10.3324/haematol.2009.018853
30. Storek, J.; Saxon, A. Reconstitution of B Cell Immunity Following Bone Marrow Transplantation. *Bone Marrow Transplant* **1992,** *9* (6), 395–408.
31. Storek, J.; Ferrara, S.; Ku, N.; Giorgi, J. V.; Champlin, R. E.; Saxon, A. B Cell Reconstitution After Human Bone Marrow Transplantation: Recapitulation of Ontogeny? *Bone Marrow Transplant* **1993,** *12* (4), 387–398.
32. Moins-Teisserenc, H.; Busson, M.; Herda, A.; Apete, S.; Peffault de Latour, R.; Robin, M. et al. CD19+CD5+ B Cells and B1-Like Cells Following Allogeneic Hematopoietic Stem Cell Transplantation. *Biol. Blood Marrow Transplant* **2013,** *19* (6), 988–991. DOI: 10.1016/j.bbmt.2013.03.006
33. Lee, J.; Kuchen, S.; Fischer, R.; Chang, S.; Lipsky, P. E. Identification and Characterization of a Human CD5+ Pre-Naive B Cell Population. *J. Immunol.* **2009,** *182* (7), 4116–4126. DOI: 10.4049/jimmunol.0803391
34. Glas, A. M.; van Montfort, E. H.; Storek, J.; Green, E. G.; Drissen, R. P.; Bechtold, V. J. et al. B-Cell-Autonomous Somatic Mutation Deficit Following Bone Marrow Transplant. *Blood* **2000,** *96* (3), 1064–1069.
35. Omazic, B.; Lundkvist, I.; Mattsson, J.; Permert, J.; Nasman-Bjork, I. Memory B Lymphocytes Determine Repertoire Oligoclonality Early After Haematopoietic Stem Cell Transplantation. *Clin. Exp. Immunol.* **2003,** *134* (1), 159–166. DOI: 10.1046/j.1365-2249.2003.02260.x
36. Small, T. N.; Robinson, W. H.; Miklos, D. B. B Cells and Transplantation: An Educational Resource. *Biol. Blood Marrow Transplant* **2009,** *15* (1 Suppl), 104–113. DOI: 10.1016/j.bbmt.2008.10.016
37. Bemark, M.; Holmqvist, J.; Abrahamsson, J.; Mellgren, K. Translational Mini-Review Series on B Cell Subsets in Disease: Reconstitution After Haematopoietic Stem Cell Transplantation—Revelation of B Cell Developmental Pathways and Lineage Phenotypes. *Clin. Exp. Immunol.* **2012,** *167* (1), 15–25. DOI: 10.1111/j.1365-2249.2011.04469.x
38. Aucouturier, P.; Barra, A.; Intrator, L.; Cordonnier, C.; Schulz, D.; Duarte, F. et al. Long Lasting IgG Subclass and Antibacterial Polysaccharide Antibody Deficiency After Allogeneic Bone Marrow Transplantation. *Blood* **1987,** *70* (3), 779–785.
39. Nasman-Bjork, I.; Lundkvist, I. Oligoclonal Dominance of immunoglobulin VH3 Rearrangements Following Allogeneic Bone Marrow Transplantation. *Bone Marrow Transplant* **1998,** *21* (12), 1223–1230. DOI: 10.1038/sj.bmt.1701261
40. Roll, P.; Muhammad, K.; Stuhler, G.; Grigoleit, U.; Einsele, H.; Tony, H. P. Effect of ATG-F on B-Cell Reconstitution After Hematopoietic Stem Cell Transplantation. *Eur. J. Haematol.* **2015,** *95* (6), 514–523. DOI: 10.1111/ejh.12524
41. Fuji, S.; Kim, S. W.; Yano, S.; Hagiwara, S.; Nakamae, H.; Hidaka, M. et al. A Prospective Multicenter Study of Unrelated Bone Marrow Transplants Using a Reduced-Intensity Conditioning Regimen with Low-Dose ATG-F. *Bone Marrow Transplant* **2016,** *51*, 451–453.
42. Zand, M. S. B-Cell Activity of Polyclonal Antithymocyte Globulins. *Transplantation* **2006,** *82* (11), 1387–1395.
43. Petersen, S. L.; Ryder, L. P.; Bjork, P.; Madsen, H. O.; Heilmann, C.; Jacobsen, N. et al. A Comparison of T-, B- and NK-Cell Reconstitution Following Conventional or

Nonmyeloablative Conditioning and Transplantation with Bone Marrow or Peripheral Blood Stem Cells from Human Leucocyte Antigen Identical Sibling Donors. *Bone Marrow Transplant* **2003,** *32* (1), 65–72.
44. Storek, J.; Wells, D.; Dawson, M. A.; Storer, B.; Maloney, D. G. Factors Influencing B Lymphopoiesis After Allogeneic Hematopoietic Cell Transplantation. *Blood* **2001,** *98* (2), 489–491.
45. Greinix, H. T.; Pohlreich, D.; Kouba, M.; Kormoczi, U.; Lohmann, I.; Feldmann, K. et al. Elevated Numbers of Immature/Transitional CD21- B Lymphocytes and Deficiency of Memory CD27+ B Cells Identify Patients with Active Chronic Graft-Versus-Host Disease. *Biol. Blood Marrow Transplant* **2008,** *14* (2), 208–219.
46. Greinix, H. T.; Kuzmina, Z.; Weigl, R.; Kormoczi, U.; Rottal, A.; Wolff, D. et al. CD19+CD21low B cells and CD4+CD45RA+CD31+ T Cells Correlate with First Diagnosis of Chronic Graft-Versus-Host Disease. *Biol. Blood Marrow Transplant* **2015,** *21* (2), 250–258.
47. Storek, J.; Witherspoon, R. P.; Maloney, D. G.; Chauncey, T. R.; Storb, R. Improved Reconstitution of CD4 T Cells and B Cells But Worsened Reconstitution of Serum IgG Levels After Allogeneic Transplantation of Blood Stem Cells Instead of Marrow. *Blood* **1997,** *89* (10), 3891–3893.
48. Storek, J.; Dawson, M. A.; Storer, B.; Stevens-Ayers, T.; Maloney, D. G.; Marr, K. A. et al. Immune Reconstitution After Allogeneic Marrow Transplantation Compared with Blood Stem Cell Transplantation. *Blood* **2001,** *97* (11), 3380–3389.
49. Fontenot, J. D.; Rudensky, A. Y. A Well Adapted Regulatory Contrivance: Regulatory T Cell Development and the Fork Head Family Transcription Factor Foxp3. *Nat. Immunol.* **2005,** *6* (4), 331–337.
50. Feuerer, M.; Hill, J. A.; Mathis, D.; Benoist, C. Foxp3+ Regulatory T Cells: Differentiation, Specification, Subphenotypes. *Nat Immunol* **2009,** *10* (7), 689–695. DOI: 10.1038/ni.1760
51. Sakaguchi, S.; Yamaguchi, T.; Nomura, T.; Ono, M. Regulatory T Cells and Immune Tolerance. *Cell* **2008,** *133* (5), 775–787.
52. Hori, S.; Nomura, T.; Sakaguchi, S. Control of Regulatory T Cell Development by the Transcription Factor Foxp3. *Science* **2003,** *299* (5609), 1057–1061.
53. Cobbold, S. P.; Castejon, R.; Adams, E.; Zelenika, D.; Graca, L.; Humm, S. et al. Induction of foxP3+ Regulatory T Cells in the Periphery of T Cell Receptor Transgenic Mice Tolerized to Transplants. *J. Immunol.* **2004,** *172* (10), 6003–6010.
54. Baron, U.; Floess, S.; Wieczorek, G.; Baumann, K.; Grützkau, A.; Dong, J. et al. DNA Demethylation in the Human FOXP3 Locus Discriminates Regulatory T Cells from Activated FOXP3+ Conventional T Cells. *Eur. J. Immunol.* **2007,** *37* (9), 2378–2389.
55. Zheng, Y.; Josefowicz, S.; Chaudhry, A.; Peng, X. P.; Forbush, K.; Rudensky, A. Y. Role of Conserved Non-Coding DNA Elements in the Foxp3 Gene in Regulatory T-Cell Fate. *Nature* **2010,** *463* (7282), 808–812.
56. Kolb, H.; Schattenberg, A.; Goldman, J.; Hertenstein, B.; Jacobsen, N.; Arcese, W. et al. Graft-Versus-Leukemia Effect of Donor Lymphocyte Transfusions in Marrow Grafted Patients: European Group for Blood and Marrow Transplantation Working Party Chronic Leukemia. *Blood* **1995,** *86* (5), 2041–2050.
57. Collins, R. H. Jr; Shpilberg, O; Drobyski, W. R., et al. Donor Leukocyte Infusions In 140 Patients with Relapsed Malignancy After Allogeneic Bone Marrow Transplantation. *J. Clin. Oncol.* **1997,** *15*(2), 433-444. doi:10.1200/JCO.1997.15.2.433.

58. Campbell, D. J.; Koch, M. A. Phenotypical and Functional Specialization of FOXP3+ Regulatory T Cells. *Nat. Rev. Immunol.* **2011,** *11* (2), 119–130.
59. Kehrl, J. H.; Wakefield, L. M.; Roberts, A. B.; Jakowlew, S.; Alvarez-Mon, M.; Derynck, R. et al. Production of Transforming Growth Factor Beta by Human T Lymphocytes and Its Potential Role in the Regulation of T Cell Growth. *J. Exp. Med.* **1986,** *163* (5), 1037–1050.
60. Haak-Frendscho, M.; Wynn, T.; Czuprynski, C.; Paulnock, D. Transforming Growth Factor-Beta *1* Inhibits Activation of Macrophage Cell Line RAW 264.7 for Cell Killing. *Clin. Exp. Immunol.* **1990,** *82* (2), 404.
61. Belkaid, Y.; Rouse, B. T. Natural Regulatory T Cells in Infectious Disease. *Nat. Immunol.* **2005,** *6* (4), 353–360.
62. Arpaia, N.; Campbell, C.; Fan, X.; Dikiy, S.; van der Veeken, J.; Liu, H. et al. Metabolites Produced by Commensal Bacteria Promote Peripheral Regulatory T-Cell Generation. *Nature* **2013,** *504* (7480), 451–455.
63. Mathewson, N. D.; Jenq, R.; Mathew, A. V.; Koenigsknecht, M.; Hanash, A.; Toubai, T. et al. Gut Microbiome-Derived Metabolites Modulate Intestinal Epithelial Cell Damage and Mitigate Graft-Versus-Host Disease. *Nat. Immunol.* **2016,** *17* (5), 505–513.
64. Shevach, E. M. CD4+ CD25+ Suppressor T Cells: More Questions Than Answers. *Nat. Rev. Immunol.* **2002,** *2* (6), 389–400.
65. Kingsley, C. I.; Karim, M.; Bushell, A. R.; Wood, K. J. CD25+ CD4+ regulatory T Cells Prevent Graft Rejection: CTLA-4-and IL-10-Dependent Immunoregulation of Alloresponses. *J. Immunol.* **2002,** *168* (3), 1080–1086.
66. Joffre, O.; Gorsse, N.; Romagnoli, P.; Hudrisier, D.; Van Meerwijk, J. P. Induction of Antigen-Specific Tolerance to Bone Marrow Allografts with CD4+ CD25+ T Lymphocytes. *Blood* **2004,** *103* (11), 4216–4221.
67. Nguyen, V. H.; Shashidhar, S.; Chang, D. S.; Ho, L.; Kambham, N.; Bachmann, M. et al. The Impact of Regulatory T cells on T-Cell Immunity Following Hematopoietic Cell Transplantation. *Blood* **2008,** *111* (2), 945–953.
68. Cohen, J. L.; Boyer, O. The Role of CD4+ CD25 Hi Regulatory T Cells in the Physiopathogeny of Graft-Versus-Host Disease. *Curr. Opin. Immunol.* **2006,** *18* (5), 580–585.
69. Ferrara, J. L.; Levine, J. E.; Reddy, P.; Holler, E. Graft-Versus-Host Disease. *Lancet* **2009,** *373* (9674), 1550–1561.
70. Taylor, P. A.; Lees, C. J.; Blazar, B. R. The Infusion of Ex Vivo Activated and Expanded CD4+CD25+ Immune Regulatory Cells Inhibits Graft-Versus-Host Disease Lethality. *Blood* **2002,** *99* (10), 3493–3499.
71. Barrett, A. J. Mechanisms of the Graft-Versus-Leukemia Reaction. *Stem Cells* **1997,** *15* (4), 248–258
72. Koenecke, C.; Czeloth, N.; Bubke, A.; Schmitz, S.; Kissenpfennig, A.; Malissen, B. et al. Alloantigen-Specific De Novo-Induced Foxp3+ Treg Revert In Vivo and Do Not Protect from Experimental GVHD. *Eur. J. Immunol.* **2009,** *39* (11), 3091–3096.
73. Laurence, A.; Amarnath, S.; Mariotti, J.; Kim, Y. C.; Foley, J.; Eckhaus, M. et al. STAT3 Transcription Factor Promotes Instability of nTreg Cells and Limits Generation of iTreg Cells During Acute Murine Graft-Versus-Host Disease. *Immunity* **2012,** *37* (2), 209–222.
74. Ligocki, A. J.; Niederkorn, J. Y. Advances on Non-CD4 + Foxp3+ T Regulatory Cells: CD8+, Type 1, and Double Negative T Regulatory Cells in Organ Transplantation. *Transplantation* **2015,** *99* (8), 1553–1559.

75. Robb, R. J.; Lineburg, K. E.; Kuns, R. D.; Wilson, Y. A.; Raffelt, N. C.; Olver, S. D. et al. Identification and Expansion of Highly Suppressive CD8+ FoxP3+ Regulatory T Cells After Experimental Allogeneic Bone Marrow Transplantation. *Blood* **2012,** *119* (24), 5898–5908.
76. Xystrakis, E.; Dejean, A. S.; Bernard, I.; Druet, P.; Liblau, R.; Gonzalez-Dunia, D. et al. Identification of a Novel Natural Regulatory CD8 T-Cell Subset and Analysis of Its Mechanism of Regulation. *Blood* **2004,** *104* (10), 3294–3301.
77. Zheng, J.; Liu, Y.; Liu, Y.; Liu, M.; Xiang, Z.; Lam K-T, et al. Human CD8+ Regulatory T Cells Inhibit GVHD and Preserve General Immunity in Humanized Mice. *Sci. Transl. Med.* **2013,** *5* (168), ra169.
78. Li, Q.; Zhai, Z.; Xu, X.; Shen, Y.; Zhang, A.; Sun, Z. et al. Decrease of CD4+ CD25+ regulatory T cells and TGF-β at Early Immune Reconstitution Is Associated to the Onset and Severity of Graft-Versus-Host Disease Following Allogeneic Haematogenesis Stem Cell Transplantation. *Leuk. Res.* **2010,** *34* (9), 1158–1168.
79. Rieger, K.; Loddenkemper, C.; Maul, J.; Fietz, T.; Wolff, D.; Terpe, H. et al. Mucosal FOXP3+Regulatory T Cells Are Numerically Deficient in Acute and Chronic GvHD. *Blood* **2006,** *107* (4), 1717–1723.
80. Lord, J. D.; Hackman, R. C.; Gooley, T. A.; Wood, B. L.; Moklebust, A. C.; Hockenbery, D. M. et al. Blood and Gastric FOXP3+ T Cells Are Not Decreased in Human Gastric Graft-Versus-Host Disease. *Biol. Blood Marrow Transplant* **2011,** *17* (4), 486–496.
81. Ratajczak, P.; Janin, A.; De Latour, R. P.; Leboeuf, C.; Desveaux, A.; Keyvanfar, K. et al. Th17/Treg Ratio in Human Graft-Versus-Host Disease. *Blood* **2010,** *116* (7), 1165–1171.
82. Imanguli, M.; Cowen, E.; Rose, J.; Dhamala, S.; Swaim, W.; Lafond, S. et al. Comparative Analysis of FoxP3 (+); Regulatory T Cells in the Target Tissues and Blood in Chronic Graft Versus Host Disease. *Leukemia* **2014,** *28* (10), 2016–2027.
83. Trzonkowski, P.; Bieniaszewska, M.; Juścińska, J.; Dobyszuk, A.; Krzystyniak, A.; Marek, N. et al. First-in-Man Clinical Results of the Treatment of Patients with Graft Versus Host Disease with Human Ex Vivo Expanded CD4+ CD25+ CD127− T Regulatory Cells. *Clin. Immunol.* **2009,** *133* (1), 22–26.
84. Hoffmann, P.; Eder, R.; Kunz-Schughart, L. A.; Andreesen, R.; Edinger, M. Large-Scale In Vitro Expansion of Polyclonal Human CD4+ CD25 High Regulatory T Cells. *Blood* **2004,** *104* (3), 895–903.
85. Strauss, L.; Whiteside, T. L.; Knights, A.; Bergmann, C.; Knuth, A.; Zippelius, A. Selective Survival of Naturally Occurring Human CD4+ CD25+ Foxp3+ Regulatory T Cells Cultured with Rapamycin. *J. Immunol.* **2007,** *178* (1), 320–329.
86. Hippen, K.; Merkel, S.; Schirm, D.; Nelson, C.; Tennis, N.; Riley, J. et al. Generation and Large-Scale Expansion of Human Inducible Regulatory T Cells That Suppress Graft-Versus-Host Disease. *Am. J. Transplant* **2011,** *11* (6), 1148–1157.
87. Matsuoka K-I, Koreth, J.; Kim, H. T.; Bascug, G.; McDonough, S.; Kawano, Y. et al. Low-Dose Interleukin-2 Therapy Restores Regulatory T Cell Homeostasis in Patients with Chronic Graft-Versus-Host Disease. *Sci. Transl. Med.* **2013,** *5* (179), ra143–ra179.
88. Koreth, J.; Matsuoka, K.; Kim, H. T.; McDonough, S. M.; Bindra, B.; Alyea EP III, et al. Interleukin-2 and Regulatory T Cells in Graft-Versus-Host Disease. *N. Engl. J. Med.* **2011,** *365* (22), 2055–2066.

89. Furusawa, Y.; Obata, Y.; Fukuda, S.; Endo, T. A.; Nakato, G.; Takahashi, D. et al. Commensal Microbe-Derived Butyrate Induces the Differentiation of Colonic Regulatory T Cells. *Nature* **2013,** *504* (7480), 446–450.
90. Mahnke, Y. D.; Brodie, T. M.; Sallusto, F.; Roederer, M.; Lugli, E. The WHO'S WHO of T-Cell Differentiation: Human Memory T-Cell Subsets. *Eur. J. Immunol.* **2013,** *43* (11), 2797–2809.
91. von Andrian, U. H.; Mempel, T. R. Homing and Cellular Traffic in Lymph Nodes. *Nat. Rev. Immunol.* **2003,** *3* (11), 867–878.
92. Smith, C. J.; Quinn, M.; Snyder, C. M. CMV-Specific CD8 T Cell Differentiation and Localization: Implications for Adoptive Therapies. *Front. Immunol.* **2016,** *7*, 352.
93. Clave, E.; Busson, M.; Douay, C.; Peffault de Latour, R.; Berrou, J.; Rabian, C. et al. Acute Graft-Versus-Host Disease Transiently Impairs Thymic Output in Young Patients After Allogeneic Hematopoietic Stem Cell Transplantation. *Blood* **2009,** *113* (25), 6477–6484.
94. Fujimaki, K.; Maruta, A.; Yoshida, M.; Kodama, F.; Matsuzaki, M.; Fujisawa, S. et al. Immune Reconstitution Assessed During Five Years After Allogeneic Bone Marrow Transplantation. *Bone Marrow Transplant* **2001,** *27* (12), 1275–1281.
95. Le, R. Q.; Melenhorst, J. J.; Battiwalla, M.; Hill, B.; Memon, S.; Savani, B. N. et al. Evolution of the Donor T-Cell Repertoire in Recipients in the Second Decade After Allogeneic Stem Cell Transplantation. *Blood* **2011,** *117* (19), 5250–5256.
96. Heining, C.; Spyridonidis, A.; Bernhardt, E.; Schulte-Monting, J.; Behringer, D.; Grullich, C. et al. Lymphocyte Reconstitution Following Allogeneic Hematopoietic Stem Cell Transplantation: A Retrospective Study Including 148 Patients. *Bone Marrow Transplant* **2007,** *39* (10), 613–622.
97. Jacobsohn, D. A.; Vogelsang, G. B. Acute Graft versus Host Disease. *Orphanet J. Rare Dis.* **2007,** 2:35.
98. Cantoni, N.; Hirsch, H. H.; Khanna, N.; Gerull, S.; Buser, A.; Bucher, C. et al. Evidence for a Bidirectional Relationship Between Cytomegalovirus Replication and Acute Graft-Versus-Host Disease. *Biol. Blood Marrow Transplant* **2010,** *16* (9), 1309–1314.
99. Auletta, J. J.; Lazarus, H. M. Immune Restoration Following Hematopoietic Stem Cell Transplantation: An Evolving Target. *Bone Marrow Transplant* **2005,** *35* (9), 835–857.
100. Blazar, B. R.; Murphy, W. J.; Abedi, M. Advances in Graft-versus-Host Disease Biology and Therapy. *Nat. Rev. Immunol.* **2012,** *12* (6), 443–458.
101. Ferrara, J. L.; Cooke, K. R.; Teshima, T. The Pathophysiology of Acute Graft-Versus-Host Disease. *Int. J. Hematol.* **2003,** *78* (3), 181–187.
102. Szyska, M.; Na, I.-K. Bone Marrow GvHD After Allogeneic Hematopoietic Stem Cell Transplantation. *Front. Immunol.* **2016,** *7*, 118.
103. Shono, Y.; Ueha, S.; Wang, Y.; Abe, J.; Kurachi, M.; Matsuno, Y. et al. Bone Marrow Graft-Versus-Host Disease: Early Destruction of Hematopoietic Niche After MHC-Mismatched Hematopoietic Stem Cell Transplantation. *Blood* **2010,** *115* (26), 5401–5411.
104. Müller, A. M. S.; Linderman, J. A.; Florek, M.; Miklos, D.; Shizuru, J. A. Allogeneic T Cells Impair Engraftment and Hematopoiesis After Stem Cell Transplantation. *Proc. Natl. Acad. Sci. USA* **2010,** *107* (33), 14721–14726.
105. McCarthy, A. L.; Malik Peiris, J. S.; Taylor, C. E.; Green, M. A.; Sviland, L.; Pearson, A. D. et al. Increase in Severity of Graft Versus Host Disease by Cytomegalovirus. *J. Clin. Pathol.* **1992,** *45* (6), 542–544.

106. Hakki, M.; Riddell, S. R.; Storek, J.; Carter, R. A.; Stevens-Ayers, T.; Sudour, P. et al. Immune Reconstitution to Cytomegalovirus After Allogeneic Hematopoietic Stem Cell Transplantation: Impact of Host Factors, Drug Therapy, and Subclinical Reactivation. *Blood* **2003,** *102* (8), 3060–3067.
107. Ozdemir, E.; St. John, L. S.; Gillespie, G.; Rowland-Jones, S.; Champlin, R. E.; Molldrem, J. J. et al. Cytomegalovirus Reactivation Following Allogeneic Stem Cell Transplantation Is Associated with the Presence of Dysfunctional Antigen-Specific CD8+ T Cells. Blood **2002,** *100* (10), 3690–3697.
108. Marek, A.; Stern, M.; Chalandon, Y.; Ansari, M.; Ozsahin, H.; Gungor, T. et al. The Impact of T-Cell Depletion Techniques on the Outcome After Haploidentical Hematopoietic SCT. *Bone Marrow Transplant* **2014,** *49* (1), 55–61.
109. Ciceri, F.; Bonini, C.; Stanghellini, M. T.; Bondanza, A.; Traversari, C.; Salomoni, M. et al. Infusion of Suicide-Gene-Engineered Donor Lymphocytes After Family Haploidentical Haemopoietic Stem-Cell Transplantation for Leukaemia (the TK007 Trial) a Non-Randomised Phase I–II Study. *Lancet Oncol.* **2009,** *10* (5), 489–500.
110. Weissinger, E. M.; Borchers, S.; Silvani, A.; Provasi, E.; Radrizzani, M.; Beckmann, I. K. et al. Long Term Follow Up of Patients After Allogeneic Stem Cell Transplantation and Transfusion of HSV-TK Transduced T-Cells. *Front. Pharmacol.* **2015,** *6*, 76.
111. Na I-K, Wittenbecher, F.; Dziubianau, M.; Herholz, A.; Mensen, A.; Kunkel, D. et al. Rabbit Antithymocyte Globulin (Thymoglobulin(®)) Impairs the Thymic Output of Both Conventional and Regulatory CD4(+) T Cells After Allogeneic Hematopoietic Stem Cell Transplantation in Adult Patients. *Haematologica* **2013,** *98* (1), 23–30. DOI: 10.3324/haematol.2012.067611
112. Juliusson, G.; Theorin, N.; Karlsson, K.; Frodin, U.; Malm, C. Subcutaneous Alemtuzumab vs ATG in Adjusted Conditioning for Allogeneic Transplantation: Influence of Campath Dose on Lymphoid Recovery, Mixed Chimerism and Survival. *Bone Marrow Transplant* **2006,** *37* (5), 503–510.
113. Cwynarski, K.; Ainsworth, J.; Cobbold, M.; Wagner, S.; Mahendra, P.; Apperley, J. et al. Direct Visualization of Cytomegalovirus-Specific T-Cell Reconstitution After Allogeneic Stem Cell Transplantation. *Blood* **2001,** *97* (5), 1232–1240.
114. Jansen, J.; Hanks, S.; Thompson, J. M.; Dugan, M. J.; Akard, L. P. Transplantation of Hematopoietic Stem Cells from the Peripheral Blood. *J. Cell Mol. Med.* **2005,** *9* (1), 37–50.
115. Fantini, M. C.; Becker, C.; Monteleone, G.; Pallone, F.; Galle, P. R.; Neurath, M. F. Cutting Edge: TGF-β Induces a Regulatory Phenotype in CD4+ CD25− T Cells Through Foxp3 induction and Down-Regulation of Smad7. *J. Immunol.* **2004,** *172* (9), 5149–5153.
116. Chakraverty, R.; Eom, H.-S.; Sachs, J.; Buchli, J.; Cotter, P.; Hsu, R. et al. Host MHC Class II+Antigen-Presenting Cells and CD4 Cells Are Required for CD8-Mediated Graft-versus-Leukemia Responses Following Delayed Donor Leukocyte Infusions. *Blood* **2006,** *108* (6), 2106–2113.
117. Nikiforow, S.; Alyea, E. P. Maximizing GVL in Allogeneic Transplantation: Role of Donor Lymphocyte Infusions. *Hematol. Am. Soc. Hematol. Educ. Program* **2014,** *2014* (1), 570–575.
118. Maloy, K. J.; Powrie, F. Regulatory T Cells in the Control of Immune Pathology. *Nat. Immunol.* **2001,** *2* (9), 816–822.

119. Kolb H-J, Schmid, C.; Barrett, A. J.; Schendel, D. J. Graft-versus-Leukemia Reactions in Allogeneic Chimeras. *Blood* **2004,** *103* (3), 767–776. DOI: 10.1182/blood-2003-02-0342
120. Meyer, R. G.; Britten, C. M.; Wehler, D.; Bender, K.; Hess, G.; Konur, A. et al. Prophylactic Transfer of CD8-Depleted Donor Lymphocytes After T-Cell-Depleted Reduced-Intensity Transplantation. *Blood* **2007,** *109* (1), 374–382. DOI: 10.1182/blood-2006-03-005769
121. Meyer, R. G.; Wagner, E. M.; Konur, A.; Bender, K.; Schmitt, T.; Hemmerling, J. et al. Donor CD4 T Cells Convert Mixed to Full Donor T-Cell Chimerism and Replenish the CD52-Positive T-Cell Pool After Alemtuzumab-Based T-Cell-Depleted Allo-Transplantation. *Bone Marrow Transplant* **2010,** *45* (4), 668–674.
122. Zhou, W.; Longmate, J.; Lacey, S. F.; Palmer, J. M.; Gallez-Hawkins, G.; Thao, L. et al. Impact of Donor CMV Status on Viral Infection and Reconstitution of Multifunction CMV-Specific T Cells in CMV-Positive Transplant Recipients. *Blood* **2009,** *113* (25), 6465–6476.
123. Ljungman, P.; Hakki, M.; Boeckh, M. Cytomegalovirus in Hematopoietic Stem Cell Transplant Recipients. *Hematol. Oncol. Clin. North Am.* **2011,** *25* (1), 151–169.
124. Boeckh, M.; Ljungman, P. How We Treat Cytomegalovirus in Hematopoietic Cell Transplant Recipients. *Blood* **2009,** *113* (23), 5711–5719.
125. Mori, T.; Mori, S.; Kanda, Y.; Yakushiji, K.; Mineishi, S.; Takaue, Y. et al. Clinical Significance of Cytomegalovirus (CMV) Antigenemia in the Prediction and Diagnosis of CMV Gastrointestinal Disease After Allogeneic Hematopoietic Stem Cell Transplantation. *Bone Marrow Transplant* **2003,** *33* (4), 431–434.
126. Eid, A. J.; Bakri, S. J.; Kijpittayarit, S.; Razonable, R. R. Clinical Features and Outcomes of Cytomegalovirus Retinitis After Transplantation. *Transpl. Infect. Dis.* **2008,** *10* (1), 13–18.
127. Travi, G.; Pergam, S. A. Cytomegalovirus Pneumonia in Hematopoietic Stem Cell Recipients. *J. Intensive Care Med.* **2014,** *29* (4), 200–212.
128. Ljungman, P.; Perez-Bercoff, L.; Jonsson, J.; Avetisyan, G.; Sparrelid, E.; Aschan, J. et al. Risk Factors for the Development of Cytomegalovirus Disease After Allogeneic Stem Cell Transplantation. *Haematologica* **2006,** *91* (1), 78–83.
129. Pourgheysari, B.; Piper, K. P.; McLarnon, A.; Arrazi, J.; Bruton, R.; Clark, F. et al. Early Reconstitution of Effector Memory CD4+ CMV-Specific T Cells Protects Against CMV Reactivation Following Allogeneic SCT. *Bone Marrow Transplant* **2008,** *43* (11), 853–861.
130. Gartner, B. C.; Schäfer, H.; Marggraff, K.; Eisele, G.; Schäfer, M.; Roemer, K. et al. Evaluation of Use of Epstein-Barr Viral Load in Patients After Allogeneic Stem Cell Transplantation to Diagnose and Monitor Posttransplant Lymphoproliferative Disease. *J. Clin. Microbiol.* **2002,** *40* (2), 351–358.
131. Peric, Z.; Cahu, X.; Chevallier, P.; Brissot, E.; Malard, F.; Guillaume, T. et al. Features of Epstein-Barr Virus (EBV) Reactivation After Reduced Intensity Conditioning Allogeneic Hematopoietic Stem Cell Transplantation. *Leukemia* **2011,** *25* (6), 932–938.
132. Uhlin, M.; Wikell, H.; Sundin, M.; Blennow, O.; Maeurer, M.; Ringden, O. et al. Risk factors for Epstein-Barr Virus-Related Post-Transplant Lymphoproliferative Disease After Allogeneic Hematopoietic Stem Cell Transplantation. *Haematologica* **2014,** *99* (2), 346–352.

133. Maschmeyer, G.; Ljungman, P. Infections in Hematopoietic Stem Cell Transplant Recipients. In: *Principles and Practice of Cancer Infectious Diseases*; Safdar, A., Ed.; Springer, 2011; pp 17–25.
134. Borchers, S.; Luther, S.; Lips, U.; Hahn, N.; Kontsendorn, J.; Stadler, M. et al. Tetramer Monitoring to Assess Risk Factors for Recurrent Cytomegalovirus Reactivation and Reconstitution of Antiviral Immunity Post Allogeneic Hematopoietic Stem Cell Transplantation. *Transpl. Infect. Dis.* **2011,** *13* (3), 222–236.
135. Chalandon, Y.; Degermann, S.; Villard, J.; Arlettaz, L.; Kaiser, L.; Vischer, S. et al. Pretransplantation CMV-Specific T Cells Protect Recipients of T-Cell-Depleted Grafts Against CMV-Related Complications. *Blood* **2006,** *107* (1), 389–396. DOI: 10.1182/blood-2005-07-2746
136. Borchers, S.; Bremm, M.; Lehrnbecher, T.; Dammann, E.; Pabst, B.; Wölk, B. et al. Sequential Anti-Cytomegalovirus Response Monitoring May Allow Prediction of Cytomegalovirus Reactivation After Allogeneic Stem Cell Transplantation. *PLoS One* **2012,** *7* (12), e50248.
137. Borchers, S.; Ogonek, J.; Varanasi, P. R.; Tischer, S.; Bremm, M.; Eiz-Vesper, B. et al. Multimer Monitoring of CMV-Specific T Cells in Research and in Clinical Applications. *Diagn. Microbiol. Infect. Dis.* **2014,** *78* (3), 201–212.
138. Odendahl, M.; Grigoleit, G. U.; Bönig, H.; Neuenhahn, M.; Albrecht, J.; Anderl, F. et al. Clinical-Scale Isolation of 'Minimally Manipulated' Cytomegalovirus-Specific Donor Lymphocytes for the Treatment of Refractory Cytomegalovirus Disease. *Cytotherapy* **2014,** *16* (9), 1245–1256. DOI: 10.1016/j.jcyt.2014.05.023
139. Kumaresan, P.; Figliola, M.; Moyes, J. S.; Huls, M. H.; Tewari, P.; Shpall, E. J. et al. Automated Cell Enrichment of Cytomegalovirus-Specific T Cells for Clinical Applications Using the Cytokine-Capture System. *J. Vis. Exp.* **2015,** (104), e52808. DOI: 10.3791/52808
140. Lacey, S. F.; Diamond, D. J.; Zaia, J. A. Assessment of Cellular Immunity to Human Cytomegalovirus in Recipients of Allogeneic Stem Cell Transplants. *Biol. Blood Marrow Transplant* **2004,** *10* (7), 433–447.
141. Tuthill, M.; Chen, F.; Paston, S.; De La Peña, H.; Rusakiewicz, S.; Madrigal, A. The Prevention and Treatment of Cytomegalovirus Infection in Haematopoietic Stem Cell Transplantation. *Cancer Immunol. Immunother.* **2009,** *58* (9), 1481–1488.
142. Gratama, J. W.; van Esser, J. W. J.; Lamers, C. H. J.; Tournay, C.; Löwenberg, B.; Bolhuis, R. L. H. et al. Tetramer-Based Quantification of Cytomegalovirus (CMV)-Specific CD8+ T Lymphocytes in T-Cell-Depleted Stem Cell Grafts and After Transplantation May Identify Patients at Risk for Progressive CMV Infection. *Blood* **2001,** *98* (5), 1358–1364.
143. Lilleri, D.; Fornara, C.; Chiesa, A.; Caldera, D.; Alessandrino, E. P.; Gerna, G. Human Cytomegalovirus-Specific CD4+ and CD8+ T-Cell Reconstitution in Adult Allogeneic Hematopoietic Stem Cell Transplant Recipients and Immune Control of Viral Infection. *Haematologica* **2008,** *93* (2), 248–256.
144. Imai, C.; Iwamoto, S.; Campana, D. Genetic Modification of Primary Natural Killer Cells Overcomes Inhibitory Signals and Induces Specific Killing of Leukemic Cells. *Blood* **2005,** *106* (1), 376–383.
145. Foster, A. E.; Gottlieb, D. J.; Sartor, M.; Hertzberg, M. S.; Bradstock, K. F. Cytomegalovirus-Specific CD4+ and CD8+ T-Cells Follow A Similar Reconstitution

Pattern After Allogeneic Stem Cell Transplantation. *Biol. Blood Marrow Transplant* **2002,** *8* (9), 501–511.
146. Altman, J. D.; Moss, P. A.; Goulder, P. J.; Barouch, D. H.; McHeyzer-Williams, M. G.; Bell, J. I. et al. Phenotypic Analysis of Antigen-Specific T Lymphocytes. *Science* **1996,** *274* (5284), 94–96.
147. Aubert, G.; Hassan-Walker, A. F.; Madrigal, J. A.; Emery, V. C.; Morte, C.; Grace, S. et al. Cytomegalovirus-Specific Cellular Immune Responses and Viremia in Recipients of Allogeneic Stem Cell Transplants. *J. Infect. Dis.* **2001,** *184* (8), 955–963.
148. Chen, F. E.; Aubert, G.; Travers, P.; Dodi, I. A.; Madrigal, J. A. HLA Tetramers and Anti-CMV Immune Responses: From Epitope to Immunotherapy. *Cytotherapy* **2002,** *4* (1), 41–48.
149. Lacey, S. F.; Villacres, M. C.; La Rosa, C.; Wang, Z.; Longmate, J.; Martinez, J. et al. Relative Dominance of HLA-B*07 Restricted CD8+ T-Lymphocyte Immune Responses to Human Cytomegalovirus pp65 in Persons Sharing HLA-A*02 and HLA-B*07 Alleles. *Hum. Immunol.* **2003,** *64* (4), 440–452.
150. Akiyama, Y.; Kuzushima, K.; Tsurumi, T.; Yamaguchi, K. Analysis of HLA-A24-Restricted CMVpp65 Peptide-Specific CTL with HLA-A*2402-CMVpp65 Tetramer. *Immunol. Lett.* **2004,** *95* (2), 199–205.
151. Bunde, T.; Kirchner, A.; Hoffmeister, B.; Habedank, D.; Hetzer, R.; Cherepnev, G. et al. Protection from Cytomegalovirus After Transplantation Is Correlated with Immediate Early 1-Specific CD8 T Cells. *J. Exp. Med.* **2005,** *201* (7), 1031–1036.
152. Lilleri, D.; Gerna, G.; Fornara, C.; Lozza, L.; Maccario, R.; Locatelli, F. Prospective Simultaneous Quantification of Human Cytomegalovirus-Specific CD4+ and CD8+ T-Cell Reconstitution in Young Recipients of Allogeneic Hematopoietic Stem Cell Transplants. *Blood* **2006,** *108* (4), 1406–1412.
153. Gratama, J. W.; Brooimans, R. A.; van der Holt, B.; Sintnicolaas, K.; van Doornum, G.; Niesters, H. G. et al. Monitoring Cytomegalovirus IE-1 and pp65-Specific CD4+ and CD8+ T-Cell Responses After Allogeneic Stem Cell Transplantation May Identify Patients at Risk for Recurrent CMV Reactivations. *Cytometry B Clin. Cytom.* **2008,** *74* (4), 211–220.

CHAPTER 2

Resistance Mechanisms in Cancer Immunology

ABSTRACT

Over the last decade, studies using animal models have helped illuminate the physiological barriers that inhibit an efficient antitumor immune response. These barriers consist of a complex set of processes that are all interconnected with the tumor and its surroundings as well as with several arms of the adaptive and innate immune systems. In this chapter, the author will look at the many ways tumors might evade an immune response, with a focus on therapeutically relevant strategies that aim to do so by targeting specific immunological checkpoints.

2.1 INTRODUCTION

The regulatory and organizational hurdles to building efficient antitumor immunotherapy have gotten increased attention in recent years, but there are still many biological barriers to overcome before immunotherapy approaches may expect to be very successful in people.[1–3] These immunologic obstacles were just recently recognized, and novel compensatory treatment was introduced, thanks to ground-breaking investigations in animal models. It is possible for many checkpoints to function simultaneously to impede the progression of anticancer immune responses.[4] Tumors may evade the immune system in two main ways: by becoming resistant to killing by immune effector cells and by inducing tolerance in the host as it grows.[5] Several processes in each of these classes have just recently begun to have their molecular basis clarified. The immune system may "ignore" a tumor if several factors come into play, including when the tumor itself suppresses

Advances in Immunological Treatment of Cancer: Cellular Mechanism and Regulation.
Rishabha Malviya, Arun Kumar Singh, and Deepika Yadav (Authors)

regulatory T cells or eliminates tumor-specific T cells; this is called tumor tolerance. Therapeutic tumor vaccination by itself seems like it should have far better results in clinical studies. There are few exceptions, but the great majority of mouse tumor investigations have shown indications of an active type of tumor-induced resistance.[6] Although large numbers of tumor-specific T cells stay in a dormant or repressed state, there is an indication that malignant cells T cells are being destroyed. To what extent this anergic/inactive state may be restored by certain drugs is a key question in the advancement of immunotherapy.[7]

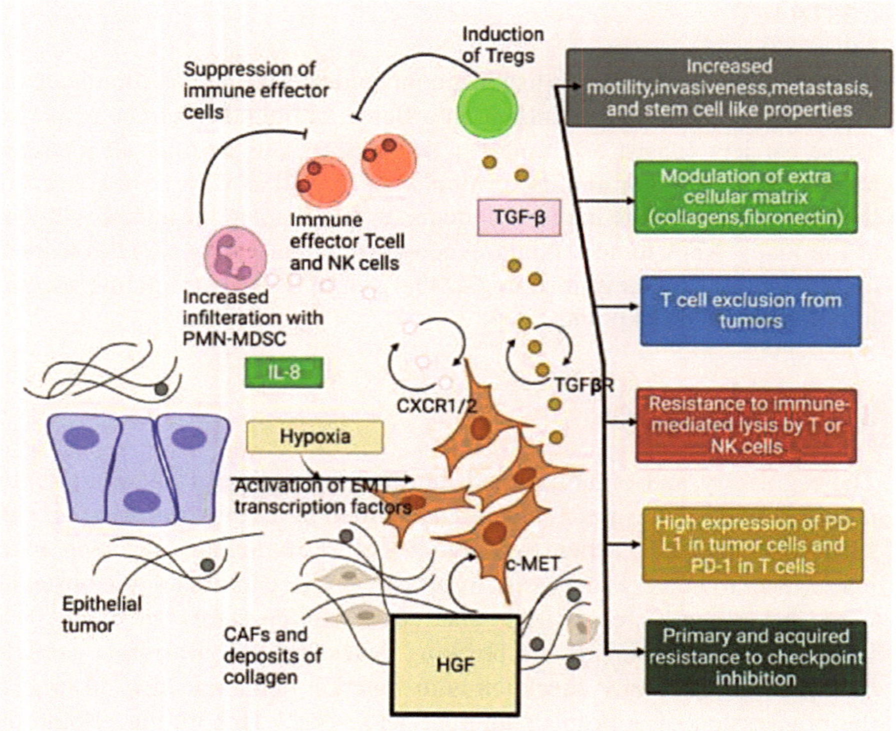

FIGURE 2.1 Schematic diagram of mechanism of action of supression of immune effector cells.

In this chapter, the author discusses the recently discovered T-cell-specific checkpoints to inherent immunity that serve as biological hurdles to efficient immunotherapy.

2.2 MUTATIONAL DOWNREGULATION OF ANTIGEN PRESENTATION IN TUMORS

The capacity of T cells to identify cancer-specific or tumor-related antigens is critical for the immune system to differentiate malignant tissues from healthy ones. It is generally known that cancer cells have a different antigenic makeup than normal cells.[8] Genetic instability is a characteristic of cancer development. Loss of function of genome protectors like p53 and/or inactivation of these proteins by mutations is the primary cause of genetic instability in malignancies. Tumors, as the malignant process develops, constantly create new antigens. In contrast, the antigenic profile of normal, nontransformed tissues is very constant.[9–11]

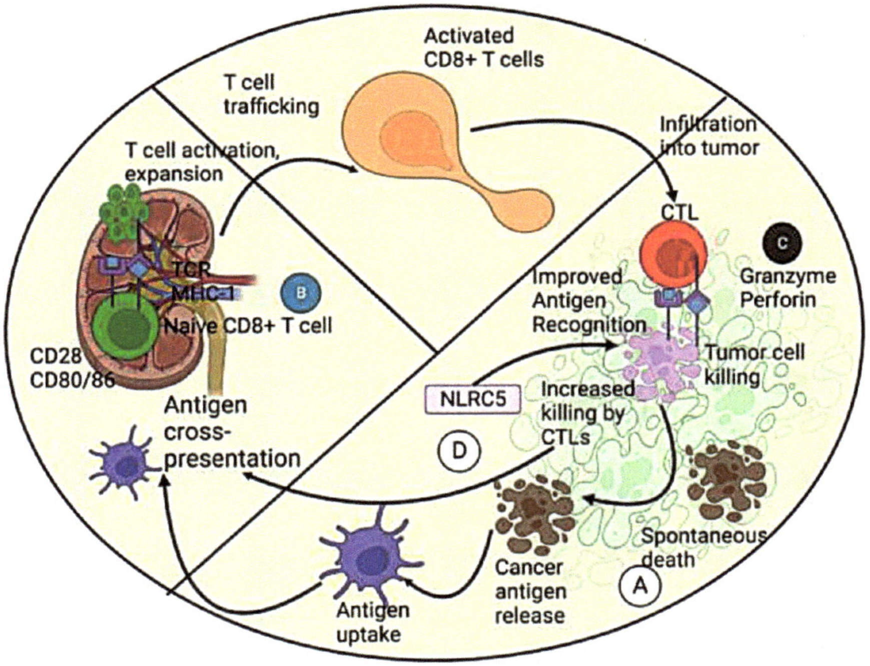

FIGURE 2.2 Diagramatic representation of mechanism of action of development of T-cell survival against the cancer.

Epigenetic abnormalities including DNA methylation changes (hyper- and hypomethylation) are common in cancer and have a significant impact on gene expression.[12] Epigenetic alterations have a substantial influence on the antigenicity of a tumor, even if they do not directly yield tumor-specific neoantigens.

Tumor innate immunity suppression was the first method of immune evasion found and researched.[13–15] Lack of MHC class I pathway or antigen processing machinery expression is diagnostic of several tumor types. Reduced expression of MHC class I molecules has been associated with many types of human cancer. Loss of specific HLA alleles[16] is hypothesized to be followed by a reduction in the generation of immunogenic tumor antigens, albeit this has never been shown. Decreased expression of major histocompatibility complex class I has also been associated with cancer (MHC). Most occurrences of MHC class I loss may be traced back to mutations that result in the deletion of b2 microglobulin genes. The deletion of one copy of the b2 microglobulin gene is the most common cause of mutations in this gene.[17,18] The reduced expression of MHC class I genes may have many transcriptional controllers. Genes encoding TAP and other immunoproteasome components, such as LMP2 and LMP7, seem to be downregulated in a broad range of tumors.[19] Since given quickly increases the MHC class I processing machinery, downregulated transcription is likely epigenetic and reversible.[20] Compared to the normal tissue they developed from, cancers often include more MHC class I molecules or processing equipment. While renal tumors express high quantities of surfaces MHC class I and low levels of TAP, normal renal epithelium does not unless prompted by gamma interferon (gIFN).[21–24]

Studies that have attempted to correlate MHC protein expression with either human clinical outcomes or tumor progression in animals have shown contradictory findings.[25] However, this might be because scientists have been studying so many different types of cancer and organs. According to human research, patients whose tumors generate MHC molecules tend to have worse results. According to the evidence, researchers have increased levels of HLA molecule expression which are associated with a favorable outcome. The progression of cervical cancer from precancerous lesions to malignant tumors is accompanied by a continuous downregulation of MHC class I. Since premalignant HPV lesions are cleared spontaneously in immunocompetent people but not in immunosuppressed people, it has been suggested that now the human papillomavirus (HPV) is essential for the genesis of cancer.[26]

Since NK cells pay more attention to and destroy cells with low MHC class I levels,[27] a tumor's downregulation of the MHC class I digesting mechanism would not necessarily signal that the tumor has succeeded in evading the immune system's identification. Suppressing MHC expression has been shown to inhibit the growth of several mouse tumors in vivo.[28–30] This may be because NK cells are better able to recognize the tumors once this has occurred. Still, increasing MHC expression has been hypothesized to slow the development of tumors in other species of animals. In conclusion, it is still unclear whether the observed change in MHC levels or antigen mechanical means throughout cancer is attributable to immune resistance development in response to a robust immune surveillance system. It is also debatable whether or not the elimination of tumor-associated antigens (TAA) is a way for tumors to evade immune monitoring. It is well known that TAA expression is very variable in animal tumor models, and researchers have tried to establish a connection between TAA loss and the transplantation of immunogenic tumors or vaccination.[31]

The researcher who treated patients with adoptive transfer of CD8+ T lymphocytes specific to melanoma antigens, also showed that melanoma antigens are lost in tumors that return after the first therapy. Patients with melanoma who get treatment with peptide vaccinations have anecdotal evidence showing a decrease in a certain antigen.[32] In conclusion, our findings support the concept that tumor antigen escape (TAA) is a powerful mechanism by which tumors may elude anticancer responses produced by immunotherapy. No convincing evidence that TAA loss represents a cancer breakout reaction to innate immunity in the whole host has been discovered, despite efforts to connect TAA loss with uncontrolled tumor formation in people (especially in melanoma).

2.2.1 TUMOR-INDUCED IMMUNE CHECKPOINT BLOCKADES

Although uncontrolled growth is a feature of all malignancies, metastasis and the ability to infiltrate neighboring tissues and spread are the fundamental pathophysiologic properties of cancer critical for morbidity and mortality.[33] Both these characteristics are indicative of substantial disruption in the tissue architecture and are never seen in normal tissues or benign tumors. Whether or not an infection is the source of tissue damage, the body will nonetheless respond with proinflammatory signals. Both the innate and adaptive immune responses rely on these signals, which may be produced as cytokines and chemokines.[34]

Since malignancies, unlike normal tissues, infiltrate and spread throughout the body, they are constantly confronted by a variety of inflammatory responses. How successfully they control and modify these responses[35] will define the nature of their interaction with the host immune system. Tissue damage brought on by invasive tumors may trigger an inflammatory response or an immune response. Malignancies, on the other hand, are frequently tolerated by the immune system after they have been established. Chronic inflammatory responses in the tumor microenvironment may promote cancer morphological change and even tumor formation, although they are not usually associated with tumor eradication.[36] There is substantial evidence that NFkB signaling in hematopoietic cells promotes carcinogenesis. In a carcinogen-induced colon cancer model, the growth and spread of tumors were stymied by the selective deletion of IKKb in macrophages and neutrophils. Nuclear factor kappa B (NFkB) is activated by hematopoietic invasion, leading to the production of a variety of proinflammatory cytokines and chemokines (such as tumor necrosis factor [TNF], interleukin [IL]-1, IL-6, CSF-1, type I interferons [IFNs],[37] and numerous chemokines [CSFs], such as IL-8). Even though numerous ideas have been suggested to explain the link between proinflammatory cytokines and chemokines and carcinogenesis, the specific processes by which these factors contribute to carcinogenesis remain unclear. Tumor necrosis factor (TNF) and other cytokines may cause an NFkB cascade to be activated in neighboring epithelial cells.[38] Extremely rarely, IL-6 and CSF-1 could promote expansion. Hematological cells like macrophages are stimulated in an NFkB-dependent manner, which in turn activates iNOS and the formation of reactive nitrogen species (i.e., NO). Utilizing the cytochrome oxidase and NADPH oxidase pathway, activated neutrophils produce reactive oxygen species (ROS).[39–41] When reactive oxygen species and nitrogen species combine, peroxynitrites form as a consequence. Exposure to any of these highly reactive chemicals may cause oxidative stress in tumor cells. However, whether or not the DNA damage response is operational, ROS and RNS may have conflicting impacts on cancer growth.[42]

A growing body of research suggests that inflammation inside the tumor microenvironment, as defined by invading hematopoietically derived cells, may be detrimental to tumor genesis and progression. A tumor's inflammatory response, however, does not seem to induce adaptive antitumor immunity,[43] in contrast to an inflammatory reaction in the context of an infection. In the next section, the author will utilize information from trials using TCR transgenic mice to show that tumor cells may become immune to their antigens.

Although there is strong evidence that antibody responses occur in animals with malignancies and in human patients with malignancies, T-cell tolerance seems to be the predominant mechanism of the action. T cells have been proven in many adoptive transfer tests to have a high capacity to eliminate growing tumors.[44] Damage of this kind may be directly induced by CTL activity, or other CD4-dependent effector pathways. For the tumor to survive, it seems natural that it would prioritize training T cells to ignore certain antigens. There is mounting evidence that tumors deliberately dampen danger signal production and sensing to infiltrate tissues and infect adjacent tissues without activating antitumor innate immunity, which would otherwise limit their growth.[45] This may cause alterations in inflammatory responses that aid tumor growth (Fig. 2.2). Oncogenic signaling pathways shape the tumor's immune milieu, which may give the tumor an inherent boost to growth and survival. Because of its capacity to suppress immune responses to tumors, the Stat3 pathway has received more attention than any other oncogenic signaling route.[46] It has been shown that Stat3 is constitutively activated in many different types of cancer, especially those of the hematological and epithelial lineages.[47] Tumor cells with Stat3 that are permanently active proliferate more quickly and are resistant to apoptosis. Dysregulated inflammatory responses, reduced DC activity, and poor T-cell immunity are all linked to cancers with chronic Stat3 activation.[48,49] Expression levels of cell cycle control, proangiogenic, and proapoptotic genes crucial to the process of transformation have been linked to the upregulation of Stat3 in tumor cells.

Studies have shown that activation of innate and adaptive immune surveillance components must be suppressed via gene control for aggressive, metastatic cancer to develop.[50] It appears that the Stat3 signaling pathway in tumor cells achieves this by reducing the production of proinflammatory surface receptors and increasing the expression of factors that inhibit the functional growth and development and multiplication of dendritic cells (DCs) as well as other cancer-derived hematopoietic-derived cells involved in immune function.[51] Tumor cells that had their Stat3 signaling inhibited produced much more of the proinflammatory cytokine genes, such as TNFa and IFNb, as well as the proinflammatory chemokine genes, IP10 and RANTES. The fact that blocking Stat3 led to the production of these chemicals even in the absence of triggering events demonstrates that Stat3 signaling restricts a tumor's innate propensity to generate them. The release of cytokines and chemokines by 3T3 cells in response to TLR agonists was downregulated by the activation of the Stat3 signaling pathway.

In addition, tumor microenvironment immunological tolerance promotion by bone marrow-derived DCs has been observed. DCs in animals with tumors and people with cancer are immature and unable to carry out their usual duties for similar reasons. Both the absence of "danger" signals required for DC activation and elements in the tumor microenvironment that hinder DC functional maturation may contribute to DC dysfunction in tumor-bearing mice. There is evidence that certain medications, including vascular endothelial growth factor (VEGF) and interleukin (IL) 10, inhibit functional DC production in culture by boosting Stat3 signaling. Reducing the number of immature DCs in the cancer microenvironment has also been linked to downregulating Stat3 activation in DC cell progenitors. Tumors do not generate novel physiological processes but rather dysregulate preexisting ones. Malfunctions in the Stat3 pathway have been associated with tumors and an abnormal wound-healing response.[52] It is generally known that when cells are injured, they release inflammatory cytokines that interfere with normal cell–cell communication and tissue organization. Wound healing is slowed by Stat3 inactivation. By specifically disrupting Stat3 in keratinocytes, the authors found that wound healing was retarded and inflammatory infiltrates were increased in animals. Activation of Stat3 serves a physiological role in the body and may explain why Stat3-activated cells can evade the immune system and other aspects of cancer biology. In addition to its action in tumor cells,[53] activation of Stat3 in tumor-infiltrating hematopoietic components seems to provide a key checkpoint for innate and adaptive anticancer responses.

To further understand Stat3 signaling in hematopoiesis, researchers used inducible hematopoietic deletion of Stat3 and detection for phosphor-Stat3 in cancer infiltrating cells. Phospho-Stat3 flow cytometric study in tumor-infiltrating NK cells and neutrophils reveals that Stat3 is also constitutively active in these cell types.[54] Inducing Stat3 deletion in the spleen decreased Stat3 signaling in neutrophils and increased their cytolytic activity toward target tumor cells. NK cells isolated from tumor-free control mice showed much less cytolytic activity than NK cells invading tumors from hematopoietic Stat3 ablated animals.[55] Stat3 may be activated by a wide variety of tumor-associated compounds, not only IL-10 from tumor-associated macrophages. A few more include interleukin-6 and vascular endothelial growth factor.[56,57]

The increased Stat3 activity within these populations in the tumor, as well as the improved function of Stat3-/- NK macrophages and cells from tumor-bearing animals, demonstrates that Stat3 signaling is involved in the downregulating purpose of these cell types at least in part due to cell-intrinsic factors.[58] Inherited hematopoietic cells invading tumors are unable to activate tumor-specific T lymphocytes because the Stat3 signaling pathway mediates their inhibition of innate responses or DC maturation.[59] Mice with Stat3 knocked out have T cells that are more responsive to tumor antigens. More T lymphocytes enter tumor sites and there are fewer regulatory T cells in mice with their Stat3 gene knocked off.[60] Blocking Stat3 signaling results in anticancer responses in vivo and has far-reaching repercussions for both innate and adaptive immune responses. Lack of Stat3 in hematopoietic cells before the tumor challenge dramatically decreased tumor growth, whereas subsequent removal of Stat3 from hematopoietic cells either slowed the progression of tumors or caused them to regress completely.[61] Since Stat3 signaling is crucial in both malignant and healthy hematopoietic cells, inhibiting it may enhance antitumor immunity. Inhibiting Stat3 with the small-molecule CPA7 was just discovered in 2016.[62] In vitro, treatment of cells with CPA7 causes a decrease in phosphoStat3 protein levels and Stat3 DNA binding activity within only a few hours. In mice treated with CPA7, phosphoStat3 levels were considerably lower in tumor-infiltrating DCs compared to DCs from vehicle-treated mice. It was demonstrated in vivo[63] that CPA7 treatment inhibited the growth of preexisting tumors, and that this effect was dependent on T cells and NK cells.

The indoleamine 2,3-dioxygenase that is produced is another regulator of T-cell irresponsibility in carcinogenesis (IDO). Both DCs inside tumors and those found in the LNs that drain tumors have been shown to generate IDO. B71/2 have been demonstrated to activate IDO in DCs through a reverse signaling route after binding to CTLA4. However, it appears that the majority of IDO in DCs comes from B220 plasmacytoid DCs and PDC-related cells. IDO's tryptophan breakdown appears to dampen T-cell reactions. A lack of tryptophan has been shown to increase the susceptibility of activated T cells to metabolic death.[64–66] Based on this observation, Munn and Mellor postulated a mechanism in which neighboring DCs deplete tryptophan by IDO overexpression, ultimately leading to the metabolic death of locally stimulated T cells. Many types of cells, including cancerous ones, secrete a molecule called transforming growth factor beta (TGFb). TGFb inhibits the body's antitumor immune responses, which is only one of its many pleiotropic physiological effects. When exposed to TGFb, a potent growth inhibitor, almost all normal epithelial cells stop dividing. However, many cancer cells develop a resistance to TGFb, which permits them to multiply uncontrolled. Cancer cells generate more TGFb, which may aid invasion by stimulating matrix metalloprotease activity.[67–69] Anti-TGFb antibodies may be able to block TGFb-induced angiogenesis in vivo because TGFb binds to its receptor. Transgenic animal models in which TGFb was overexpressed in keratinocytes showed the protein's dual role in carcinogenesis.

These animals show early signs of being resistant to developing superficial skin malignancies of a low or intermediate grade. Tumors are immediately recognizable as having an aggressive spindle cell nature once they have begun growing. Although increased blood TGFb levels are linked to a poor prognosis in several cancers,[70] identifying this specific bimodal pattern of activity in a clinical setting is more difficult.

Animals that lack TGFb are more susceptible to fatal inflammatory diseases and have a higher mortality rate because of the immune suppression that TGFb provides. Similar autoimmune symptoms are seen after targeted interruption of T-cell TGFb signaling,[71] showing that T cells are likely mediating the majority of these effects. New and persuasive experiments were conducted by the researcher, linking TGFb with Treg-mediated suppression of CD8 T-cell antitumor responses. The author discovered that anticancer CD8(+) T-cell effector response was suppressed by adoptive transfer of CD4(+) CD25(-) regulatory T cells,[72] but this suppression was much alleviated when the CD8(+) T cells were produced from mice with a dominant negative TGF(b1) receptor. A similar approach was used by Zhang et al. (2005), who transduced CD8 T cells ex vivo with a dominant negative TGFb receptor and then reintroduced them, by adoptive transfer, into mice with progressing endogenous prostate tumors. Tumor size and mortality are decreased in treated rats, which is consistent with the idea that TGFb contributes to tumor development by suppressing endogenous CD8 T-cell activity.[73] Furthermore, these findings are supported by CD8 T cells from melanoma patients who received an in vitro gp100 targeted vaccination. Similar to melanoma-targeting T-cell clones, the presence of TGFb in early antigen activation cultures of these cells dampened their ability to elicit an immune response. These results suggest that inhibiting TGFb with antibodies or medicines will be an important part of future combinatorial immunotherapy strategies.[74]

2.2.2 REDUCED ACTIVITY OF ANTIGEN-PRESENTING CELLS

Tumor immune evasion research is still unsure of how tumors establish antigen-specific T-cell tolerance.

Although it is known that mechanisms like Stat3 signaling regularly interact, IDO, ROS, RNS, TGFb, etc. inhibit T-cell response increasing pressure and/or tumor killing by enabled effector T cells, it is unclear which mechanisms actively enhance the antigen-specific T-cell tolerance observed in transgenic models.[75] It is thought that self-tolerance for antigens generated by peripheral tissues relies on the distribution of tissue-specific antigens to mature T lymphocytes in the absence of appropriate costimulatory signals. Tumor-induced tolerance most likely involves the same mechanisms. Costimulatory signals were formerly thought to be supplied by DC-produced molecules of the B7 class.

Proinflammatory cytokines, such as interferons, IL-12, tumor necrosis factor (TNF), etc. are necessary to distinguish between effector T-cell stimulation and tolerance induction.[76]

It is believed that self-antigen presentation by immature or incompletely formed DCs is essential for T-cell tolerance in the absence of TLR-mediated surface receptors associated with infection.

Tumor-resident DCs are immature and inactivated because they produce few inflammatory cytokines, display less CD86, and have little MHC class II on their surfaces.[77] It has been discovered that the Stat3 pathway strongly inhibits TLR and CD40-driven DC activation in tumor-infiltrating dendritic cells (DC). There is evidence that tumor-derived IL-10, IL-6, and VEGF may activate Stat3 in DCs (which is partially produced by Stat3 signaling in the tumor cell). It has been shown that Stat3 activation directly inhibits NFkB signaling in DCs, either at the IKK level or farther downstream in the nucleus.[78] Tumor microenvironments may be able to dampen DC activation if tumors need more oncogenic signaling. Melanoma cells, which are caused by activated mutations of the BRAF gene and account for approximately 60% of melanomas, have recently been identified to secrete chemicals that prevent the activation of scavenger DCs (sDCs). These "activation-inhibited" immature splenic DCs may play a role in fostering tolerance in tumor-specific T cells.[79] Although iMC/MSC may offer tolerogenic antigens to T cells, it is still debatable whether or not they represent a distinct population of intertumoral cells. Adoptive transfer of antigen-loaded iMC into mice has been demonstrated in recent studies to have the capacity to establish antigen-specific T-cell tolerance. Therefore, IDO-expressing DC could establish antigen-specific T-cell tolerance by causing the selective death or reduction of proliferation of activated T cells through IDO-mediated tryptophan. Antigen-presenting DCs expressing IDO are hypothesized to convey antigen to T lymphocytes, which then activate and undergo activation-associated cell death due to a shortage of local tryptophan reserves. T-cell tolerance to tumor antigens is discussed further in the next section, which highlights the role of regulatory T cells in achieving and sustaining this state. It is not entirely apparent how Treg cells and immature or toleragenic APCs are related to one another[80]; for example, whether antigen-specific T cells are induced to adopt a Treg phenotype by toleragenic DCs or if Treg cells operate upon DCs to enhance their toleragenic potential.

2.2.3 RESISTANCE OF CD4 T CELLS

Over time, the immune system has refined a number of mechanisms for blocking CD4+ T-cell-mediated autoimmune responses. The CD4+CD8+ double-positive stage is where the development of autoreactive T cells has gotten the most study. Bogen et al. demonstrated comparable results in a myeloma animal model. Myeloma immunotherapy in mice was achieved by genetic engineering by introducing a T-cell receptor that recognizes a myeloma protein. Mice genetically modified to express the T-cell receptor are very effective in responding to and rejecting tumors as small as a few millimeters in size.[81]

However, tumor formation is seen when the initial implant is greater than 0.5 cm. Loss of T cells happened in these animals during the twofold positive ontogenetic stage, which is consistent with results from conventional thymic deletion models. A lack of sufficient costimulatory signals prevents the T-cell receptor from reviving after its interaction with the MHC antigen–peptide complex, leading to energy.[82] Tolerance is achieved via this second, perhaps more nuanced method. A number of in vitro studies have shown that T-cell energy may lead to tolerance of malignancies. To investigate its potential as a new tumor antigen, influenza hemagglutinin (HA) peptide was stably transfected into A20 lymphoma cells in one model. The authors predicted strong immunological rejection when HA-specific T cells were transplanted into mice with expanding A20 (HA) lymphomas (and hoped for). In fact, after being transplanted into a recipient, specific T cells quickly became anergic and lost their ability to multiply in vitro in response to a specific antigen.[83] Interestingly enough, the kind of adoptively transplanted T cells had no effect on the rate at which A20 (HA) tumors formed in mice. Antigen-specific T lymphocytes may be primed in one of two ways: either by the target cell type directly, via the production of Class I antigen complexes, or by expert antigen-presenting cells generated from bone marrow. The majority of tumor forms, it seems, present primarily by cross presentation. However, as has been pointed out by a number of study groups, this cross priming process is not particularly effective, which might result in the host immune system "ignoring" growing tumors despite their microscopic size. The host may have already launched an anticancer response by the time cancer had progressed to the lymph nodes draining the initial tumor.[84]

2.2.4 REGULATORY T CELLS

Characterizing regulatory T cells and expanding families of coinhibitory and costimulatory molecules has received a lot of interest in the fields of tumor immunity and viral immunology (Treg). Our interest in this field was aroused by the finding by the researcher that CD4 T cells lacking the CD25 protein may prevent the onset of autoimmunity after neonatal thymectomy. Ultimately, a number of models of autoimmunity, infection, and cancer were used to validate the regulatory capability of thymic CD4+CD25 T cells. Studies examining the presence of CD4+CD25+ regulatory T cells in the ascites fluid of women with advanced ovarian cancer and its connection with shorter survival have provided the strongest evidence to date for a role for these cells in human cancer. Ex vivo research further verified the regulatory role of Tregs. Treg have now been identified in cancer tissue samples from several sites, including the lung, pancreas, ovary, and breast.[85]

At now, there is no foolproof way to isolate Treg by the use of a cell surface marker. Since CD25 (the alpha chain of the interleukin-2 receptor) is produced by both engaged and regulatory T cells, CD4 CD25 T cells serve as a good marker for differentiating between the two types of T cells. Glucocorticoid-inducible TNF receptor (GITR), neuropilin, and monocytes activation gene 3 are all possible markers of regulatory T cells (Lag-3). Without a distinct cell surface marker in model systems, it is challenging to conceive of Treg-specific therapeutic techniques.[86] Absolute specificity may not be necessary, as shown by the observation that the systemic inhibition of CD4 CD25 T cells using antibodies specific (PC61 in the animal) or a CD25 targeted toxin (Ontac) enhances a specific immune response. Robert North undertook pioneering work in the late 1970s when he discovered that low doses of the popular chemotherapeutic medication cyclophosphamide appeared to selectively reduce cells with regulatory activity. As a result of this discovery, a new approach to inhibiting Treg's unique roles has become possible. The significance of this effect has long been recognized in clinical settings, and new studies using murine models of prostate and breast cancer have helped understand the mechanism behind this effect. This may be the most fascinating findings so far on cyclophosphamide's participation in a tumor model, since it was discovered by Ercolini et al. that immunoregulatory doses of cyclophosphamide seemed to selectively expose a high affinity population of tumor-specific CD8 T cells. Low-dose cyclophosphamide has been found to be successful in the past and provides no risk of lymphopenia, so it may be a valuable complement to anticancer immunotherapy strategies.[87]

2.2.5 DEGRADATION OF CD8 T CELLS

Endogenous tumor-specific CD8 T cells have been exposed to cancer antigens for a long time since tumors tend to be present for years before becoming clinically apparent. Chronic viral models have been used to study several long-lived antigens.[88]

Repeated analyses of this phenomenon demonstrate that CD8 T-cell function is compromised in these states. Antigen-specific CD8 T cells are not destroyed but instead they become stuck in a limbo between the CD62L low effector remember phenotype and the CD62L high usually consists phenotype. Fatigue may further reduce the CD8 T-cell response to lingering tumor antigens. Insight into CD8 T-cell exhaustion provides hope for immunotherapy since it suggests that tumor-specific CD8 T cells may be present and partly primed even in the tumor-bearing host. Researchers found that CD8 T cells isolated from tumor-bearing hosts might potentially restore function in tumor-free mice, providing protection against tumor challenge, suggesting that the removal of persistent antigen could have far-reaching implications. These results are consistent with the groundbreaking research from over 20 years ago by Ramsdell and Fowlkes,[89] which demonstrated that removing a persistent antigen might restore T-cell activity in a live creature. Although these findings may be challenging to implement in clinical practise, they provide support for the hypothesis that individuals with minimal lingering symptoms may benefit most from immunotherapy therapies. Patients with chemosensitive tumors, such as hormone-responsive prostate cancer and small cell lung cancer (SCLC), offer promising groups for therapeutic research due to their low residual disease state. The in vivo response of tumor-specific CD8 T lymphocytes has been studied using animal models of insulinoma. Researchers showed that injecting tumor-specific CD8 T cells and enabling them to develop into an effector phenotype might induce protection in vivo. Adoptively transplanted CD8 T cells were shown to grow and acquire cytolytic activity in vivo by Lyman et al. using a CD8 T-cell transgenic donor. Our results imply that CD8 T-cell tolerance of tumors may be less strong than CD4 T-cell tolerance, and that low-affinity tumor-specific T cells may be able to compensate for a decline in their in vivo cytolytic efficiency with the help of CD4 T-cells, if provided adequate support.[90]

After being activated by antigen presentation cells, CD8 T lymphocytes rapidly gain full effector function by lysing target cells via granzyme- and perforin-mediated processes. Evidence suggests that CD8 T lymphocytes are more effective in killing off their targets than antigen-presenting cells. Serine protease inhibitor (SPI)6, a serpin family member that selectively inactivates granzyme B and suppresses CTL-induced apoptosis, is a recently discovered mechanism for this protection offered by activated antigen presentation cells. Recent studies demonstrate that SPI6 is expressed in colon cancer cell lines that are resistant to CTL-mediated death, in addition to the crucial serpin SPICI (serine protease inhibitor implicated in cytotoxicity inhibition).[91] Despite their recency, these results suggest a unique strategy cancers may use to evade death at the hands of activated CD8 T cells. The degree to which this pathway suppresses an anticancer immune response in practise may be elucidated by future investigations comparing the relative expression of serpins in different tumor types.

2.3 CONCLUSION

There has been a recent upsurge in the molecular identification of immunologic checkpoints that may impede anticancer immune responses, albeit the relative impact of overcoming any single checkpoint in vivo is yet unclear. Using model systems where a very little change may have a major influence on the result of an otherwise ineffectual treatment approach is becoming the norm. Determining which checkpoint intervention strategies are most effective will be an important research priority over the next years, requiring studies on both laboratory models and actual patients.

Combinatorial approaches for assessing the blocking of several checkpoints bring new obstacles in clinical development.

In this regard, hormone treatment, surgery, radiation, or conventional chemotherapy may constitute the first step of a combined approach for reducing tumor antigen levels. Then, tumor antigen processing and presentation regulator proteins including Stat3 and transforming growth factor beta (TGFb) could be muted. Afterward, a Phase III trial will most likely be conducted on one of the numerous viable vaccination methods. The priming phase of a T-cell response is regulated by checkpoints such as CTLA4, which vaccines may interfere with. Checkpoints like B7-H1 that affect CD8 T-cell activation in the presence of a tumor may now be amenable to targeted inhibition. For a lab-based immunologist, a sequential and combinatorial strategy like this may seem inconceivable, but in reality, far more complicated therapy regimens are frequently used to combat human cancer. Immunological treatments may need to reach this degree of intricacy to be really effective.

KEYWORDS

- **cancer immunology**
- **resistance pathway**
- **checkpoint blockers**
- **T cell**
- **anticancer**

REFERENCES

1. Ansari, M. J.; Salama, A. D.; Chitnis, T.; Smith, R. N.; Yagita, H.; Akiba, H.; Yamazaki, T.; Azuma, M.; Iwai, H.; Khoury, S. J.; Auchincloss, H.; Jr.; Sayegh, M. H. The Programmed Death-1 (PD-1) Pathway Regulates Autoimmune Diabetes in Non-obese Diabetic (NOD) Mice. *J. Exp. Med.* 2003, *198* (1), 63–69.
2. Baban, B.; Hansen, A. M.; Chandler, P. R.; Manlapat, A.; Bingaman, A.; Kahler, D. J.; Munn, D. H.; Mellor, A. L. A Minor Population of Splenic Dendritic Cells Expressing CD19 Mediates IDO-Dependent T Cell Suppression via Type I IFN Signaling Following B7 Ligation. *Int. Immunol.* 2005, *17* (7), 909–919.
3. Bakkenist, C. J.; Kastan, M. B. Initiating Cellular Stress Responses. *Cell* **2004,** *118* (1), 9–17.
4. Benitez, R.; Godelaine, D.; Lopez-Nevot, M. A.; Brasseur, F.; Jimenez, P.; Marchand, M.; Oliva, M. R.; van Baren, N.; Cabrera, T.; Andry, G.; Landry, C.; Ruiz-Cabello, F. et al. Mutations of the Beta2-Microglobulin Gene Result in a Lack of HLA Class I Molecules on Melanoma Cells of Two Patients Immunized with MAGE Peptides. *Tissue Antigens* **1998,** *52* (6), 520–529.
5. Berd, D.; Mastrangelo, M. J.; Engstrom, P. F.; Paul, A.; Maguire, H. Augmentation of the Human Immune Response by Cyclophosphamide. *Cancer Res.* 1982, *42* (11), 4862–4866.

6. Berd, D.; Maguire, H. C., Jr.; Mastrangelo, M. J. Induction of Cell-Mediated Immunity to Autologous Melanoma Cells and Regression of Metastases After Treatment with a Melanoma Cell Vaccine Preceded by Cyclophosphamide. *Cancer Res.* 1986, *46* (5), 2572–2577.
7. Blackman, M. A.; Lund, F. E.; Surman, S.; Corley, R. B.; Woodland, D. L. Major histocompatibility Complex-Restricted Recognition of Retroviral Superantigens by V Beta 17þ T Cells. *J. Exp. Med.* 1992, *176* (1), 275–280.
8. Blanchet, O.; Bourge, J. F.; Zinszner, H.; Israel, A.; Kourilsky, P.; Dausset, J.; Degos, L.; Paul, P. Altered Binding of Regulatory Factors to HLA Class I Enhancer Sequence in Human Tumor Cell Lines Lacking Class I Antigen Expression. *Proc. Natl. Acad. Sci. USA* **1992,** *89* (8), 3488–3492.
9. Blobe, G. C.; Schiemann, W. P.; Lodish, H. F. Role of Transforming Growth Factor Beta in Human Disease. *N. Engl. J. Med.* 2000, *342* (18), 1350–1358.
10. Bogen, B.; Munthe, L.; Sollien, A.; Hofgaard, P.; Omholt, H.; Dagnaes, F.; Dembic, Z.; Lauritzsen, G. F. Naive CD4þ T Cells Confer Idiotype-Specific Tumor Resistance in the Absence of Antibodies. *Eur. J. Immunol.* 1995, *25* (11), 3079–3086.
11. Salama, A. D.; Chitnis, T.; Imitola, J.; Ansari, M. J.; Akiba, H.; Tushima, F.; Azuma, M.; Yagita, H.; Sayegh, M. H.; Khoury, S. J. Critical Role of the programmed Death-1 (PD-1) Pathway in Regulation of Experimental Autoimmune Encephalomyelitis. *J. Exp. Med.* 2003, *198* (1), 71–78.
12. Bonifaz, L.; Bonnyay, D.; Mahnke, K.; Rivera, M.; Nussenzweig, M. C.; Steinman, R. M. Efficient Targeting of Protein Antigen to the Dendritic Cell Receptor DEC-205 in the Steady State Leads to Antigen Presentation on Major Histocompatibility Complex Class I Products and Peripheral CD8þ T Cell Tolerance. *J. Exp. Med.* 2002, *196* (12), 1627–1638.
13. Bots, M.; Kolfschoten, I. G.; Bres, S. A.; Rademaker, M. T.; de Roo, G. M.; Kruse, M.; Franken, K. L.; Hahne, M.; Froelich, C. J.; Melief, C. J.; Offringa, R.; Medema, J. P. SPI-CI and SPI-6 Cooperate in the Protection from Effector Cell-Mediated Cytotoxicity. *Blood* **2005,** *105* (3), 1153–1161.
14. Bowman, T.; Broome, M. A.; Sinibaldi, D.;Wharton, W.; Pledger, W. J.; Sedivy, J.M.; Irby, R.; Yeatman, T.; Courtneidge, S. A.; Jove, R. Stat3-Mediated Myc Expression Is Required for Src Transformation and PDGF-Induced Mitogenesis. *Proc. Natl. Acad. Sci. USA* **2001,** *98* (13), 7319–7324.
15. (a) Bretschen, P.; Cohn, M. A Theory of Self-Nonself Discrimination. *Science* **1970,** *169*, 1042–1049. (b) Bromberg, J. F.; Wrzeszczynska, M. H.; Devgan, G.; Zhao, Y.; Pestell, R. G.; Albanese, C.; Darnell, Bronte, J. E., Jr. Stat3 as an Oncogene. *Cell* **1999,** *98* (3), 295–303.
16. Bronte, V.; Apolloni, E.; Cabrelle, A.; Ronca, R.; Serafini, P.; Zamboni, P.; Restifo, N. P.; Zanovello, P. Identification of a CD11b (þ)/Gr-1 (þ)/CD31 (þ) Myeloid Progenitor Capable of Activating or Suppressing CD8 (þ) T Cells. *Blood* **2000,** *96* (12), 3838–3846.
17. Bronte, V.; Serafini, P.; De Santo, C.; Marigo, I.; Tosello, V.; Mazzoni, A.; Segal, D. M.; Staib, C.; Lowel, M.; Sutter, G.; Colombo, M. P.; Zanovello, P. IL-4-Induced Arginase 1 Suppresses Alloreactive T Cells in Tumor-Bearing Mice. *J. Immunol.* 2003, *170* (1), 270–278.
18. Bruder, D.; Probst-Kepper, M.; Westendorf, A. M.; Geffers, R.; Beissert, S.; Loser, K.; von Boehmer, H.; Buer, J.; Hansen, W. Neuropilin-1: A Surface Marker of Regulatory T Cells. Eur. *J. Immunol.* 2004, *34* (3), 623–630.

19. Dong, H.; Zhu, G.; Tamada, K.; Chen, L. B7-H1, a Third Member of the B7 Family, Co- Stimulates T-Cell Proliferation and Interleukin-10 Secretion. *Nat. Med.* 1999, *5* (12), 1365–1369.
20. Dong, H.; Strome, S. E.; Salomao, D. R.; Tamura, H.; Hirano, F.; Flies, D. B.; Roche, P. C.; Lu, J.; Zhu, G.; Tamada, K.; Lennon, V. A.; Celis, E. et al. Tumor-Associated B7-H1 Promotes T-Cell Apoptosis: A Potential Mechanism of Immune Evasion. *Nat. Med.* 2002, *8* (8), 793–800.
21. Doyle, A.; Martin, W. J.; Funa, K.; Gazdar, A.; Carney, D.; Martin, S. E.; Linnoila, I.; Cuttitta, F.; Mulshine, J.; Bunn, P.; Minna, J. Markedly Decreased Expression of Class I Histocompatibility Antigens, Protein, and mRNA in Human Small-Cell Lung Cancer. *J. Exp. Med.* 1985, *161* (5), 1135–1151.
22. D'Urso, C. M.; Wang, Z. G.; Cao, Y.; Tatake, R.; Zeff, R. A.; Ferrone, S. Lack of HLA Class I Antigen Expression by Cultured Melanoma Cells FO-1 Due to a Defect in B2m Gene Expression. *J. Clin. Invest.* 1991, *87* (1), 284–292.
23. Young, M. R.; Petruzzelli, G. J.; Kolesiak, K.; Achille, N.; Lathers, D. M.; Gabrilovich, D. I. Human Squamous Cell Carcinomas of the Head and Neck Chemoattract Immune Suppressive CD34 (þ) Progenitor Cells. *Hum. Immunol.* 2001, *62* (4), 332–341.
24. Esteban, F.; Concha, A.; Huelin, C.; Perez-Ayala, M.; Pedrinaci, S.; Ruiz-Cabello, F.; Garrido, F. Histocompatibility Antigens in Primary and Metastatic Squamous Cell Carcinoma of the Larynx. *Int. J. Cancer* **1989,** *43* (3), 436–442.
25. Esteban, F.; Concha, A.; Delgado, M.; Perez-Ayala, M.; Ruiz-Cabello, F.; Garrido, F. Lack of MHC Class I Antigens and Tumour Aggressiveness of the Squamous Cell Carcinoma of the Larynx. *Br. J. Cancer* **1990,** *62* (6), 1047–1051.
26. Esteban, F.; Redondo, M.; Delgado, M.; Garrido, F.; Ruiz-Cabello, F. MHC Class I Antigens and Tumour-Infiltrating Leucocytes in Laryngeal Cancer: Long-Term Follow-Up. *Br. J. Cancer* **1996,** *74* (11), 1801–1804.
27. Fearon, E. R.; Vogelstein, B. A Genetic Model for Colorectal Tumorigenesis. *Cell* **1990,** *61* (5), 759–767.
28. Ferrone, S.; Marincola, F. M. Loss of HLA Class I Antigens by Melanoma Cells: Molecular Mechanisms, Functional Significance and Clinical Relevance. *Immunol. Today* **1995,** *16* (10), 487–494.
29. (a) Finn, O. J. Cancer Vaccines: Between the Idea and the Reality. *Nat. Rev. Immunol.* **2003,** *3* (8), 630–641. Gorelik, L.; Flavell, R. A. Abrogation of TGFbeta Signaling in T Cells Leads to Spontaneous T Cell Differentiation and Autoimmune Disease. *Immunity* **2000,** *12* (2), 171–181.
30. Greten, F. R.; Eckmann, L.; Greten, T. F.; Park, J. M.; Li, Z. W.; Egan, L. J.; Kagnoff, M. F.; Karin, M. IKKbeta Links Inflammation and Tumorigenesis in a Mouse Model of Colitis- Associated Cancer. *Cell* **2004,** *118* (3), 285–296.
31. Halak, B. K.; Maguire, H. C.; Jr.; Lattime, E. C. Tumor-Induced Interleukin-10 Inhibits Type 1 Immune Responses Directed at a Tumor Antigen as Well as a Non-Tumor Antigen Present at the Tumor Site. *Cancer Res.* 1999, *59*, 911–917.
32. Hasegawa, Y.; Takanashi, S.; Kanehira, Y.; Tsushima, T.; Imai, T.; Okumura, K. Transforming Growth Factor-Beta1 Level Correlates with Angiogenesis, Tumor Progression, and Prognosis in Patients with Nonsmall Cell Lung Carcinoma. *Cancer* **2001,** *91* (5), 964–971.
33. Hatachi, S.; Iwai, Y.; Kawano, S.; Morinobu, S.; Kobayashi, M.; Koshiba, M.; Saura, R.; Kurosaka, M.; Honjo, T.; Kumagai, S. CD4þ PD-1þ T Cells Accumulate as Unique

Anergic Cells in Rheumatoid Arthritis Synovial Fluid. *J. Rheumatol.* 2003, *30* (7), 1410–1419.

34. Hawiger, D.; Inaba, K.; Dorsett, Y.; Guo, M.; Mahnke, K.; Rivera, M.; Ravetch, J. V.; Steinman, R. M.; Nussenzweig, M. C. Dendritic Cells Induce Peripheral T Cell Unresponsiveness Under Steady State Conditions In Vivo. *J. Exp. Med.* 2001, *194* (6), 769–779.
35. Ishida, Y.; Agata, Y.; Shibahara, K.; Honjo, T. Induced Expression of PD-1, a Novel Member of the Immunoglobulin Gene Superfamily, Upon Programmed Cell Death. *EMBO J.* 1992, *11* (11), 3887–3895.
36. Jager, E.; Ringhoffer, M.; Altmannsberger, M.; Arand, M.; Karbach, J.; Jager, D.; Oesch, F.; Knuth, A. Immunoselection In Vivo: Independent Loss of MHC class I and Melanocyte Differentiation Antigen Expression in Metastatic Melanoma. *Int. J. Cancer* **1997,** *71* (2), 142–147.
37. (a) Jones, P. A.; Baylin, S. B. The Fundamental Role of Epigenetic Events in Cancer. *Nat. Rev. Genet.* 2002, *3* (6), 415–428. (b) Kappler, J. W.; Roehm, N.; Marrack, P. T Cell Tolerance by Clonal Elimination in the Thymus. *Cell* **2002,** *49* (2), 273–280.
38. (a) Karin, M.; Greten, F. R. NF-kappaB: Linking Inflammation and Immunity to Cancer Development and Progression. *Nat. Rev. Immunol.* 2005, *5* (10), 749–759. (b) Karre, K.; Ljunggren, H. G.; Piontek, G.; Kiessling, R. Selective Rejection of H-2- Deficient Lymphoma Variants Suggests Alternative Immune Defence Strategy. *Nature* **1986,** *319* (6055), 675–678.
39. Klein, S. C.; Ljunggren, H. G.; Piontek, G.; Kiessling, R. IL6 and IL6 Receptor Expression in Burkitt's Lymphoma and Lymphoblastoid Cell Lines: Promotion of IL6 Receptor Expression by EBV. *Hematol. Oncol.* 1995, *13* (3), 121–130.
40. Koopman, L. A.; Corver, W. E.; van der Slik, A. R.; Giphart, M. J.; Fleuren, G. J. Multiple Genetic Alterations Cause Frequent and Heterogeneous Human Histocompatibility Leukocyte Antigen Class I Loss in Cervical Cancer. *J. Exp. Med.* 2000, *191* (6), 961–976.
41. Kortylewski, M.; Kujawski, M.; Wang, T.; Wei, S.; Zhang, S.; Pilon-Thomas, S.; Niu, G.; Kay, H.; Mule, J.; Kerr, W. G.; Jove, R.; Pardoll, D. et al. Inhibiting Stat3 Signalling in the Hematopoietic System Elicits Multicomponent Antitumor Immunity. *Nature Med.* 2005, *11*, 1314–1321.
42. Haywood, G. R.; McKhann, C. F. Antigenic Specificities on Murine Sarcoma Cells. Reciprocal Relationship Between Normal Transplantation Antigens (H-2) and Tumor-Specific Immunogenicity. *J. Exp. Med.* 1971, *133* (6), 1171–1187.
43. Hilders, C. G.; Houbiers, J. G.; Krul, E. J.; Fleuren, G. J. The Expression of Histocompatibility-Related Leukocyte Antigens in the Pathway to Cervical Carcinoma. *Am. J. Clin. Pathol.* 1994, *101* (1), 5–12.
44. Hirano, F.; Kaneko, K.; Tamura, H.; Dong, H.; Wang, S.; Ichikawa, M.; Rietz, C.; Flies, D. B.; Lau, J. S.; Zhu, G.; Tamada, K.; Chen, L. Blockade of B7-H1 and PD-1 by Monoclonal Antibodies Potentiates Cancer Therapeutic Immunity. *Cancer Res.* 2005, *65* (3), 1089–1096.
45. Huang, A. Y.; Golumbek, P.; Ahmadzadeh, M.; Jaffee, E.; Pardoll, D.; Levitsky, H. Role of Bone Marrow-Derived Cells in Presenting MHC Class I-Restricted Tumor Antigens. *Science* **1994,** *264* (5161), 961–965.
46. Huang, C. T.; Workman, C. J.; Flies, D.; Pan, X.; Marson, A. L.; Zhou, G.; Hipkiss, E. L.; Ravi, S.; Kowalski, J.; Levitsky, H. I.; Powell, J. D.; Pardoll, D. M. et al. Role of LAG-3 in Regulatory T Cells. *Immunity* **2004,** *21* (4), 503–513.

47. Hugo, P.; Boyd, R. L.; Waanders, G. A.; Petrie, H. T.; Scollay, R. Timing of Deletion of Autoreactive V Beta 6þ Cells and Down-Modulation of Either CD4 or CD8 on Phenotypically Distinct CD4þ8þ Subsets of Thymocytes Expressing Intermediate or High Levels of T Cell Receptor. *Int. Immunol.* 1991, *3* (3), 265–272.
48. Hui, K.; Grosveld, F.; Festenstein, H. Rejection of Transplantable AKR Leukaemia Cells Following MHC DNA-Mediated Cell Transformation. *Nature* **1984,** *311* (5988), 750–752.
49. Hurchla, M. A.; Sedy, J. R.; Gavrieli, M.; Drake, C. G.; Murphy, T. L.; Murphy, K. M. B and T Lymphocyte Attenuator Exhibits Structural and Expression Polymorphisms and Is Highly Induced in Anergic CD4þ T Cells. *J. Immunol.* 2005, *174* (6), 3377–3385.
50. Lanier, L. L.; Phillips, J. H. Inhibitory MHC Class I Receptors on NK Cells and T Cells. *Immunol. Today* **1996,** *17* (2), 86–91.
51. Laping, N. J.; Grygielko, E.; Mathur, A.; Butter, S.; Bomberger, J.; Tweed, C.; Martin, W.; Fornwald, J.; Lehr, R.; Harling, J.; Gaster, L.; Callahan, J. F. et al. Inhibition of Transforming Growth Factor (TGF)-Beta1-Induced Extracellular Matrix with a Novel Inhibitor of the TGF-Beta Type I Receptor Kinase Activity: SB-431542. *Mol. Pharmacol.* 2002, *62* (1), 58–64.
52. Lauritzsen, G. F.; Hofgaard, P. O.; Schenck, K.; Bogen, B. Clonal Deletion of Thymocytes as a Tumor Escape Mechanism. *Int. J. Cancer* **1998,** *78* (2), 216–222. Lechner, O.; Lauber, J.; Franzke, A.; Sarukhan, A.; von Boehmer, H.; Buer, J. Fingerprints of Anergic T Cells. *Curr. Biol.* 2001, *11* (8), 587–595.
53. Lethe, B.; van den, E. B.; Van Pel, A.; Corradin, G.; Boon, T. Mouse Tumor Rejection Antigens P815A and P815B: Two Epitopes Carried by a Single Peptide. *Eur. J. Immunol.* 1992, *22* (9), 2283–2288.
54. (a) Letterio, J. J.; Roberts, A. B. Regulation of Immune Responses by TGF-BETA. *Annu. Rev. Immunol.* 1998, *16*, 137–161. (b). Lin, E. Y.; Gouon-Evans, V.; Nguyen, A. V.; Pollard, J. W. The Macrophage Growth Factor CSF-1 in Mammary Gland Development and Tumor Progression. *J. Mammary Gland Biol. Neoplasia* **2002,** *7* (2), 147–162.
55. Lind, M. H.; Rozell, B.; Wallin, R. P.; van Hogerlinden, M.; Ljunggren, H. G.; Toftgard, R.; Sur, I. Tumor Necrosis Factor Receptor 1-Mediated Signaling Is Required for Skin Cancer Development Induced by NF-kappaB Inhibition. *Proc. Natl. Acad. Sci. USA* **2004,** *101* (14), 4972–4977.
56. Liyanage, U. K.; Moore, T. T.; Joo, H. G.; Tanaka, Y.; Herrmann, V.; Doherty, G.; Drebin, J. A.; Strasberg, S. M.; Eberlein, T. J.; Goedegebuure, P. S.; Linehan, D. C. Prevalence of Regulatory T Cells Is Increased in Peripheral Blood and Tumor Microenvironment of Patients with Pancreas or Breast Adenocarcinoma. *J. Immunol.* 2002, *169* (5), 2756–2761.
57. Ljunggren, H. G.; Karre, K. Host Resistance Directed Selectively Against H-2-Deficient Lymphoma Variants: Analysis of the Mechanism. *J. Exp. Med.* 1985, *162* (6), 1745–1759.
58. Lu, X.; Lane, D. P. Differential Induction of Transcriptionally Active p53 Following UV or Ionizing Radiation: Defects in Chromosome Instability Syndromes? *Cell* **1993,** *75* (4), 765–778.
59. Lucas, C.; Bald, L. N.; Fendly, B. M.; Mora-Worms, M.; Figari, I. S.; Patzer, E. J.; Palladino, M. A. The Autocrine Production of Transforming Growth Factor-Beta 1 During Lymphocyte Activation: A Study with a Monoclonal Antibody-Based ELISA. *J. Immunol.* 1990, *145* (5), 1415–1422.

60. Ramsdell, F.; Fowlkes, B. J. Maintenance of In Vivo Tolerance by Persistence of Antigen. *Science* **1992,** *257* (5073), 1130–1134.
61. MacDonald, H. R.; Schneider, R.; Lees, R. K.; Howe, R. C.; Acha-Orbea, H.; Festenstein, H.; Zinkernagel, R. M.; Hengartner, H. T-Cell Receptor V Beta Use Predicts Reactivity and Tolerance to Mlsa-Encoded Antigens. *Nature* **1988,** *332* (6159), 40–45.
62. Machiels, J. P.; Reilly, R. T.; Emens, L. A.; Ercolini, A. M.; Lei, R. Y.; Weintraub, D.; Okoye, F. I.; Jaffee, E. M. Cyclophosphamide, Doxorubicin, and Paclitaxel Enhance the Antitumor Immune Response of Granulocyte/Macrophage-Colony Stimulating Factor-Secreting Whole-Cell Vaccines in HER-2/neu Tolerized Mice. *Cancer Res.* 2001, *61* (9), 3689–3697.
63. Nefedova, Y.; Huang, M.; Kusmartsev, S.; Bhattacharya, R.; Cheng, P.; Salup, R.; Jove, R.; Gabrilovich, D. Hyperactivation of Stat3 is Involved in Abnormal Differentiation of Dendritic Cells in Cancer. *J. Immunol.* 2004, *172* (1), 464–474.
64. Nefedova, Y.; Nagaraj, S.; Rosenbauer, A.; Muro-Cacho, C.; Sebti, S. M.; Gabrilovich, D. I. Regulation of Dendritic Cell Differentiation and Antitumor Immune Response in Cancer by Pharmacologic-Selective Inhibition of the Janus-Activated Kinase 2/Signal Transducers and Activators of Transcription 3 Pathway. *Cancer Res.* 2005, *65* (20), 9525–9535.
65. Nguyen, L. T.; Radhakrishnan, S.; Ciric, B.; Tamada, K.; Shin, T.; Pardoll, D. M.; Chen, L.; Rodriguez, M.; Pease, L. R. Cross-Linking the B7 Family Molecule B7-DC Directly Activates Immune Functions of Dendritic Cells. *J. Exp. Med.* 2002, *196* (10), 1393–1398.
66. Nguyen, L. T.; Elford, A. R.; Murakami, K.; Garza, K. M.; Schoenberger, S. P.; Odermatt, B.; Speiser, D. E.; Ohashi, P. S. Tumor Growth Enhances Cross-Presentation Leading to Limited T Cell Activation Without Tolerance. *J. Exp. Med.* 2002, *195* (4), 423–435.
67. Nigam, A.; Yacavone, R. F.; Zahurak, M. L.; Johns, C. M. S.; Pardoll, D. M.; Piantadosi, S.; Levitsky, H. I.; Nelson, W. G. Immunomodulatory Properties of Antineoplastic Drugs Administered in Conjunction with GM-CSF-Secreting Cancer Cell Vaccines. *Int. J. Oncol.* 1998, *12* (1), 161–170.
68. Nishimura, H.; Nose, M.; Hiai, H.; Minato, N.; Honjo, T. Development of Lupus-Like Autoimmune Diseases by Disruption of the PD-1 Gene Encoding an ITIM Motif-Carrying Immunoreceptor. *Immunity* **1999,** *11* (2), 141–151.
69. Nishimura, H.; Okazaki, T.; Tanaka, Y.; Nakatani, K.; Hara, M.; Matsumori, A.; Sasayama, S.; Mizoguchi, A.; Hiai, H.; Minato, N.; Honjo, T. Autoimmune Dilated Cardiomyopathy in PD-1 Receptor-Deficient Mice. *Science* **2001,** *291* (5502), 319–322.
70. Niu, G.; Wright, K. L.; Huang, M.; Song, L.; Haura, E.; Turkson, J.; Zhang, S.; Wang, T.; Sinibaldi, D.; Coppola, D.; Heller, R.; Ellis, L. M. et al. Constitutive Stat3 Activity Up-Regulates VEGF Expression and Tumor Angiogenesis. *Oncogene* **2002,** *21* (13), 2000–2008.
71. (a) North, R. J. Cyclophosphamide-Facilitated Adoptive Immunotherapy of an Established Tumor Depends on Elimination of Tumor-Induced Suppressor T Cells. *J. Exp. Med.* 1982, *155* (4), 1063–1074. (b). Ochsenbein, A. F. Immunological Ignorance of Solid Tumors. *Springer Semin Immunopathol.* 2005, *27*, 19–35.
72. Quill, H.; Schwartz, R. H. Stimulation of Normal Inducer T Cell Clones with Antigen Presented by Purified Ia Molecules in Planar Lipid Membranes: Specific Induction of a Long-Lived State of Proliferative Nonresponsiveness. *J. Immunol.* 1987, *138* (11), 3704–3712.

73. Zhang, Q.; Yang, X.; Pins, M.; Javonovic, B.; Kuzel, T.; Kim, S. J.; Parijs, L. V.; Greenberg, N. M.; Liu, V.; Guo, Y.; Lee, C. Adoptive Transfer of Tumor-Reactive Transforming Growth Factor-Beta-Insensitive CD8þ T Cells: Eradication of Autologous Mouse Prostate Cancer. *Cancer Res.* 2005, *65* (5), 1761–1769.
74. Restifo, N. P.; Marincola, F. M.; Kawakami, Y.; Taubenberger, J.; Yannelli, J. R.; Rosenberg, S. A. Loss of Functional Beta 2-Microglobulin in Metastatic Melanomas from Five Patients Receiving Immunotherapy. *J. Natl. Cancer Inst.* 1996, *88* (2), 100–108.
75. Riker, A.; Cormier, J.; Panelli, M.; Kammula, U.; Wang, E.; Abati, A.; Fetsch, P.; Lee, K. H.; Steinberg, S.; Rosenberg, S.; Marincola, F. Immune Selection After Antigen-Specific Immunotherapy Of Melanoma. *Surgery* **1999,** *126* (2), 112–120.
76. Rowe, M.; Khanna, R.; Jacob, C. A.; Argaet, V.; Kelly, A.; Powis, S.; Belich, M.; Croom-Carter, D.; Lee, S.; Burrows, S. R.; Trowsdale, J.; Moss, D. J. et al. Restoration of Endogenous Antigen Processing in Burkitt's Lymphoma Cells by Epstein-Barr Virus Latent Membrane Protein-1: Coordinate Up-Regulation of Peptide Transporters and HLA-Class I Antigen Expression. *Eur. J. Immunol.* 1995, *25* (5), 1374–1384.
77. Ruiz-Cabello, F.; Lopez Nevot, M. A.; Gutierrez, J.; Oliva, M. R.; Romero, C.; Ferron, A.; Esteban, F.; Huelin, C.; Piris, M. A.; Rivas, C. Phenotypic Expression of Histocompatibility Antigens in Human Primary Tumours and Metastases. *Clin. Exp. Metastasis* **1989,** *7* (2), 213–226.
78. Ruiz-Cabello, F.; Perez-Ayala, M.; Gomez, O.; Redondo, M.; Concha, A.; Cabrera, T.; Garrido, F. Molecular Analysis of MHC-Class-I Alterations in Human Tumor Cell Lines. *Int. J. Cancer Suppl.* 1991, *6*, 123–130.
79. Saito, H.; Tsujitani, S.; Oka, S.; Kondo, A.; Ikeguchi, M.; Maeta, M.; Kaibara, N. The Expression of Transforming Growth Factor-Beta1 Is Significantly Correlated with the Expression of Vascular Endothelial Growth Factor and Poor Prognosis of Patients with Advanced Gastric Carcinoma. *Cancer* **1999,** *86* (8), 1455–1462.
80. Sakaguchi, S.; Sakaguchi, N.; Asano, M.; Itoh, M.; Toda, M. Immunologic Self-Tolerance Maintained by Activated T Cells Expressing IL-2 Receptor Alpha-Chains (CD25): Breakdown of a Single Mechanism of Self-Tolerance Causes Various Autoimmune Diseases. *J. Immunol.* 1995, *155* (3), 1151–1164.
81. Bogen, B.; Schenck, K.; Munthe, L. A.; Dembic, Z. Deletion of Idiotype (Id)-Specific T Cells In Multiple Myeloma. *Acta Oncol.* 2000, *39* (7), 783–788.
82. (a) Sanda, M. G.; Restifo, N. P.; Walsh, J. C.; Kawakami, Y.; Nelson, W. G.; Pardoll, D. M.; Simons, J. W. Molecular Characterization of Defective Antigen Processing in Human Prostate Cancer. *J. Natl. Cancer Inst.* 1995, *87* (4), 280–285. (b) Sano, S.; Itami, S.; Takeda, K.; Tarutani, M.; Yamaguchi, Y.; Miura, H.; Yoshikawa, K.; Akira, S.; Takeda, J. Keratinocyte-Specific Ablation of Stat3 Exhibits Impaired Skin Remodeling, But Does Not Affect Skin Morphogenesis. *EMBO J.* 1999, *18* (17), 4657–4668.
83. Scheibenbogen, C.; Weyers, I.; Ruiter, D.; Willhauck, M.; Bittinger, A.; Keilholz, U. Expression of gp100 in Melanoma Metastases Resected Before or After Treatment with IFN Alpha and IL-2. J. *Immunother. Emphasis Tumor Immunol.* 1996, *19* (5), 375–380.
84. Schmid, P.; Itin, P.; Rufli, T. In Situ Analysis of Transforming Growth Factor-Beta s (TGF- Beta 1, TGF-Beta 2, TGF-Beta 3), and TGF-Beta Type II Receptor Expression in Malignant Melanoma. *Carcinogenesis* **1995,** *16* (7), 1499–1503.

85. Schmielau, J.; Finn, O. J. Activated Granulocytes and Granulocyte-Derived Hydrogen Peroxide Are the Underlying Mechanism of Suppression of T-Cell Function in Advanced Cancer Patients. *Cancer Res.* 2001, *61* (12), 4756
86. Yee, C.; Thompson, J. A.; Byrd, D.; Riddell, S. R.; Roche, P.; Celis, E.; Greenberg, P. D. Adoptive T Cell Therapy Using Antigen-Specific CD8þ T Cell Clones for the Treatment of Patients with Metastatic Melanoma: In Vivo Persistence, Migration, and Antitumor Effect of Transferred T Cells. *Proc. Natl. Acad. Sci. USA* **2002,** *99* (25), 16168–16173.
87. Ercolini, A. M.; Ladle, B. H.; Manning, E. A.; Pfannenstiel, L. W.; Armstrong, T. D.; Machiels, J. P.; Bieler, J. G.; Emens, L. A.; Reilly, R. T.; Jaffee, E.M. Recruitment of Latent Pools of High- Avidity CD8 (þ) T Cells to the Antitumor Immune Response. *J. Exp. Med.* 2005, *201* (10), 1591–1602
88. Zajac, A. J.; Blattman, J. N.; Murali-Krishna, K.; Sourdive, D. J.; Suresh, M.; Altman, J. D.; Ahmed, R. Viral Immune Evasion Due to Persistence of Activated T Cells Without Effector Function. *J. Exp. Med.* 1998, *188* (12), 2205–2213.
89. Zea, A. H.; Rodriguez, P. C.; Atkins, M. B.; Hernandez, C.; Signoretti, S.; Zabaleta, J.; McDermott, D.; Quiceno, D.; Youmans, A.; O'Neill, A.; Mier, J.; Ochoa, A. C. Arginase-Producing Myeloid Suppressor Cells in Renal Cell Carcinoma Patients: A Mechanism of Tumor Evasion. *Cancer Res.* 2005, *65* (8), 3044–3048.
90. Lyman, M. A.; Aung, S.; Biggs, J. A.; Sherman, L. A. A Spontaneously Arising Pancreatic Tumor Does Not Promote the Differentiation of Naive CD8þ T lymphocytes into Effector CTL. *J. Immunol.* 2004, *172* (11), 6558–6567.
91. Zuk, J. A.; Walker, R. A. Immunohistochemical Analysis of HLA Antigens and Mononuclear Infiltrates of Benign and Malignant Breast. *J. Pathol.* 1987, *152*, 275–275.4760.

CHAPTER 3

Cancer Immunoediting: A New Hope for Survival

ABSTRACT

The field of cancer immunoediting offers a framework for understanding the immune system's host-protective and tumor-sculpting actions against cancer and for creating cutting-edge personalized cancer immunotherapies. Tumor immunoediting is a process in which the host immune system directs a malignant growth by using both innate and adaptive immunological mechanisms. The first step, called Elimination, involves a healthy immune system eliminating the damaged cells. The immune system can only eliminate so many cancer cells before the rest of the population dies. The Escape phase is the last stage of tumor growth, during which immunologically generated tumors begin to show clinical signs of progression while simultaneously forming an immunosuppressive tumor microenvironment. This article examines how the development of innovative and objectively successful cancer immunotherapies and the discovery of new predictive and prognostic biomarkers have contributed to our knowledge of cancer immunoediting.

Tumors that spread despite immune monitoring have likely been "immunosculpted." It is hypothesized that following prolonged exposure to immune effectors, tumors lose their immunogenicity (i.e., undergo immunoediting) and become more dangerous. Immunosculpting may have permanent and irreversible consequences in certain circumstances. However, cancer immune resistance may in certain instances be due to reversible epigenetic processes. Even on the other hand, these immune-induced alterations could aid tumor development. When it comes to cancer, prevention, and therapy, knowing how tumors hide from the immune system is crucial.

Advances in Immunological Treatment of Cancer: Cellular Mechanism and Regulation.
Rishabha Malviya, Arun Kumar Singh, and Deepika Yadav (Authors)

3.1 INTRODUCTION

To what extent the immune system supports, inhibits, or has no effect on tumor development has been resolved during the last 20 years. Several different labs have shown in recent years that immunology may foster cellular plasticity, stop or reduce tumor progression, and shape tumor immunogenicity. Important distinctions exist between these seemingly competing immune functions due to factors such as the timing of these processes during tumor formation, the nature of the transforming event, the specific immune components involved in each process, and the nature of tumor-specific antigens expressed in the transformed cell.[1,2] However, our research team and others have demonstrated that a well-functioning immune system may achieve all of these objectives and more via a process the researcher calls "Cancer Immunoediting".[3] It is well acknowledged that chronic inflammation leads to the formation and spread of cancer.[4] Thanks to the evolution of the cancer immunoediting concept from the earlier, maybe more contentious "cancer immunosurveillance" hypothesis, researchers have a better knowledge of the predictive and prognostic relevance of immune infiltrates into tumors.

In a phrase that has stood the test of time, "immune surveillance" was first used by Burnet and Thomas over 50 years ago. The immune system would act as a sentinel, keeping an eye out for and killing off any abnormally growing cells, according to this theory. Over the last 15 years, it has been clear that the immune system's role in surveillance is just a small part of the story. To better describe the many facets of immune system-tumor interactions,[5] the researcher has refined and enlarged the theory into one author call "cancer immunoediting." The "Elimination," "Equilibrium," and "Escape" phases of this dynamic process have been addressed at length elsewhere[6] (in which the immune system not only protects against cancer but also influences how new tumors develop). Research in mice and humans has provided evidence for the equilibration and escape phases of malignancy, whereas studies in animal cancer models have provided the majority of the data for the elimination phase.[7] The ability of cancer cells to hide from the body's defences is now widely recognized as a fundamental characteristic of the disease. The cancer immunoediting literature has grown exponentially in recent years, mostly as a result of its paradigmatic usage in the research of the immune system's involvement in cancer. The current state of knowledge on cancer immunoediting is discussed, along with the potential implications for future treatment strategies.[8]

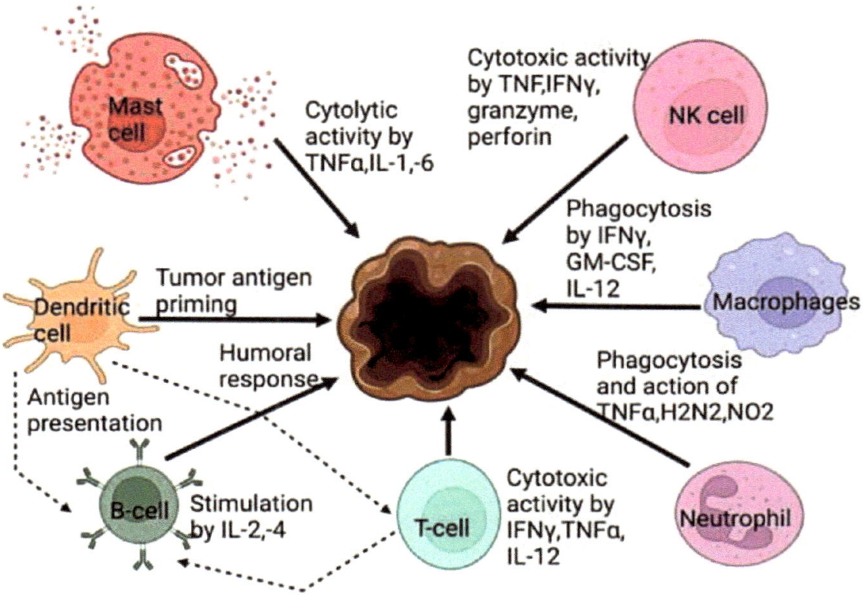

FIGURE 3.1 Schematic representation of immunoediting to fight cancer through interleukins.

3.2 ELIMINATION

Immunodeficient mice are more susceptible to carcinogen-induced and spontaneous malignancies than their wild-type counterparts, and the tumors developed in these mice are more immunogenic than tumors generated in competent animals.[9] As previously mentioned,[10,11] protective antitumor immunity is dependent on a working lymphocyte compartment, recognition markers like NKG2D, and host effector molecules including IFN-, perforin, Fas/FasL, and TRAIL. SECOND PART Cancer immunoediting is a field in which both type I (IFN-/) and type II (IFN-) interferons play distinct but essential roles.[12] As opposed to IFN-, which primarily targets tumor and hematopoietic cells, IFN-/ acts mostly on host cells. Type I IFNs are crucial for the initiation of the early antitumor response because they function on CD8+/CD103+ DCs to enhance cross-presentation of tumor antigens to CD8+ T cells.[13,14] Although granulocyte, macrophage, and NK lineage cells have type I interferon receptors, tumor rejection may still take place in their absence.[15]

An exciting new study demonstrates that tumor cells with increased tetraploidization (and, by extension, aneuploidy) are more immunogenic, increasing the chance of their elimination by the immune system. Hyperploid cells have a higher immunogenicity because their endoplasmic reticulum stress response is hyperactive, exposing calreticulin on their surface.[16] Ploidy control through the immunosurveillance/elimination mechanism was discovered to need CD4+ and CD8+ T lymphocytes, IFN-, and the IFN-receptor (reviewed in[17]). These results highlight a fascinating new discovery: that hyperploidy in carcinogen-induced and oncogene-induced cancer acts as a recognition function for this immunological eradication of tumors.[18]

Many people have trouble understanding Elimination research because they do not use tumors as a read-out (i.e., a negative result). As a result, it is not often clear if a tumor was there from birth and was recently found during surgery, or whether it was always there and was overlooked. There were previously no models that provided compelling evidence that eradication was possible. Tumorigenic early-stage B cells were first identified by Croxford et al.[19] between weeks 6 and 9 in E-myc mice, coinciding with a profound regression of the periphery. This model suggests that sustained myc expression may lead to DNA damage, which may in turn activate ligands that enhance immune recognition by activating the serine/threonine protein kinase ATM.[20] Reversion of B cells to their progenitor state was delayed by eliminating T cells and NK1.1+ cells or by inhibiting DNAM-1, a lymphocyte receptor for one such ligand (CD155). There is no description in the scientific literature of the rate of tumor development in E-myc mice under these conditions.

Cells from the innate immune system have been found to play an important role in the newly revealed techniques of elimination. Evidence from a mouse model of liver carcinoma[21] demonstrates that NK cells may eliminate senescent tumor cells, and that this elimination is dependent on the intrinsic expression of p53 in the tumor cells. Tumor cells that had expressed p53 went into senescence and produced chemokines (CCL2) and interleukins (IL-6, IL-12, and IL-15), which attracted in natural killer (NK) cells.[22] Natural killer (NK) cells were successful in curing tumors when NKG2D ligand was expressed by tumor cells. NKG2D ligands on tumor cells might be activated by a number of different mechanisms, such as Ras signaling and the DNA damage response.[23,24] In a model of K-ras driven hepatocellular carcinoma,[25] it was revealed that K-ras-specific CD4+ T cells cooperate with macrophages to eliminate premalignant senescent hepatocytes from the liver. There was a failure to monitor senescent tissue, which contributed to the emergence of liver cancer.[26]

There is indirect evidence that NK cells serve an integral role in the Elimination process as well. When cancer cells have telomeres that have been mutated, the cells undergo a cascade of events that eventually leads to growth arrest, halting the disease's progress. Overexpression of TRF2, an essential telomere-preserving protein, is shown in a wide range of human cancers. Recent studies[27] demonstrate that tumor cells with high TRF2 levels are less likely to attract and activate natural killer (NK) cells. Through a search for TRF2-bound genes, the author identified heparan sulfate (glucosamine) 3-O-sulphotransferase 4 (HS3ST4), which was shown to be under the control of TRF2 and to restrict NK cell recruitment through an epistatic interaction with TRF2. Collectively, our results show that a tumor-cell-intrinsic TRF2-dependent mechanism controls NK cell host-protective functions during carcinogenesis.[28]

It has also been proposed that macrophages may play a druggable role in innate tumor clearance. To avoid being digested by macrophages, tumor cells continually secrete a protein called CD47, which binds to a receptor called SIRP inhibitory. Phagocytes consumed and killed tumor cells when a monoclonal antibody was employed to suppress CD47.[29] Because erythrocytes express CD47 as well, this medicine has the potential to produce a potentially fatal anemia. It has been demonstrated to be successful in a number of animal cancer models to use a soluble human SIRP variant with strong binding affinity to CD47 on tumor cells. When combined with antitumor antibodies, these variants showed safety and efficacy in lymphoma and HER2+ breast cancer mice models (through antibody-dependent cellular cytotoxicity).

3.3 EQUILIBRIUM

Since the Equilibrium stage of immune-mediated tumor dormancy has not been reliably replicated in mice, and has only been reported in people anecdotally, little is known about the molecular pathways that begin during this period. The importance of adaptive Th1-like immunity in immune-mediated dormancy of fibrosarcoma was shown in a meta-analysis of large-scale investigations[30] published in 2007. Another study using the MCA-induced fibrosarcoma animal model provided evidence for immune-mediated tumor dormancy lasting for a long time.[31] The fine balance between IL-12 (which promotes elimination) and IL-23 (which promotes persistence) (and has a similar component with IL-12 (p40))[32] keeps malignancies in control. Even while IL-10 was revealed to have just a minor role in tumor promotion, it was shown that many other pathways (including IL-4, IL-17A, TNF, and IFN-) were redundant. Recent research comparing the cellular environment of stable and evasive tumors showed that low numbers of NKT cells, Foxp3+ Treg cells, and MDSCs were linked with immune-mediated homeostasis in cancer. To what degree tumor cells stay in immunologically mediated dormancy is strongly correlated with the ratio of immunosuppressive cells to immune cells able to demonstrate antitumor effector activities in the cancer microenvironment, as shown by this study.[33] Through a process of continuous editing, tumor cells are constantly generating variants that, depending on the tumor type, lack either important tumor-specific antigens or the ability to distribute them, playing a major part in the decision to avoid the immune system or be removed. Additional evidence for the occurrence of the Equilibrium phase comes from the discovery that tumor antigen-specific T cells may halt the development of Tag-induced multistage carcinogenesis (an experimental model of pancreatic cancer) in mice by coordinating the production of IFN- and TNF.[34] Without TNFR and IFN-, these same T cells promoted angiogenesis and multistage carcinogenesis. Tag-expressing tumors subjected to interferon (IFN)- and tumor necrosis factor (TNF) treatment undergo senescence, as shown by a prolonged growth arrest in G1/G0, activation of p16INK4a, and downstream Rb hypophosphorylation at Ser795.[35,36] Cytokines signal via p16INK4a, STAT1, and TNFR1 to cause senescence. Numerous tumors in both murines and humans enter senescence in response to IFN- and TNF treatment, indicating that this approach to treating cancer may be effective across species. Despite potential imaging advancements with suitable antigen markers that would allow for a more in-depth analysis of circulating tumor cells and habitats, detecting hidden cancer in equilibrium in mice and humans remains a big technological barrier.[37]

3.4 IMMUNE ESCAPE-TARGETED IMMUNOTHERAPIES

Immunotherapy is a cutting-edge method that might dramatically improve cancer care. However, the adaptability of cells in the cancer microenvironment often causes therapeutic intervention to fail. Thus, it is typically possible to generate a systemically effective memory response by using a series of tactics that use combinations that target several pathways, some of which may be synergistic.[38,39] There have been a number of published systematic

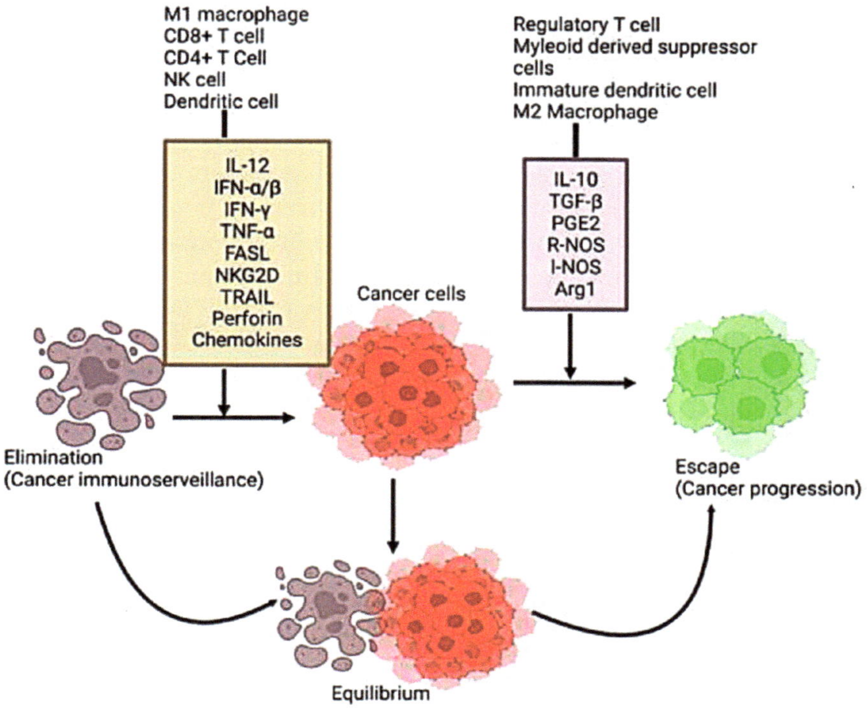

FIGURE 3.2 Schematic representation of cancer immunosurveillance to survive from cancer.

reviews of preclinical therapeutics, and some promising approaches are on the edge of being tried in people.[40] The innovative therapeutic combination of ipilimumab (anti-CTLA-4)[41] and nivolumab (anti-PD-1)[42] may have caused a rethinking of the treatment approach for melanoma patients.[43] Metastatic melanoma patients who had previously failed conventional therapy had significant improvement after receiving this combination.[44] Combination medication usage raises the stakes for both positive and negative immunopathology responses in the treatment of disease. Finding a happy medium between tumor immunity and inflammatory pathophysiology is, therefore, crucial.[45] As a result, preclinical models that can unravel the processes underlying this double-edged sword of combination medicines are necessary to discover whether combinations produce a superior therapeutic index.

Cancer's chances and prognosis in light of immune cells Advantages of being able to quantify the antitumor effects of immunotherapies in patients include the growth of genomic studies of tumors and the host, the definition of key parameters that differentiate between responders and non-responders, and the creation of novel methods of stratifying patients and their tumors.[46,47] Recent years have seen a precipitous surge in the discovery of immunological markers for a wide range of human cancers.[48] There are shared indicators between cancers that have a better prognosis and those that react to treatment or go into full remission. The density of CD3+, CD8+, and CD45RO+ T cells in the tumor core and invasive margin, as well as the quality of the tertiary lymphoid islets in the afflicted organ, are all indicators of the tumor's immunological contexture (the "Immunoscore").[49] IFN-, STAT1, IL-12, IRF1, T-bet, perforin, granzymes, CXCR3 ligand chemokines CXCL9 and CXCL10 and CCR5 ligand chemokines CXCL5 and adhesion molecules are all characteristics of an active Th1 profile (MAD- CAM1, ICAM1, and VCAM1). Several recent case studies[50,51] have shown the use of tumor infiltrates as a predictive biomarker for predicting treatment effectiveness. Additionally, these characteristics have shown to be a more accurate predictor of cancer stage than traditional pathological criteria for a number of different malignancies. Similar lymphocyte signatures have been identified in individuals treated with IL-2,[52] MAGE-A3 immunization,[53] and Ipilimumab,[54] in addition to being present in various targeted and conventional therapies.[55–57]

3.5 IMMUNOLOGICAL SCULPTING

The fundamental role of the innate immune system is to protect the body against foreign infections and malfunctioning self-cells, such as those seen in cancerous tissues. Patients with poor NK cell numbers or nonfunctional NK cells have a poor prognosis for beating their malignancy. Proficient in locating and destroying sick, "stressed," or diseased cells, NK cells play a crucial role in the immune system.[58]

NK cells are able to tell the difference between healthy and stressed cells thanks to the activating and inhibiting signals they get from a variety of germline-imprinted receptors.[59] Somatic rearrangement of variable diversity or joining (VDJ) gene segments with or without a recombinase activating gene allows NK-cell receptors to build an endless number of antigen-specific receptors, but T-cell receptors may only construct a finite number of such receptors (RAG). However, whether or not NK cells can respond to a target cell is determined by the expression of genes encoding specialized receptors for self-antigens.[60] Because of the random nature of their expression, various NK clones emerge with variable ratios of inhibitory to activating receptors.

The kind of ligand they bind to determines whether or not the expressed receptors trigger the target cell to increase its production of stimulatory "death" impulses or inhibitory "no-kill" signals. MHC class I (MHC I) molecules are ubiquitously expressed in normal tissues and act as ligands for receptors that instruct NK cells to retreat.[61,62] Without MHC I, inhibitory signals to the NK cell are blocked, enabling the activation signal to be potentiated indirectly. Moreover, proteins produced in response to stress bind to receptors on NK cells, sending them a "death" signal. These receptors are normally absent in healthy cells. When an NK cell comes into contact with its target, it activates a potent cocktail of MHC I molecules and stress-induced ligands. The intensity of the signals arising from these exchanges determines the NK-reaction cell's response to the stressed cell. Several NK-cell receptors and the ligands that bind to them are summarized in this chapter to offer a foundational understanding of NK-cell receptors and signaling.[63]

NK cells engage in "receptor-ligand interactions" to locate and kill damaged cells. Tolerance of healthy tissues and the prevention of autoimmunity are two benefits that result from NK cells recognizing and interacting with self-MHC I molecules.[64] The binding of an MHC I molecule to an MHC I-specific receptor produced on the surface of an NK cell is an example of a cis interaction that provides mice with this information.[65] The immunotyrosine-based inhibitory motif (ITIM) in the cytoplasmic tail must be phosphorylated at tyrosine (pTyr) for major histocompatibility complex class I binding.[66] NK cells cannot acquire the "license" to kill stressed cells unless they possess at least one MHC I-specific receptor or engage at least one MHC I molecule in such cis interactions. This is why self-MHC tolerance is essential for NK cells to acquire prior to the killing. Experiments back up the widely held belief that NK cells are armed, even if the precise licensing procedure for NK cells has yet to be defined in full. A lack of inhibitory ITIM makes natural killer cells ineffective[67]; hence, its presence is seen to constitute a positive signal according to the arming idea. It was previously believed that the two concepts were incompatible. Microscopic analysis, however, has shown that NK cells get their license to operate by binding MHC I in cis (on the surface of the NK cell). These variants may bind to cis MHC I, preventing the inhibitory receptors from working.

Active and inactive receptors are incapable of exchanging information with one another. When this barrier is gone, activating receptors may get close to their ligands and start sending the signals that mobilize lytic granules. This suggests that physical separation from inhibitory chemicals is essential for the positive regulation of NK-cell cytolysis.

Activation of receptors correctly is crucial.[68] The tyrosine phosphatases recruited by ITIM signaling are necessary for activation by both MHC I-specific and non-MHC I-specific receptors, indicating that their signaling pathways are comparable and conserved.

An essential step in setting off the lytic cascade is the action of dephosphorylating kinases. This research clarified previously unexplained phenomena and provided evidence that inhibitory ITIM signaling might lead to advantageous control.[69]

The ability to acquire a functional license is shared by both growing NK cells in the bone marrow and fully mature NK cells in the periphery. The notion that NK production is dynamic and ever-evolving is further shown by the possibility that NKs lose their killing power without MHC I. The variety of the host's major histocompatibility complex haplotypes, the frequency of MHC I-specific alleles, and other variables all influence the licensing signal.

In mice[70] and humans, cytokine activation may circumvent MHC I-dependent licensing, allowing unlicensed NK cells to destroy stressed cells in inflammatory conditions.

Cytokines are a class of proteins that stimulate immune cell growth. Type I interferons, interleukin (IL)-2, IL-15, IL-12, IL-18, and IL-21 are all essential cytokines for NK cell development and function (IFN). The vast majority of cytokines are produced by immune cells such as T cells, activated macrophages, and dendritic cells. Increased activation of surface killing receptors[71,72] may be responsible for the beneficial effects of the interleukin-2 and interleukin-15 systems, which are produced by T cells and dendritic cells, respectively, on natural killer (NK) cell survival, proliferation, cytokine production, and cytotoxicity.[73,74] IL-18, generated by macrophages and dendritic cells, increases the likelihood of survival,[75] whereas IL-12 and IL-21, produced by activated macrophages and T cells, increase cell proliferation, cytokine production, and cytotoxicity.[76] T cells and macrophages produce IL-6, which encourages the production of IFNg and IL-17 and, in turn, boosts cytotoxicity.

Viral type I interferons a and b have comparable effects on promoting robust cytokine production and cytotoxicity.[77,78] In other words, NK cells' communication skills with both other immune cells and their potential targets are crucial to NK-mediated host defence.

The fate of NK cells is determined by the balance between stimulating and repressive ligands that they encounter when they probe foreign cells. An NK cell will lyse its target cell if the activating impulses are powerful enough. Although there are certain ligands that do not need MHC, the vast majority of inhibitory signals are transmitted by MHC I ligands.[79] The MHC I-specific inhibitory receptor present on all functionally competent NK cells would instruct the NK cell to disregard the target since MHC I is generated by all normal cells (apart from erythrocytes and platelets). A common response of stressed cells is the downregulation of MHC I, which prevents the transmission of the inhibitory signal and allows NK cell activation. As a result of a process known as "missing self," NK cells will ignore targets that lack the MHC class I antigen.[80] Stress proteins generated in response to viral infections or malignant transformation are potentially possible activating ligands for NK-cell receptors, which may cause MHC I dysregulation and subsequently stimulate NK-cell-mediated cytolysis. Later chapters will go into further depth about each receptor and its corresponding ligand. However, it seems that a stronger signal is required to cause cytotoxicity of the target than is sufficient to elicit inflammatory cytokines like IFNg, which in turn mobilize components of the adaptive immune response to evaluate the suspected invader (determined by either the strength or quantity of the receptor-ligand interactions).[81] According to prior research, major histocompatibility complex class I (MHC I)-specific receptors on mouse NK cells may bind to MHC I molecules on the plasma membrane of NK cells. Enhancing the NK-mediated cytolytic activity of target cells, these cis connections reduce the requirement for trans-binding or the binding of NK-expressed MHC I-specific inhibitory receptors to MHC I on target cells.[82,83] This fine-tuning demonstrates how the MHC repertoire present inside NK cells may play a role in regulating the strength of the signal required to kill the target cell, suggesting that NK cells with a wide variety of MHC alleles may be more "active" than those with a restricted repertoire.

3.6 SIGNALING

The surface receptors of an NK cell operate as an external mirror of the cell's internal communication when it recognizes a possible target. Major receptor-activating and -inhibiting signaling pathways are substantially conserved. Active receptors have short, non-signaling cytoplasmic tails. Phosphorylation of the ITAM follows the binding of an activating receptor to an adaptor molecule and initiates a cascade of events that results in the production of lytic granules and/or cytokines.[84] ITAM domain-containing adaptor proteins connect to the positively charged transmembrane amino acids and form associations with the short cytoplasmic tails. DAP12, CD3z,

FcRg, and DAP10 are only some of the most often used adapter molecules.[85] After the tyrosine in the adaptor's intracellular tyrosine-based activation motif (ITAM) is phosphorylated in response to receptor-ligand interaction by Src family kinases like lymphocyte-specific protein tyrosine kinase (Lck) and Fyn, DAP12 associates with NK activating receptors via charged residues in the receptor's transmembrane region of the cytoplasmic tail. Syk[86] is a tyrosine kinase exclusive to the spleen.

By binding to the ITAM phosphotyrosine, zeta-chain-associated protein kinase 70 (Zap70)[87] activates phosphatidylinositol-3 kinase (PI3K), which in turn activates Rac1, p38 mitogen-activated protein kinase 1, and mitogen-activated protein kinase/extracellular signal-regulated kinase/extracellular effector. Recent studies[88,89] show that protein kinase C- (PKC-) is required for prolonged ITAM signaling and IFNg release, but not for cytolysis. A number of other adaptors, including CD3z and FcRg, use the same ITAM as DAP12 and hence perform indistinguishably like DAP12.[90]

Although DAP10 has a PI3K-binding domain, it lacks an ITAM. Granules are mobilized in a similar fashion as DAP12 after tyrosine phosphorylation in this region by Fyn/Lck.[91,92] More specifically, CD3z (FcRg) interacts with FcgRIII (CD16) to trigger antibody-dependent NK cell death. While the initial signaling molecules triggered upon ligand binding to NK cell receptors may differ between receptor types, they all eventually lead to the same conserved common cascade that culminates in ERK activation and NK cell proliferation regardless of the ligand that initially activated the cell.[93]

Then, the author set off the lytic granules. However, inhibitory receptors may be identified by the presence of one or more immune tyrosine inhibitory motifs in their rather long cytoplasmic tails (ITIM). These ITIMs are phosphorylated by Lck and Fyn at the region where the receptor and ligand engage. SHP-1 and SHP-2 bind to the phosphotyrosine in the ITIM via their Src homology-2 (SH2) domains.[94] When engaged, SHP-1 and SHP-2 hinder the lytic cascade in distinct NK activation calls for the dephosphorylation[95] of the signaling proteins Syk and Zap70.

3.7 KILLER IMMUNOGLOBULIN-LIKE RECEPTORS (KIR)

MHC class I molecules are particularly crucial for NK cell signaling. In order to determine the safety of their immediate surroundings, natural killer cells employ a protein called the major histocompatibility complex (MHC).

These interactions are critical to NK function because they transmit both inhibitory and activating signals.[96] Killer immunoglobulin-like receptors (KIR) are a large family of molecules expressed on human NK cells that recognize major histocompatibility complex class I (MHC I) on their targets. Haplotypes A and B are two examples of receptors that are passed down via generations. Depending on the length of their cytoplasmic tails, killer immunoglobulin-like receptors (KIRs) may either stimulate or dampen NK cell target signaling.[97] KIRs have either two (KIR2D) or three (KIR3D) extracellular immunoglobulin (Ig)-like domains.

In this discussion, the researcher will use the abbreviations aKIR and iKIR to distinguish between activation-inducing and -suppressing KIR gene products, respectively. aKIR is responsible for stimulating NK cells once they have made contact with their target, as indicated above.

Depending on the ligand, cells may release immune-stimulating cytokines or cell-killing lytic granules. Although these proteins are ineffective as signaling components due to their short cytoplasmic tails, adding a charged lysine or arginine permits them to interact with DAP12. Allele expression of the MHC I chain varies among six aKIR with two ECDs.98

It is well known that person X loves eating Y. KIR2DS1, KIR2DS2, and KIR2DS3 bind human leukocyte antigen (HLA)-C, MHC I alleles similarly to their inhibitory homologs. This additional layer of complexity further confounds NK cell responses to HLA-C-expressing targets, since the same ligand may convey both inhibitory and activating signals.

Furthermore, this suggests that the balance between activating and inhibitory KIR on the surface of the target cell may determine the cell's ultimate destiny. KIR2DS4 is exceptional since it is the only KIR that is overexpressed in Haplotype A individuals. Several more compounds have been shown to have their own distinct ligands, joining the list of those already known to bind to the major histocompatibility complex (MHC). KIR2DS4 gene expression is increased when HLA classes C1, C2, and A11 are present.[99]

This aKIR may have a role in antitumoral defence via its interactions with unknown proteins generated by melanoma cells and chronic lymphocytic leukemia (CLL) cells. The specific nature of the interactions between KIR2DS5 and KIR2DS3 is unknown at this time.

When bound to their ligand, iKIR binds SHP-1 and SHP-2 by their long cytoplasmic tails, while aKIR binds SHP-1 and SHP-2 through their short cytoplasmic tails and so creates negative intracellular signals. The eight distinct iKIR have varying degrees of affinity for MHC alleles. HLA-C alleles are bound by many KIR genes (KIR2DL1, KIR2DL2, and

KIR2DL3), HLA-B alleles are bound by KIR3DL1,[100,101] and HLA-A3, A11, and additional KIR genes are bound by KIR3DL2.102 To yet, KIR2DL5 has no identified ligands. KIR2DL4, unlike the overwhelming majority of other genes, is expressed everywhere in the human body and has been shown to convey both activating and inhibitory signals in experimental conditions. KIR2DL4 is able to recognize soluble HLA-G, which is produced by the placenta in the absence of conventional HLA molecules, and so serves as a secondary line of defence against spontaneous abortion. An epigenetic mechanism, other than variations in promoter sequence, that regulates KIR expression is the methylation of cytosines in CpG dinucleotides.[103] Over several cell divisions, these DNA methylation patterns have been demonstrated to remain, leading to a broader range of gene expression in NK clones.[104] Due to the wide variety of KIR expression, NK clones may have varying ratios of activating to inhibitory KIR. Genotyping and epidemiology studies have shown two main categories of KIR expression. Carrier individuals with the haplotype A allele have five inhibitory killer immune receptors (iKIRs) and one activating killer immune receptor (aKIR), namely KIR2DS4 and KIR2DL4. Haplotype B individuals, on the other hand, show a far broader variation in gene expression, with several permutations of aKIR and iKIR.[105] KIR haplotypes have lately been linked to an increased risk of certain cancers and infections, although their prevalence varies widely between ethnic groups.

3.8 NATURAL CYTOTOXICITY RECEPTORS

In order to properly identify stressed cells from healthy ones, NK cells must rely on protein interactions that occur independently of major histocompatibility complexes. Researchers concluded that there must be ligands unrelated to MHC since NK cells were able to efficiently remove MHC I-deficient tumor cells. Receptors that detect molecules other than MHC I are activated in the absence of MHC I-mediated inhibition, a phenomenon known as the "missing self" phenomenon, as has been proven in recent research to serve this purpose. It has been shown that the natural cytotoxicity receptor (NCR) family of receptors plays a crucial role in MHC-independent target lysis. Loss of even a single NCR reduces NK cells' ability to kill tumor targets in vivo,106 since tumor-generated transforming growth factor b-1 (TGFb) downregulates members of the NCR family to restrict NK cytolysis.[107] In addition, NCRs may improve interactions with antigen-presenting cells like dendritic cells, which aid in cellular maturation and recruitment of the adaptive immune system.[108]

NKp30, NKp44, NKp46, and NKp80 are the four proteins that make up the NCR, and their names are derived from their molecular masses.[109–111] NKp44 expression is upregulated by interleukin-2 (IL-2), although NKp30, NKp46, and Nkp80 are expressed by both active and resting NK cells. Multiple ligands have been proposed for NKp46, including influenza and parainfluenza hemagglutinin and hemagglutinin-neuraminidase[112] and the heparin sulfate proteoglycans,[113] both of which bind NKp30.[114] As well as B7-H6,[115] CMV pp65,[116] and BAT3,[114] there are other ligands for NKp30. Tumor cells release BAT3, a nuclear factor HLA-B-associated transcript.

On the other hand, only NKp44 of the NCRs has been shown to interact with DAP12.[117] Together, NKp30 and NK46 bind to CD3z and FcRg to create heterotrimeric adaptor complexes.[118,119] By binding to activation-induced C-type lectin (AICL), the genetically similar activating receptor NKp80 increases NK cytotoxicity against malignant myeloid cells.[120] Signals are sent from NKp80 directly to Syk[121] and, presumably, the PI3K pathway, bypassing an adaptor protein through a hemi-ITAM in the cytoplasmic tail.

3.9 CONCLUSION

NK cells are an important element of the first immune response, especially when malignancy is present. Unlike T cells, which undergo negative selection and clonal elimination in the thymus if they exhibit self-reactivity, NK cells are genetically programmed to tolerate self-MHC while still detecting and eliminating malignant tissue via the detection of aberrantly produced self-proteins.

The outcome of a putative malignancy depends on the ratio of inhibitory to activating signals exchanged between stressed NK cells and the cells they mistake for cancer. Tumors may avoid NK-cell-mediated apoptosis by releasing immunosuppressive chemicals or losing soluble forms of activating receptor ligands. Tumors tend to make natural killer cells unhappy (NK-cells). Intratumoral NK cells may be reprogrammed toward an aggressive and active phenotype, one that is ready to identify and destroy invading malignancies, by elucidating the mechanisms that sustain the expression of stimulatory receptors.

KEYWORDS

- **immunoediting**
- **immunological sculpting**
- **signaling**
- **cytotoxicity receptors**
- **tumors**

REFERENCES

1. Shankaran, V.; Ikeda, H.; Bruce, A. T.; White, J. M.; Swanson, P. E.; Old, L. J.; Schreiber, R. D. IFNgamma and lymphocytes prevent primary tumour development and shape tumour immunogenicity. *Nature* **2001,** *410*, 1107–1111.
2. van den Broek, M. E.; Kagi, D.; Ossendorp, F.; Toes, R.; Vamvakas, S.; Lutz, W. K.; Melief, C. J.; Zinkernagel, R. M.; Hengartner, H. Decreased Tumor Surveillance in Perforin-Deficient Mice. *J. Exp. Med.* **1996,** *184*, 1781–1790.
3. Smyth, M. J.; Thia, K. Y.; Street, S. E.; MacGregor, D.; Godfrey, D. I.; Trapani, J. A. Perforin-Mediated Cytotoxicity Is Critical for Surveillance of Spontaneous Lymphoma. *J. Exp. Med.* **2000,** *192*, 755–760.
4. Zerafa, N.; Westwood, J. A.; Cretney, E.; Mitchell, S.; Waring, P.; Iezzi, M.; Smyth, M. J. Cutting edge, TRAIL Deficiency Accelerates Hematological Malignancies. *J. Immunol.* **2005,** *175*, 5586–5590.
5. Takeda, K.; Smyth, M. J.; Cretney, E.; Hayakawa, Y.; Kayagaki, N.; Yagita, H.; Okumura, K. Critical Role for Tumor Necrosis Factor-Related Apoptosis-Inducing Ligand in Immune Surveillance Against Tumor Development. *J. Exp. Med.* **2002,** *195*, 161–169.
6. Takeda, K.; Smyth, M. J.; Cretney, E.; Hayakawa, Y.; Yamaguchi, N.; Yagita, H.; Okumura, K. Involvement of Tumor Necrosis Factor-Related Apoptosis-Inducing Ligand in NK Cell-Mediated and IFN-Gamma-Dependent Suppression of Subcutaneous Tumor Growth. *Cell Immunol.* **2001,** *214*, 194–200.
7. Takeda, K.; Hayakawa, Y.; Smyth, M. J.; Kayagaki, N.; Yamaguchi, N.; Kakuta, S.; Iwakura, Y.; Yagita, H.; Okumura, K. Involvement of Tumor Necrosis Factor-Related Apoptosis-Inducing Ligand in Surveillance of Tumor Metastasis by Liver Natural Killer Cells. *Nat Med.* **2001,** *7*, 94–100.
8. Smyth, M. J.; Cretney, E.; Takeda, K.; Wiltrout, R. H.; Sedger, L. M.; Kayagaki, N.; Yagita, H.; Okumura, K. Tumor Necrosis Factor-Related Apoptosis-Inducing Ligand (TRAIL) Contributes to Interferon Gamma-Dependent Natural Killer Cell Protection from Tumor Metastasis. *J Exp. Med.* **2001,** *193*, 661–670.
9. Seki, N.; Hayakawa, Y.; Brooks, A. D.; Wine, J.; Wiltrout, R. H.; Yagita, H.; Tanner, J. E.; Smyth, M. J.; Sayers, T. J. Tumor Necrosis Factor-Related Apoptosis-Inducing

Ligand-Mediated Apoptosis Is an Important Endogenous Mechanism for Resistance to Liver Metastases in Murine Renal Cancer. *Cancer Res.* **2003,** *63*, 207–213.
10. Cretney, E.; Takeda, K.; Yagita, H.; Glaccum, M.; Peschon, J. J.; Smyth, M. J. Increased Susceptibility to Tumor Initiation and Metastasis in TNF-Related Apoptosis-Inducing Ligand-Deficient Mice. *J. Immunol.* **2002,** *168*, 1356–1361.
11. Wiley, S. R.; Schooley, K.; Smolak, P. J.; Din, W. S.; Huang, C. P.; Nicholl, J. K.; Sutherland, G. R.; Smith, T. D.; Rauch, C.; Smith, C. A. et al. Identification and Characterization of a New Member of the TNF Family That Induces Apoptosis. *Immunity* **1995,** *3*, 673–682.
12. Smyth, M. J.; Thia, K. Y.; Street, S. E.; Cretney, E.; Trapani, J. A.; Taniguchi, M.; Kawano, T.; Pelikan, S. B.; Crowe, N. Y.; Godfrey, D. I. Differential Tumor Surveillance by Natural Killer (NK) and NKT Cells. *J. Exp. Med.* **2000,** *191*, 661–668.
13. Smyth, M. J.; Crowe, N. Y.; Godfrey, D. I. NK Cells and NKT Cells Collaborate in Host Protection from Methylcholanthrene-Induced Fibrosarcoma. Int Immunol. **2001,** *13*, 459–463.
14. Gao, Y.; Yang, W.; Pan, M.; Scully, E.; Girardi, M.; Augenlicht, L. H.; Craft, J.; Yin, Z. Gamma Delta T Cells Provide an Early Source of Interferon Gamma in Tumor *Immunity J. Exp. Med.* **2003,** *198*, 433–442.
15. Girardi, M.; Oppenheim, D. E.; Steele, C. R.; Lewis, J. M.; Glusac, E.; Filler, R.; Hobby, P.; Sutton, B.; Tigelaar, R. E.; Hayday, A. C. Regulation of Cutaneous Malignancy by Gamma Delta T Cells. *Science* **2001,** *294*, 605–609.
16. Noguchi, Y.; Jungbluth, A.; Richards, E. C.; Old, L. J. Effect of Interleukin 12 on Tumor Induction by 3-Methylcholanthrene. *Proc. Natl. Acad. Sci. USA* **1996,** *93*, 11798–11801.
17. Enzler, T.; Gillessen, S.; Manis, J. P.; Ferguson, D.; Fleming, J.; Alt, F. W.; Mihm, M.; Dranoff, G. Deficiencies of GM-CSF and Interferon Gamma Link Inflammation and Cancer. *J. Exp. Med.* **2003,** *197*, 1213–1219.
18. Hussein, M. R.; Hassan, H. I. Analysis of the Mononuclear Inflammatory Cell Infiltrate in the Normal Breast, Benign Proliferative Breast Disease, In Situ and Infiltrating Ductal Breast Carcinomas: Preliminary Observations. *J. Clin. Pathol.* **2006,** *59*, 972–977.
19. 19.Croxford, J. L.; Tang, M. L.; Pan, M. F.; Huang, C. W.; Kamran, N.; Phua, C. M.; Chng, W. J.; Ng, S. B.; Raulet, D. H.; Gasser, S. ATM-Dependent Spontaneous Regression of Early Emu-Myc-Induced Murine B-Cell Leukemia Depends on Natural Killer and T Cells. *Blood* **2013,** *121*, 2512–2521.
20. Zhang, L.; Conejo-Garcia, J. R.; Katsaros, D.; Gimotty, P. A.; Massobrio, M.; Regnani, G.; Makrigiannakis, A.; Gray, H.; Schlienger, K.; Liebman, M. N.; Rubin, S. C.; Coukos, G. Intratumoral T Cells, Recurrence, and Survival in Epithelial Ovarian Cancer. *N. Engl. J. Med.* **2003,** *348*, 203–213.
21. Sato, E.; Olson, S. H.; Ahn, J.; Bundy, B.; Nishikawa, H.; Qian, F.; Jungbluth, A. A.; Frosina, D.; Gnjatic, S.; Ambrosone, C.; Kepner, J.; Odunsi, T.; Ritter, G.; Lele, S.; Chen, Y. T.; Ohtani, H.; Old, L. J.; Odunsi, K. Intraepithelial CD8+ Tumor-Infiltrating Lymphocytes and a High CD8+/Regulatory T Cell Ratio Are Associated with Favorable Prognosis in Ovarian Cancer. *Proc. Natl. Acad. Sci. USA* **2005,** *102*, 18538–18543.
22. Pages, F.; Berger, A.; Camus, M.; Sanchez-Cabo, F.; Costes, A.; Molidor, R.; Mlecnik, B.; Kirilovsky, A.; Nilsson, M.; Damotte, D.; Meatchi, T.; Bruneval, P.; Cugnenc, P. H.; Trajanoski, Z.; Fridman, W. H.; Galon, J. Effector Memory T Cells, Early Metastasis, and Survival in Colorectal Cancer. *N. Engl. J. Med.* **2005,** *353*, 2654–2666.

23. Galon, J.; Costes, A.; Sanchez-Cabo, F.; Kirilovsky, A.; Mlecnik, B.; Lagorce-Pages, C.; Tosolini, M.; Camus, M.; Berger, A.; Wind, P.; Zinzindohoue, F.; Bruneval, P.; Cugnenc, P. H.; Trajanoski, Z.; Fridman, W. H.; Pages, F. Type, Density, and Location of Immune Cells Within Human Colorectal Tumors Predict Clinical Outcome. *Science* **2006,** *313*, 1960–1964.
24. Sabharanjak, S.; Mayor, S. Folate Receptor Endocytosis and Trafficking. *Adv. Drug Deliv. Rev.* **2004,** *56*, 1099–1109.
25. Li, P. Y.; Del Vecchio, S.; Fonti, R.; Carriero, M. V.; Potena, M. I.; Botti, G.; Miotti, S.; Lastoria, S.; Menard, S.; Colnaghi, M. I.; Salvatore, M. Local concentration of folate binding protein GP38 in sections of human ovarian carcinoma by in vitro quantitative autoradiography. *J. Nucl. Med.* **1996,** *37*, 665–672.
26. Rentzsch, C.; Kayser, S.; Stumm, S.; Watermann, I.; Walter, S.; Stevanovic, S.; Wallwiener, D.; Guckel, B. Evaluation of Pre-Existent Immunity in Patients with Primary Breast Cancer: Molecular and Cellular Assays to Quantify Antigen-Specific T Lymphocytes in Peripheral Blood Mononuclear Cells. *Clin. Cancer Res.* **2003,** *9*, 4376–4386.
27. Disis, M. L.; Knutson, K. L.; Schiffman, K.; Rinn, K.; McNeel, D. G. Pre-Existent Immunity to the HER-2/neu Oncogenic Protein in Patients with HER-2/neu Overexpressing Breast and Ovarian Cancer. *Breast Cancer Res. Treat.* **2000,** *62*, 245–252.
28. Beckhove, P.; Feuerer, M.; Dolenc, M.; Schuetz, F.; Choi, C.; Sommerfeldt, N.; Schwendemann, J.; Ehlert, K.; Altevogt, P.; Bastert, G.; Schirrmacher, V.; Umansky, V. Specifically Activated Memory T Cell Subsets from Cancer Patients Recognize and Reject Xenotransplanted Autologous Tumors. *J. Clin. Invest.* **2004,** *114*, 67–76.
29. Goodell, V.; Salazar, L. G.; Urban, N.; Drescher, C. W.; Gray, H.; Swensen, R. E.; McIntosh, M. W.; Disis, M. L. Antibody Immunity to the p53 Oncogenic Protein Is a Prognostic Indicator in Ovarian Cancer. *J. Clin. Oncol.* **2006,** *24*, 762–768.
30. Vial, T.; Descotes, J. Immunosuppressive Drugs and Cancer. *Toxicology* **2003,** *185*, 229–240.
31. de Visser, K. E.; Eichten, A.; Coussens, L. M. Paradoxical Roles of the Immune System During Cancer Development. *Nat. Rev. Cancer* **2006,** *6*, 24–37.
32. Vajdic, C. M.; McDonald, S. P.; McCredie, M. R.; van Leeuwen, M. T.; Stewart, J. H.; Law, M.; Chapman, J. R.; Webster, A. C.; Kaldor, J. M.; Grulich, A. E. Cancer Incidence Before and After Kidney Transplantation. *JAMA* **2006,** *296*, 2823–2831.
33. Bui, J. D.; Schreiber, R. D. Cancer Immunosurveillance, Immunoediting and Inflammation: Independent or Interdependent Processes? *Curr. Opin. Immunol.* **2007,** *19*, 203–208.
34. Chambers, W. M.; Warren, B. F.; Jewell, D. P.; Mortensen, N. J. Cancer Surveillance in Ulcerative Colitis. *Br. J. Surg.* **2005,** *92*, 928–936.
35. Friedman, S. Cancer in Crohn Disease. *Gastroenterol. Clin. North Am.* **2006,** *35*, 621–639.
36. Clevers, H. At the Crossroads of Inflammation and Cancer. *Cell* **2004,** *118*, 671–674.
37. Itzkowitz, S. H.; Yio, X. Inflammation and Cancer, I. V. Colorectal Cancer in Inflammatory Bowel Disease: The Role of Inflammation. *Am. J. Physiol. Gastrointest. Liver Physiol.* **2004,** *287*, G7–G17.

38. Seril, D. N.; Liao, J.; Yang, G. Y.; Yang, C. S. Oxidative Stress and Ulcerative Colitis-Associated Carcinogenesis: Studies in Humans and Animal Models. *Carcinogenesis* **2003,** *24*, 353–362.
39. Ballaz, S.; Mulshine, J. L. The Potential Contributions of Chronic Inflammation to Lung Carcinogenesis. *Clin. Lung Cancer* **2003,** *5*, 46–62.
40. Shacter, E.; Weitzman, S. A. Chronic Inflammation and Cancer. *Oncology* **2002,** *16*, 217–226; 29, discussion 30–32.
41. Prinz, C.; Schwendy, S.; Voland, P. H Pylori and Gastric Cancer: Shifting the Global Burden. *World J. Gastroenterol.* **2006,** *12*, 5458–5464.
42. Karin, M. Nuclear Factor-kappa B in Cancer Development and Progression. *Nature* **2006,** *441*, 431–436.
43. Karin, M.; Greten, F. R. NF-kappa B: Linking Inflammation and Immunity to Cancer Development and Progression. *Nat. Rev. Immunol.* **2005,** *5*, 749–759.
44. Karin, M. NF-kappaB and Cancer: Mechanisms and Targets. *Mol. Carcinog.* **2006,** *45*, 355–361.
45. Greten, F. R.; Eckmann, L.; Greten, T. F.; Park, J. M.; Li, Z. W.; Egan, L. J.; Kagnoff, M. F.; Karin, M. IKKbeta Links Inflammation and Tumorigenesis in a Mouse Model of Colitis-Associated Cancer. *Cell* **2004,** *118*, 285–296.
46. de Visser, K. E.; Korets, L. V.; Coussens, L. M. De novo Carcinogenesis Promoted by Chronic Inflammation Is B Lymphocyte Dependent. *Cancer Cell* **2005,** *7*, 411–423.
47. Dunn, G. P.; Ikeda, H.; Bruce, A. T.; Koebel, C.; Uppaluri, R.; Bui, J.; Chan, R.; Diamond, M.; White, J. M.; Sheehan, K. C.; Schreiber, R. D. Interferon-Gamma and Cancer Immunoediting. *Immunol. Res.* **2005,** *32*, 231–245.
48. Pollard, J. W. Tumour-Educated Macrophages Promote Tumour Progression and Metastasis. *Nat. Rev. Cancer* **2004,** *4*, 71–78.
49. Rollins, B. J. Inflammatory Chemokines in Cancer Growth and Progression. *Eur. J. Cancer* **2006,** *42*, 760–767.
50. Leek, R. D.; Lewis, C. E.; Whitehouse, R.; Greenall, M.; Clarke, J.; Harris, A. L. Association of Macrophage Infiltration with Angiogenesis and Prognosis in Invasive Breast Carcinoma. *Cancer Res.* **1996,** *56*, 4625–4629.
51. Tsutsui, S.; Yasuda, K.; Suzuki, K.; Tahara, K.; Higashi, H.; Era, S. Macrophage Infiltration and Its Prognostic Implications in Breast Cancer: The Relationship with VEGF Expression and Microvessel Density. *Oncol. Rep.* **2005,** *14*, 425–431.
52. Lin, E. Y.; Pollard, J. W. Macrophages: Modulators of Breast Cancer Progression. *Novartis Found Symp.* **2004,** *256*, 158–168; Discussion 68–72, 259–269.
53. Grivennikov, S. I.; Greten, F. R.; Karin, M. Immunity, Inflammation, and Cancer. *Cell* **2010,** *140*, 883–899.
54. Shankaran, V.; Ikeda, H.; Bruce, A. T.; White, J. M.; Swanson, P. E.; Old, L. J.; Schreiber, R. D. IFNgamma and Lymphocytes Prevent Primary Tumor Development and Shape Tumor Immunogenicity. *Nature* **2001,** *410*, 1107–1111.
55. Dunn, G. P.; Koebel, C. M.; Schreiber, R. D. Interferons, Immunity and Cancer Immunoediting. *Nat. Rev. Immunol.* **2006,** *6*, 836–848.
56. Burnet, M. Cancer; A Biological Approach: I—The Processes of Control. *Br. Med. J.* **1957,** *1*, 779
57. Schreiber, R. D.; Old, L. J.; Smyth, M. J. Cancer Immunoediting: Integrating Immunity Roles in Cancer Suppression and Promotion. *Science* **2011,** *331*, 1565–1570.

58. Dunn, G. P.; Old, L. J.; Schreiber, R. D. The Immunobiology of Cancer Immunosurveillance and Immunoediting. *Immunity* **2004,** *21*, 137–148.
59. Hanahan, D.; Weinberg, R. A. Hallmarks of Cancer: The Next Generation. *Cell* **2011,** *144*, 646–674.
60. Matsushita, H.; Vesely, M. D.; Koboldt, D. C.; Rickert, C. G.; Uppaluri, R.; Magrini, V. J.; Arthur, C. D.; White, J. M.; Chen, Y. S.; Shea, L. K. et al. Cancer Exome Analysis Reveals a T-Cell-Dependent Mechanism of Cancer Immunoediting. *Nature* **2012,** *482*, 400–404.
61. DuPage, M.; Mazumdar, C.; Schmidt, L. M.; Cheung, A. F.; Jacks, T. Expression of Tumor-Specific Antigens Underlies Cancer Immunoediting. *Nature* **2012,** *482*, 405–409.
62. DuPage, M.; Cheung, A. F.; Mazumdar, C.; Winslow, M. M.; Bronson, R.; Schmidt, L. M.; Crowley, D.; Chen, J.; Jacks, T. Endogenous T Cell Responses to antigens expressed in Lung Adenocarcinomas Delay Malignant Tumor Progression. *Cancer Cell* **2011,** *19*, 72–85.
63. Castle, J. C.; Kreiter, S.; Diekmann, J.; Lower, M.; van de Roemer, N.; de Graaf, J.; Selmi, A.; Diken, M.; Boegel, S.; Paret, C. et al. Exploiting the Mutanome for Tumor Vaccination. *Cancer Res.* **2012,** *72*, 1081–1091.
64. Robbins, P. F.; Lu, Y. C.; El-Gamil, M.; Li, Y. F.; Gross, C.; Gartner, J.; Lin, J. C.; Teer, J. K.; Cliften, P.; Tycksen, E. et al. Mining Exomic Sequencing Data to Identify Mutated Antigens Recognized by Adoptively Transferred Tumor-Reactive T Cells. *Nat. Med.* **2013,** *19*, 747–752.
65. van Rooij, N.; van Buuren, M. M.; Philips, D.; Velds, A.; Toebes, M.; Heemskerk, B.; van Dijk, L. J.; Behjati, S.; Hilkmann, H.; El Atmioui, D. et al. Tumor Exome Analysis Reveals Neoantigen-Specific T-Cell Reactivity in an Ipilimumab-Responsive Melanoma. *J. Clin. Oncol.* **2013,** *10*, e439–e442.
66. von Boehmer, L.; Mattle, M.; Bode, P.; Landshammer, A.; Schafer, C.; Nuber, N.; Ritter, G.; Old, L.; Moch, H.; Schafer, N. et al. NY-ESO-1-Specific Immunological Pressure and Escape in a Patient with Metastatic Melanoma. *Cancer Immunol.* **2013,** *13*, 12.
67. Nicholaou, T.; Chen, W.; Davis, I. D.; Jackson, H. M.; Dimopoulos, N.; Barrow, C.; Browning, J.; Macgregor, D.; Williams, D.; Hopkins, W. et al. Immunoediting and Persistence of Antigen-Specific Immunity in Patients Who Have Previously Been Vaccinated with NY-ESO-1 Protein Formulated in ISCOMATRIX. *Cancer Immunol. Immunother.* **2011,** *60*, 1625–1637.
68. O and Sullivan, T.; Saddawi-Konefka, R.; Vermi, W.; Koebel, C. M.; Arthur, C.; White, J. M.; Uppaluri, R.; Andrews, D. M.; Ngiow, S. F.; Teng, M. W. et al. Cancer Immunoediting by the Innate Immune System in the Absence of Adaptive Immunity. *J. Exp. Med.* **2012,** *209*, 1869–1882.
69. Vesely, M. D.; Kershaw, M. H.; Schreiber, R. D.; Smyth, M. J. Natural Innate and Adaptive Immunity to Cancer. *Annu. Rev. Immunol.* **2011,** *29*, 235–271.
70. Diamond, M. S.; Kinder, M.; Matsushita, H.; Mashayekhi, M.; Dunn, G. P.; Archambault, J. M.; Lee, H.; Arthur, C. D.; White, J. M.; Kalinke, U. et al. Type I Interferon Is Selectively Required by Dendritic Cells for Immune Rejection of Tumors. *J. Exp. Med.* **2011,** *208*, 1989–2003.
71. Fuertes, M. B.; Kacha, A. K.; Kline, J.; Woo, S. R.; Kranz, D. M.; Murphy, K. M.; Gajewski, T. F. Host Type I IFN Signals Are Required for Antitumor CD8+ T Cell Responses Through CD8{alpha}+ Dendritic Cells. *J. Exp. Med.* **2011,** *208*, 2005–2016.

72. Senovilla, L.; Vitale, I.; Martins, I.; Tailler, M.; Pailleret, C.; Michaud, M.; Galluzzi, L.; Adjemian, S.; Kepp, O.; Niso-Santano, M. et al. An Immunosurveillance Mechanism Controls Cancer Cell Ploidy. *Science* **2012,** *337*, 1678–1684.
73. 73.Hayakawa, Y.; Rovero, S.; Forni, G.; Smyth, M. J. Alpha-Galactosylceramide (KRN7000) Suppression of Chemical- and Oncogene-Dependent Carcinogenesis. *Proc. Natl. Acad. Sci. USA* **2003,** *100*, 9464–9469.
74. Iannello, A.; Thompson, T. W.; Ardolino, M.; Lowe, S. W.; Raulet, D. H. p53-Dependent Chemokine Production by Senescent Tumor Cells Supports NKG2D-Dependent Tumor Elimination by Natural Killer Cells. *J. Exp. Med.* **2013,** *210*, 2057–2069.
75. Gasser, S.; Orsulic, S.; Brown, E. J.; Raulet, D. H. The DNA Damage Pathway Regulates Innate Immune System Ligands of the NKG2D Receptor. *Nature* **2005,** *436*, 1186–1190.
76. Liu, X. V.; Ho, S. S.; Tan, J. J.; Kamran, N.; Gasser, S. Ras Activation Induces Expression of Raet1 Family NK Receptor Ligands. *J. Immunol.* **2012,** *189*, 1826–1834.
77. Kang, T. W.; Yevsa, T.; Woller, N.; Hoenicke, L.; Wuestefeld, T.; Dauch, D.; Hohmeyer, A.; Gereke, M.; Rudalska, R.; Potapova, A. et al. Senescence Surveillance of Pre-Malignant Hepatocytes Limits Liver Cancer Development. *Nature* **2011,** *479*, 547–551.
78. Biroccio, A.; Cherfils-Vicini, J.; Augereau, A.; Pinte, S.; Bauwens, S.; Ye, J.; Simonet, T.; Horard, B.; Jamet, K.; Cervera, L. et al. TRF2 Inhibits a Cell-Extrinsic Pathway Through Which Natural Killer Cells Eliminate Cancer Cells. *Nat. Cell. Biol.* **2013,** *15*, 818–828.
79. Willingham, S. B.; Volkmer, J. P.; Gentles, A. J.; Sahoo, D.; Dalerba, P.; Mitra, S. S.; Wang, J.; Contreras-Trujillo, H.; Martin, R.; Cohen, J. D. et al. The CD47-Signal Regulatory Protein Alpha (SIRPa) Interaction Is a Therapeutic Target for Human Solid Tumors. *Proc. Natl. Acad. Sci. USA* **2012,** *109*, 6662–6667.
80. Weiskopf, K.; Ring, A. M.; Ho, C. C.; Volkmer, J. P.; Levin, A. M.; Volkmer, A. K.; Ozkan, E.; Fernhoff, N. B.; van de Rijn, M.; Weissman, I. L. et al. Engineered SIRPalpha Variants as Immunotherapeutic Adjuvants to Anticancer Antibodies. *Science* **2013,** *341*, 88–91.
81. Koebel, C. M.; Vermi, W.; Swann, J. B.; Zerafa, N.; Rodig, S. J.; Old, L. J.; Smyth, M. J.; Schreiber, R. D. Adaptive Immunity Maintains Occult Cancer in an Equilibrium State. *Nature* **2007,** *450*, 903–907.
82. Teng, M. W.; Vesely, M. D.; Duret, H.; McLaughlin, N.; Towne, J. E.; Schreiber, R. D.; Smyth, M. J. Opposing Roles for IL-23 and IL-12 in Maintaining Occult Cancer in an Equilibrium State. *Cancer Res.* **2012,** *72*, 3987–3996.
83. Wu, X.; Peng, M.; Huang, B.; Zhang, H.; Wang, H.; Xue, Z.; Zhang, L.; Da, Y.; Yang, D.; Yao, Z. et al. Immune Microenvironment Profiles of Tumor Immune Equilibrium and Immune Escape States of Mouse Sarcoma. *Cancer Lett.* **2013,** *340*, 124–133.
84. Muller-Hermelink, N.; Braumuller, H.; Pichler, B.; Wieder, T.; Mailhammer, R.; Schaak, K.; Ghoreschi, K.; Yazdi, A.; Haubner, R.; Sander, C. A. et al. TNFR1 Signaling and IFN-Gamma Signaling Determine Whether T Cells Induce Tumor Dormancy or Promote Multistage Carcinogenesis. *Cancer Cell* **2008,** *13*, 507–518.
85. Braumuller, H.; Wieder, T.; Brenner, E.; Assmann, S.; Hahn, M.; Alkhaled, M.; Schilbach, K.; Essmann, F.; Kneilling, M.; Griessinger, C. et al. T-Helper-1-Cell Cytokines Drive Cancer into Senescence. *Nature* **2013,** *494*, 361–365.
86. Mellman, I.; Coukos, G.; Dranoff, G. Cancer Immunotherapy Comes of Age. *Nature* **2011,** *480*, 480–489.

87. Gajewski, T. F.; Fuertes, M.; Spaapen, R.; Zheng, Y.; Kline, J. Molecular Profiling to Identify Relevant Immune Resistance Mechanisms in the Tumor Microenvironment. *Curr. Opin. Immunol.* **2011,** *23*, 286–292.
88. Mantovani, A.; Sica, A. Macrophages, Innate Immunity and Cancer: Balance, Tolerance, and Diversity. *Curr. Opin. Immunol.* **2010,** *22*, 231–237.
89. Beavis, P. A.; Stagg, J.; Darcy, P. K.; Smyth, M. J. CD73: A Potent Suppressor of Antitumor Immune Responses. *Trends Immunol.* **2012,** *33*, 231–237.
90. Beavis, P. A.; Divisekera, U.; Paget, C.; Chow, M. T.; John, L. B.; Devaud, C.; Dwyer, K.; Stagg, J.; Smyth, M. J.; Darcy, P. K. Blockade of A2A Receptors Potently Suppresses the Metastasis of CD73+Tumors. *Proc. Natl. Acad. Sci. USA* **2013,** *110*, 14711–14716.
91. Wang, L.; Rubinstein, R.; Lines, J. L.; Wasiuk, A.; Ahonen, C.; Guo, Y.; Lu, L. F.; Gondek, D.; Wang, Y.; Fava, R. A. et al. VISTA, a Novel Mouse Ig Superfamily Ligand That Negatively Regulates T Cell Responses. *J. Exp. Med.* **2011,** *208*, 577–592.
92. Ceeraz, S.; Nowak, E. C.; Noelle, R. J. B7 Family Checkpoint Regulators in Immune Regulation and Disease. *Trends Immunol.* **2013,** *34*, 556–563.
93. Hobo, W.; Norde, W. J.; Schaap, N.; Fredrix, H.; Maas, F.; Schellens, K.; Falkenburg, J. H.; Korman, A. J.; Olive, D.; van der Voort, R. et al. B and T Lymphocyte Attenuator Mediates Inhibition of Tumor-Reactive CD8+ T Cells in Patients After Allogeneic Stem Cell Transplantation. *J. Immunol.* **2012,** *189*, 39–49.
94. Fourcade, J.; Sun, Z.; Pagliano, O.; Guillaume, P.; Luescher, I. F.; Sander, C.; Kirkwood, J. M.; Olive, D.; Kuchroo, V.; Zarour, H. M. CD8(+) T Cells Specific for Tumor Antigens Can Be Rendered Dysfunctional by the Tumor Microenvironment Through Upregulation of the Inhibitory Receptors BTLA and PD-1. *Cancer Res.* **2012,** *72*, 887–896.
95. Marabelle, A.; Kohrt, H.; Sagiv-Barfi, I.; Ajami, B.; Axtell, R. C.; Zhou, G.; Rajapaksa, R.; Green, M. R.; Torchia, J.; Brody, J. et al. Depleting Tumor-Specific Tregs at a Single Site Eradicates Disseminated Tumors. *J. Clin. Investig.* **2013,** *123*, 2447–2463.
96. Pardoll, D. M. The Blockade of Immune Checkpoints in Cancer Immunotherapy. *Nat. Rev. Cancer* **2012,** *12*, 252–264.
97. Vanneman, M.; Dranoff, G. Combining Immunotherapy and Targeted Therapies in Cancer Treatment. *Nat. Rev. Cancer* **2012,** *12*, 237–251.
98. Hodi, F. S.; O'Day, S. J.; McDermott, D. F.; Weber, R. W.; Sosman, J. A.; Haanen, J. B.; Gonzalez, R.; Robert, C.; Schadendorf, D.; Hassel, J. C. et al. Improved Survival with Ipilimumab in Patients with Metastatic Melanoma. *N. Engl. J. Med.* **2010,** *363*, 711–723.
99. Brahmer, J. R.; Tykodi, S. S.; Chow, L. Q.; Hwu, W. J.; Topalian, S. L.; Hwu, P.; Drake, C. G.; Camacho, L. H.; Kauh, J.; Odunsi, K. et al. Safety and Activity of Anti-PD-L1 Antibody in Patients with Advanced Cancer. *N. Engl. J. Med.* **2012,** *366*, 2455–2465.
100. Topalian, S. L.; Hodi, F. S.; Brahmer, J. R.; Gettinger, S. N.; Smith, D. C.; McDermott, D. F.; Powderly, J. D.; Carvajal, R. D.; Sosman, J. A.; Atkins, M. B. et al. Safety, activity, and Immune Correlates of Anti-PD-1 Antibody in Cancer. *N. Engl. J. Med.* **2012,** *366*, 2443–2454.
101. Wolchok, J. D.; Kluger, H.; Callahan, M. K.; Postow, M. A.; Rizvi, N. A.; Lesokhin, A. M.; Segal, N. H.; Ariyan, C. E.; Gordon, R. A.; Reed, K. et al. Nivolumab Plus Ipilimumab in Advanced Melanoma. *N. Engl. J. Med.* **2013,** *369*, 122–133.
102. Dranoff, G. Immunotherapy at Large: Balancing Tumor Immunity and Inflammatory Pathology. *Nat. Med.* **2013,** *19*, 1100–1101.

103. Angell, H.; Galon, J. From the Immune Contexture to the Immunoscore: The Role of Prognostic and Predictive Immune Markers in Cancer. *Curr. Opin. Immunol.* **2013,** *25*, 261–267.
104. Galon, J.; Angell, H. K.; Bedognetti, D.; Marincola, F. M. The Continuum of Cancer Immunosurveillance: Prognostic, Predictive, and Mechanistic Signatures. *Immunity* **2013,** *39*, 11–26.
105. Mlecnik, B.; Tosolini, M.; Charoentong, P.; Kirilovsky, A.; Bindea, G.; Berger, A.; Camus, M.; Gillard, M.; Bruneval, P.; Fridman, W. H. et al. Biomolecular Network Reconstruction Identifies T-Cell Homing Factors Associated with Survival in Colorectal Cancer. *Gastroenterology* **2010,** *138*, 1429–1440.
106. Gajewski, T. F.; Schreiber, H.; Fu, Y. X. Innate and Adaptive Immune Cells in the Tumor Microenvironment. *Nat. Immunol.* **2013,** *14*, 1014–1022.
107. Balachandran, V. P.; Cavnar, M. J.; Zeng, S.; Bamboat, Z. M.; Ocuin, L. M.; Obaid, H.; Sorenson, E. C.; Popow, R.; Ariyan, C.; Rossi, F. et al. Imatinib Potentiates Antitumor T Cell Responses in Gastrointestinal Stromal Tumor Through the Inhibition of Ido. *Nat. Med.* **2011,** *17*, 1094–1100.
108. Delahaye, N. F.; Rusakiewicz, S.; Martins, I.; Menard, C.; Roux, S.; Lyonnet, L.; Paul, P.; Sarabi, M.; Chaput, N.; Semeraro, M. et al. Alternatively Spliced NKp30 Isoforms Affect the Prognosis of Gastrointestinal Stromal Tumors. *Nat. Med.* **2011,** *17*, 700–707.
109. Weiss, G. R.; Grosh, W. W.; Chianese-Bullock, K. A.; Zhao, Y.; Liu, H.; Slingluff, C. L.; Jr, Marincola, F. M.; Wang, E. Molecular Insights on the Peripheral and Intratumoral Effects of Systemic High-Dose rIL- 2 (aldesleukin) Administration for the Treatment of Metastatic Melanoma. *Clin. Cancer Res.* **2011,** *17*, 7440–7450.
110. Ulloa-Montoya, F.; Louahed, J.; Dizier, B.; Gruselle, O.; Spiessens, B.; Lehmann, F. F.; Suciu, S.; Kruit, W. H.; Eggermont, A. M.; Vansteenkiste, J. et al. Predictive Gene Signature in MAGE-A3 Antigen-Specific Cancer Immunotherapy. *J. Clin. Oncol.* **2013,** *31*, 2388–2395.
111. Ji, R. R.; Chasalow, S. D.; Wang, L.; Hamid, O.; Schmidt, H.; Cogswell, J.; Alaparthy, S.; Berman, D.; Jure-Kunkel, M.; Siemers, N. O. et al. An Immune-Active Tumor Microenvironment Favors Clinical Response to Ipilimumab. *Cancer Immunol. Immunother.* **2012,** *61*, 1019–1031.
112. Halama, N.; Michel, S.; Kloor, M.; Zoernig, I.; Benner, A.; Spille, A.; Pommerencke, T.; von Knebel, D. M.; Folprecht, G.; Luber, B. et al. Localization and Density of Immune Cells in the Invasive Margin of Human Colorectal Cancer Liver Metastases Are Prognostic for Response to Chemotherapy. *Cancer Res.* **2011,** *71*, 5670–5677.
113. Ignatiadis, M.; Singhal, S. K.; Desmedt, C.; Haibe-Kains, B.; Criscitiello, C.; Andre, F.; Loi, S.; Piccart, M.; Michiels, S.; Sotiriou, C. Gene Modules and Response to Neoadjuvant Chemotherapy in Breast Cancer Subtypes: A Pooled Analysis. *J. Clin. Oncol.* **2012,** *30*, 1996–2004.
114. Denkert, C.; Loibl, S.; Noske, A.; Roller, M.; Muller, B. M.; Komor, M.; Budczies, J.; Darb-Esfahani, S.; Kronenwett, R.; Hanusch, C. et al. Tumor-Associated Lymphocytes as an Independent Predictor of Response to Neoadjuvant Chemotherapy in Breast Cancer. *J. Clin. Oncol.* **2010,** *28*, 105–113.
115. Tosolini, M.; Kirilovsky, A.; Mlecnik, B.; Fredriksen, T.; Mauger, S.; Bindea, G.; Berger, A.; Bruneval, P.; Fridman, W. H.; Pages, F. et al. Clinical Impact of Different Classes of Infiltrating T Cytotoxic and Helper Cells (Th1, th2, treg, th17) in Patients with Colorectal Cancer. *Cancer Res.* **2011,** *71*, 1263–1271.

116. Grivennikov, S. I.; Wang, K.; Mucida, D.; Stewart, C. A.; Schnabl, B.; Jauch, D.; Taniguchi, K.; Yu, G. Y.; Osterreicher, C. H.; Hung, K. E. et al. Adenoma-Linked Barrier Defects and Microbial Products Drive IL-23/IL-17-Mediated Tumor Growth. *Nature* **2012,** *491*, 254–258.
117. Hayashi, T.; Kobayashi, Y.; Kohsaka, S.; Sano, K. The Mutation in the ATP-Binding Region of JAK1, Identified in Human Uterine Leiomyosarcomas, Results in Defective Interferon-Gamma Inducibility of TAP1 and LMP2. *Oncogene* **2006,** *25*, 4016–4026.
118. Moretta, L.; Bottino, C.; Pende, D.; Mingari, M. C.; Biassoni, R.; Moretta, A. Human Natural Killer Cells: Their Origin, Receptors and Function. *Eur. J. Immunol.* **2002,** *32*, 1205–1211.
119. Cicinnati, V. R.; Zhang, X.; Yu, Z.; Ferencik, S.; Schmitz, K. J.; Dworacki, G.; Kaczmarek, E.; Oldhafer, K.; Frilling, A.; Baba, H. A.; Schmid, K. W.; Grosse-Wilde, H.; Broelsch, C. E.; DeLeo, A. B.; Gerken, G.; Beckebaum, S. Increased Frequencies of CD8+ T Lymphocytes Recognizing Wild-Type p53-Derived Epitopes in Peripheral Blood Correlate with Presence of Epitope Loss Tumor Variants in Patients with Hepatocellular Carcinoma. *Int. J. Cancer*. **2006,** *119*, 2851–2860.
120. Zhou, G.; Lu, Z.; McCadden, J. D.; Levitsky, H. I.; Marson, A. L. Reciprocal Changes in Tumor Antigenicity and Antigen-Specific T Cell Function During Tumor Progression. *J Exp. Med.* **2004,** *200*, 1581–1592.
121. Vainio, S.; Lin, Y. Coordinating Early Kidney Development: Lessons from Gene Targeting. *Nat. Rev. Genet.* **2002,** *3*, 533–543.

CHAPTER 4

Understanding of Antigen and Phenotypic Plasticity Molecules for the Purpose of Cancer Immunotherapy

ABSTRACT

Involving the immune system in the battle against cancer has shifted its attention to the T lymphocyte, in particular its ability for antigen-directed cytotoxicity. Checkpoint blocking, adoptive cellular therapy, and cancer vaccinology are just a few examples of the innovative approaches that have emerged because of advances in basic research into the T cell's molecular and cellular biology. Over the last half-century, this discipline of immunology has seen extraordinary bench-to-bedside clinical success. Here, the author provides a holistic biological and historical context for the development and clinical use of cancer immunotherapeutics, with a special focus on the central role of T cell control. Here, the researcher focuses on clinical studies that prove the treatment effectiveness and related toxicities of different medication classes. Finally, the author highlights the unanswered issues and untapped potential in cancer immunotherapy by summarizing developing medicines.

4.1 INTRODUCTION

In the nineteenth century, researchers first started considering ways to use the body's immune system to combat the disease. Earlier epidemiological data between immunity and cancer was initially provided by researchers like Wilhelm Busch and Friedrich Fehleisen. Patients who had erysipelas, a superficial skin infection often caused by Streptococcus pyogenes, experienced spontaneous remission of their tumors.[1,2] Later research by William

Advances in Immunological Treatment of Cancer: Cellular Mechanism and Regulation.
Rishabha Malviya, Arun Kumar Singh, and Deepika Yadav (Authors)

Coley, sometimes known as the "Father of Cancer Immunotherapy," found that erysipelas was linked to a better prognosis in sarcoma patients. Coley gave heat-inactivated S. pyogenes and Serratia marcescens extracts to cancer patients as a technique to prospectively examine his epidemiological results[3,4] Coley's toxins were utilized in successful cancer therapies. Coley's poisons were not extensively employed as a treatment because of a lack of scientific rigor and repeatability, as well as the development of radiation and chemotherapy medications.[5]

Thanks to modern medical advancements, the concept of cancer immunotherapy has been revived in the twenty-first century. Paul Ehrlich hypothesized the existence of neoplastic cells in humans in 1909,[6,7] which are subsequently destroyed by the immune system. Cancer immunosurveillance was separately proposed by Lewis Thomas and Sir Frank Macfarlane Burnet. This theory proposes that the immune system recognizes and kills tumor-associated neoantigens to prevent carcinogenesis, in a manner analogous to graft rejection. Despite the lack of a unifying mechanism, this idea has been supported by data such as productive immune responses following tumoral adoptive transfer in mice[4] and clinical reports of spontaneous regression of melanoma in patients with associated autoimmune sickness.[8–10] Thanks to advances in knockout mice models, researchers now have the tools to make the first direct experimental connection between immune insufficiency and cancer. Additional molecular and pharmacological research allowed for the discovery of tumor-specific immune responses.[11] This conclusively demonstrated the ability of T cells to destroy malignant tissue in the immune system. Major shifts have occurred in oncology with the advent of immunotherapy for cancer,[12] which has dramatically improved survival rates for individuals with especially lethal forms of the disease. As immunotherapies become the gold standard for treating many forms of cancer, the number of individuals who can benefit from them increases dramatically. It is expected that immunotherapy will become more important in the treatment of cancer over the coming several decades,[13] as new druggable targets are discovered and novel treatment combinations are developed.

In this chapter, the author discusses the role of T cells in contemporary cancer immunotherapies, including the use of immune checkpoint blockade to "unleash" potent T cell responses, adoptive cellular therapies based on the infusion of tumor-fighting immune cells into the body, and cancer vaccines that can be engineered to have either preventative or therapeutic effects. Finally, the author provides a summary of some promising recent developments in the field of cancer immunotherapy.[14]

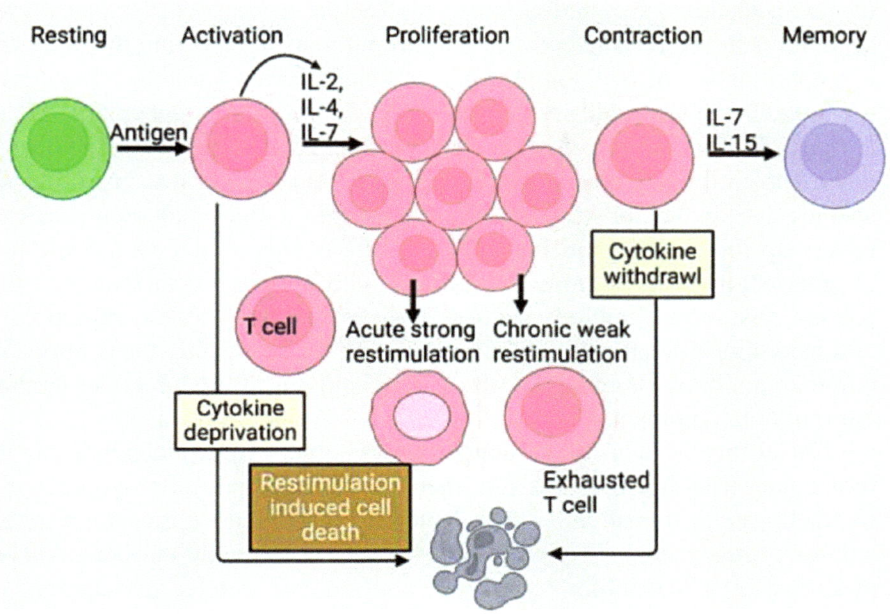

FIGURE 4.1 Schematic representation of phenotypic plasticity molecules for the purpose of cancer immunotherapy.

4.2 POTENTIALS AND CONSTRAINTS OF MONOCLONAL ANTIBODIES IN CANCER TREATMENT

The amazing evolution of monoclonal antibodies (mAbs) from research tools to strong medications in clinical usage[15] dates back to their initial use on a cancer patient in 1980. More than 90 monoclonal antibodies have made it from the lab to the clinic in the past 30 years, thanks in large part to the European Medicines Agency (EMA) and the Food and Drug Administration (FDA). Specifically for cancer treatment, the use of therapeutic monoclonal antibodies (mAbs) has risen rapidly in popularity as a promising first-line option for the treatment of immunologic illnesses. Since the first licensed antibody was released, several more have been produced (muromonab, which inhibits allograft rejection by targeting CD3). From their inception to the introduction of revolutionary new ideas in structural design and therapeutically relevant targets,[16–20] a wide range of methods have been developed to create them for the human immune system.

IgGs are the most prevalent kind of antibody, and their Y-shaped structure is composed of two dimers, each of which consists of three domains of nearly identical size and is joined by flexible linkers. The Y-shaped Fab segment is the active site because it binds antigens. Everything that they are is a combination of two components: a constant and a variable.[21] Paratopes are found in the latter and are responsible for binding the target epitope of the antigen. The Y is represented by stem Fc fragments, which engage effector cells. The antibody molecule itself may function as an antigen in a different species due to the subtype and species-specific amino acid sequence that makes up the majority of the antibody. Due to this, antibody therapy has progressed from its humble beginnings with mouse monoclonals to the present day, when methods such as transferring the variable region to a corresponding human antibody (chimeric Ab), replacing only the paratope in a human antibody (humanized Ab), and developing fully human antibodies are routinely employed.[22]

The ability of a mAb to neutralize its target antigen depends on its molecular structure. It is possible that they directly promote apoptosis by interacting with transmembrane receptors to either reduce receptor expression or disrupt receptor connections with nearby signaling partners. Abciximab is a drug that inhibits gpIIb/IIIa and reduces platelet aggregation; in this situation, the Fab fragment is excellent as it supplies the solitary required action of receptor blockage. This chimeric antibody is the first to get FDA approval and represents an additional attempt to reduce immunogenicity.[23] This chimeric antibody retains just the variable section of the original mouse IgG, whereas the constant component is human IgG1.

Antibody-dependent cellular cytotoxicity may be triggered by IgG1 and IgG3 subtypes by activating Fc receptors on cytotoxic immune cells (monocytes, macrophages, NK cells, and granulocytes). Some mAbs may induce indirect cytotoxicity by triggering the complement cascade. Many different types of cancer have responded well to monoclonal antibodies because of their specificity. After 10 years of waiting, rituximab (which targets CD20 on lymphoma cells) was finally given the green signal. In the same way, as abciximab is a chimeric, mouse/human antibody,[24–26] so is rituximab.

A few years later, toward the end of the decade, trastuzumab was authorized as the first antibody targeting HER2 in solid tumors. Because it was humanized by replacing the mouse variable area framework residues and IgG1 constant domains with those from humans, this antibody is a great example of antibody engineering. Initially, bevacizumab was hypothesized to block VEGF and reduce tumor angiogenesis; currently, humanized antibody therapy[27] targets the cancer stroma.

The next conceptual and structural step was the use of antibodies to deliver poisons or conventional chemotherapeutic drugs to tumor cells. To treat Hodgkin lymphoma, brentuximab vedotin was included in the first antibody–drug combination (ADC) licensed for usage. It works by binding to CD30 on Reed–Sternberg cells. When it comes to treating solid tumors, trastuzumab was the first ever authorized monoclonal antibody (mAb).[28]

Chimeric multipara topic antibodies have been developed and tested successfully in preclinical settings in part because of how easy it is to produce chimeric and humanized antibodies. Blinatumomab is the first to get clearance from regulators since it is a bispecific T cell engager (BiTE). This novel class of immunotherapeutic medicines, developed from synthetic antibodies, is intended to target two antigens on cancer cells (in this example, CD19 on B-ALL cells) at once.[29]

A big advancement in cancer treatment occurred in the previous decade with the discovery of T cell-targeted immunomodulators that interfere with immunological checkpoints. Ipilimumab, which blocks CTLA4 signaling, was one of the first mAb-based immune checkpoint inhibitors to be authorized by the FDA. Later, other monoclonal antibodies specifically targeting PD1 (pembrolizumab and nivolumab) and PD-L1 (advertise) were developed (atezolizumab and durvalumab). In recent years, anti-PD1/PD-L1 monoclonal antibodies have become a standard anticancer treatment option in the United States. More than seventy different types of cancer have been treated with these compounds, either alone or in combination with other therapies like chemotherapy and radiation.[30]

Despite significant progress achieved with mAbs, a subset of patients still have poor responses due to the advanced stage of their malignancy upon diagnosis or to treatment-induced drug resistance. Several mechanisms have been identified that cause resistance to monoclonal antibodies (mAbs), such as those that prevent mAbs from binding to their target antigen (tumor microenvironment [TME]/extracellular matrix [ECM] or downregulation of the target), those that induce resistance by activating alternative signaling pathways in the tumor, and those that have immunomodulatory effects on tumor cells or the TMME via checkpoint activators, exosomes, and microvesicles.[31]

The researcher has spent the past decade investigating the causes of resistance to trastuzumab, the most common HER2-specific monoclonal antibody. To dampen effector function, the author found that overexpressing ECM components such as MUC4 or CD44/hyaluronan is crucial.[32] These extracellular matrix (ECM) components build up a steric barrier that hides the HER2 antigen and prevents therapeutic mAbs from recognizing it. In advanced tumors with an abundant extracellular matrix, epitope masking of HER2 has been shown in vitro, in vivo, and in clinical models. The immune response known as antibody-dependent cellular cytotoxicity (ADCC) may still be able to kill tumor cells that have spread from an earlier stage after trastuzumab has bound to them. Increasing ADCC and effector cell recruitment is achieved with the combination of pertuzumab and trastuzumab, which in turn inhibits tumor development.[33] These findings provide support for the hypothesis that ADCC by cytotoxic cells might perhaps overcome early resistance and impede cancer growth in the presence of an immature extracellular matrix, even when tumor cells demonstrate their intrinsic biological resistance.

Based on these results, the author attempted to engineer a cytotoxic T cell that could recognize extracellularly generated trastuzumab. Chimeric antigen receptors (CARs)[34] are the genetically engineered names for these synthetic molecules. In the treatment of tumors that have become resistant to traditional cancer therapies, our latest study shows that chimeric antigen receptor (CAR)-modified T cells have a considerable survival advantage over antibodies.

4.3 CANCER TREATMENT WITH IMMUNE CELLS: A POTENTIAL NEW STRATEGY

The development of cutting-edge biological and immunotherapeutic therapies has been widely hailed as a major step forward in cancer treatment during the last decade. The end game is to develop personalized therapies that are both patient-friendly and highly successful against various cancers.[35]

Adoptive immune cell treatment represents a major advancement in the search for safe and effective specialized medications. Autologous T cell therapy was first used by Steven A. Rosenberg and colleagues in the early 1980s to treat cancers that had spread to other parts of the body (so-called solid tumors). This groundbreaking study[36] used a megadose of IL-2 cytokine stimulation to activate and grow T cells isolated from the peripheral blood mononuclear cells of cancer patients. Adoptive transplants of T cells isolated from peripheral blood and, later, from tumor tissue (tumor-infiltrating lymphocyte—TIL) were used to treat patients with advanced cancer in the next decade. Patients with melanoma (a malignant skin tumor) had a low rate of persistent success; over 40% of them reacted to treatment, and a sizable fraction (approximately 15%) were even totally cured. On the other hand, some solid tumors did not show any significant improvement, and in many instances, no treatment response was seen.[37]

Scientists introduced the first genetically engineered T cells in the early 1990s to enhance adoptive T cell immunotherapy. This cell type selectively targeted and eliminated many tumor cell lines by adhering to tumor-associated antigens (TAAs) on their surfaces. The promising results of the preclinical studies employing these genetically engineered T cells have sparked new interest in tailoring cancer treatments to each patient individually. Redirecting T cells to attack tumors, this cutting-edge method combines gene editing through retroviral or lentiviral transduction with adoptive immunotherapy.[38–40]

Alpha and beta subunits of the T cell receptor complex are involved in targeting tumors, specifically in immunoengineered T cells. These molecules are essential for the detection of peptide antigens released by cells regularly, whether they originate from viruses or the cellular proteome. Technological advancements have made it possible to construct and encode in DNA an endless number of TCRs and molecules specific to tumor neoantigens, which can subsequently be integrated into the T cell membrane for recognition.[41] Despite the impressive effectiveness of adoptive transfer of tumor-infiltrating lymphocytes in the treatment of malignancies expressing TAA peptides persistently, especially melanomas, designer T cells transduced with specific TCRs have demonstrated even better efficiency. Though TCR-redirected T cells have entered the early phases of clinical testing, they have not yet acquired full clinical approval. To that end, the author would like to draw your attention to a recent meta-analysis of clinical studies including transgenic TCRs,[42] which the author judged to be of extremely excellent quality.

4.4 STRUCTURE OF CHIMERIC ANTIGEN RECEPTORS

It has been shown that the first generation of chimeric antigen receptors, which are synthetic molecules, can recognize tumor-associated antigens on the surface of cells and trigger a cytotoxic T cell response. As can be seen in Figure 4.2, these CARs do not provide the first antigen-dependent signal (signal I) required for T cell activation and target death.[43]

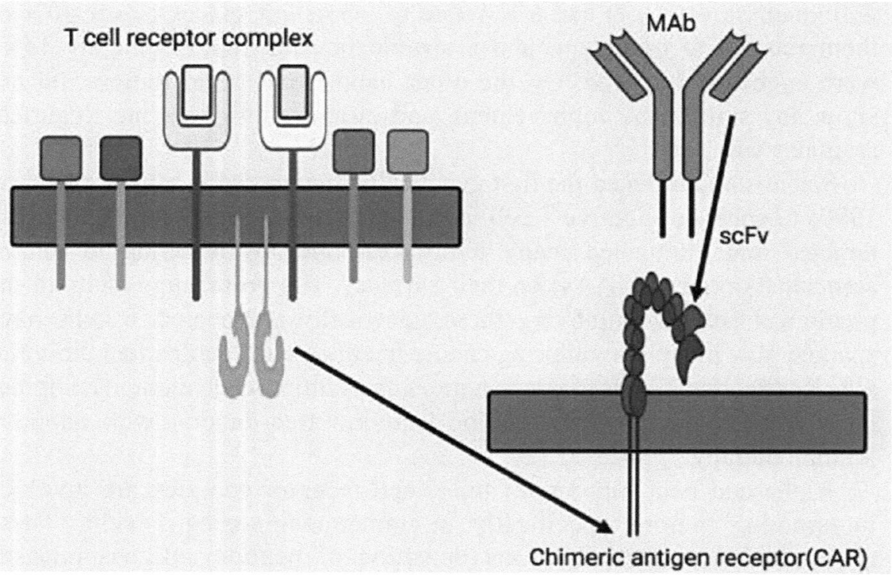

FIGURE 4.2 Diagrammatic representation of the T cell receptor complex.

The variable light (VL) and heavy (VH) chains of the parent monoclonal antibody are commonly used as the extracellular antigen recognition domain of these antibodies.[44] This yields components that are independent of more than one chain (scFvs). Native extracellular proteins may be bound efficiently in the absence of MHC thanks to their mAb-like target recognition capabilities. Even though CAR T cells' targets may downregulate HLA molecules to avoid immune surveillance, these cells are nonetheless effective. As chimeric antigen receptors recognize their targets like antibodies, they may be able to bind tumor-specific antigens that are not protein-dominant, such as polysaccharides or highly glycosylated proteoglycans.[45]

To go from the antigen recognition moiety to the intracellular effector domain, the molecule requires transmembrane segments and a hinge (or linker) region. Collectively, they constitute a relatively inconsequential but still important part of the CAR's overall architecture.[46] Usually, CD4, CD8, or CD28 provide the transmembrane domain, whereas immunoglobulin G provides the hinge region (IgG1 or IgG4). The composition and length of the hinge region regulate the mobility of the antigen recognition domain and the physical distance between the T cell and the target cell.[47–50]

The transmembrane domains of CARs are essential for membrane localization, endogenous or target-induced dimerization, and cluster formation.

The rate at which CAR synapses are formed between cells has a significant impact on the cytolytic T cell response. CD4, CD8, and CD28 sequences are used because they are thought to transmit signals just as effectively as the native proteins.[51] This is due to the transmembrane region's importance in relaying the conformational change generated by target attachment to the CAR's cytosolic component.

The intracellular effector (or signaling) region is the least conserved portion of the CAR design despite its importance in T cell activation. It is a common feature of the gamma chain of the high-affinity immunoglobulin-E receptor and the zeta subunit of the T cell receptor complex (CD3) (FcRI). Effector domains stimulate T cell activation, proliferation, and cytotoxic response[52] by activating phosphatidylinositol and tyrosine kinase-dependent signaling pathways.

One or more costimulatory endodomains are inserted intracellularly upstream of the signaling domain in second- and third-generation CARs.[53] These domains, which can originate from molecules like CD27, CD28, 4-1BB, or OX40, act as signal amplifiers by passing along the costimulatory signals required for full T cell activation (signal II). First-generation chimeric antigen receptors (CARs) were effective in killing cancer cells in animal tests, but CARs containing a defense mechanism from the costimulatory endodomain were even more effective. However, it seems that different costimulatory domains have varying degrees of impact on the effector response. As an example, the antigen-binding signaling cascade of the CD28 cassette may inhibit mitochondrial biogenesis. After showing that CD28 costimulatory CAR T cells produce most of their energy through glycolysis, Kawalekar et al. showed that 4-1BB CAR T cells generate most of their energy via oxidative phosphorylation in mitochondria. CAR T cells undergo rapid self-exhaustion and differentiation into terminal effector memory cells as a functional consequence of CD28-derived costimulation, resulting in a powerful cytotoxic response. Due to the attenuated cytotoxic response mediated by the 4-1BB costimulatory endodomain, CAR T cells persist as core memory subsets for a longer period of time.[54]

In in vitro and in vivo preclinical models, third-generation chimeric antigen receptors, including both types of costimulatory endodomains, enhanced effector functions and killing capability, but there was a significant therapeutic risk associated with their translation into clinical use.

State-of-the-art CAR T cell research focuses on reprogrammed T cells for universal cytokine-mediated death (TRUCKs). These CAR T cells have been engineered to secrete a transgenic cytokine (often IL-7, IL-12, or

IL-15) when they encounter NFB in tumor tissue. Combining the anticancer activities of CAR T cells with the immune-modulating effects of the chosen cytokine makes these TRUCKs an effective cancer therapy.[55] Transporter-like antigen-presenting cells (TRUCKs) are optimized for the delivery of monoclonal antibodies to particular organs (such as PD-1 inhibitors or bispecific T cell activator molecules). These compounds either release T cells from the possible checkpoint regulation by the tumor environment or attract bystander tumor-infiltrating immune cells, and together they result in "full T cell activation".[56]

4.5 CHIMERIC ANTIGEN RECEPTOR T CELL GENETIC ALTERATION

Before the development of virus-free DNA transfection technology, chimeric antigen receptor (CAR)-modified T cells had to be produced by hand. Despite the procedure's advantages (lack of immunogenicity and mutation likelihood), some have voiced worry about the short life cycle of transfected cells and the quick decline in CAR expression.[57]

Transduction using either lentiviral or gamma retroviral vectors is often used in the production of CAR T cells for use in human clinical trials and animal models.[58] Transfecting a plasmid expressing the chimeric receptor into primitive human T cells may be an effective and safe way to modify these cells. There is no difference in therapeutic efficacy between lentiviruses and retroviruses, even though lentiviruses may infect and genetically modify quiescent cells whereas retroviruses can only infect developing cells.[59,60]

It is challenging and costly to create new methods of manufacturing CAR T cells because of worldwide bans on the commercial use of viral systems. New, nonviral methods are being sought in the hope that they may make long-term, low-cost CAR T cell infusions possible.[61] Functional CAR T cells were generated in preclinical animals using the Sleeping Beauty (SB) and piggyBac transposase techniques, but their clinical effectiveness was limited. CAR T cells were generated by the Penn team by mRNA electroporation[62] rather than the more common practice of inserting the transgene into the DNA. These measures helped researchers maintain their experiment going smoothly throughout the changeover, accomplish transitory CAR expression, and bank a sizable number of CAR T cells for use in future clinical trials. However, there is a paucity of published information on the approach's therapeutic efficacy, and it does not provide long-term persistence or the establishment of particular memory cells. Researchers at the

Fred Hutchinson Cancer Research Center in Seattle[63] used DNA-carrying nanoparticles to efficiently alter the genome of circulating peripheral T cells with chimeric antigen receptors in an animal model of leukemia caused by xenotransplantation of a cell line, resulting in long-lasting remission.

4.6 CLINICAL STUDIES WITH CAR T CELLS

4.6.1 PRODUCTION OF CHIMERIC ANTIGEN RECEPTOR T CELL PROTEINS

The patient's blood is used in many CAR T cell therapies, which is manufactured in GMP-approved bioreactors (autologous treatment). Using human monoclonal antibodies against CD3 (signal I) and CD28 (signal II), T cells are activated after fast flow cytometric sorting of blood mononuclear cells.[64] The cell culture is given the appropriate dose of cytokines (IL-2 or IL-7 + IL-15) for maximal activation. Then, the CAR construct is introduced into the activated T cells' DNA using one of the previously discussed techniques.[65] Next, the author evaluates the efficiency and precision of CAR transduction. It will be frozen until it is ready to be used if it passes all of the rigorous release conditions. Lymph ablative chemotherapy is a beneficial therapeutic modality before CAR T cell therapy, as shown by several clinical studies.[66]

4.6.2 UTILIZING CHIMERIC ANTIGEN RECEPTOR T CELLS TO FIGHT HEMATOPOIETIC TUMORS

Patients with acute lymphoblastic leukemia who have not responded to normal therapy or a bone marrow transplant have a new source of optimism in CD19-specific CAR T cells. Clinical use of CD19-specific CAR T cells led to CR in about 70% of patients in this very poor prognosis group according to a meta-analysis of studies published within the previous 3 years.[67] Only around 10% of the improvement group showed no response to the therapy. In a clinical assessment of anti-CD19 CART at the University of Pennsylvania, 87% of adults and 75% of children had complete responses. The results from the NCI and the MSKCC were identical.[68] Lower complete remission (CR) rates were seen in patients who had extramedullary or neurological disease involvement or who had had numerous allogeneic bone marrow transplants prior to CAR T cell treatment (40–50 percent). Those with "escape" variations (those positive for markers other than CD19, such as CD22 or myeloid markers) were at a higher risk of recurrence.[69]

Similarly, anti-CD19 CAR T cell therapy is showing promising results in the treatment of chronic B cell (CD19+) lymphocytic leukemia (B-CLL) in the United States.[70] In clinical studies at the National Cancer Institute (NCI), Memorial Sloan Kettering Cancer Center (MSKCC), University of Pennsylvania (UPenn), and Baylor College of Medicine (BCM), more than forty patients were treated with a CD19-specific T cell treatment. Complete remission (CR), partial remission (PR), or stable disease (SD) occurred in 10% of individuals with leukemia.[71,72]

After poor findings from clinical trials employing first-generation chimeric antigen receptor T cells (CAR T cells) for lymphoma, later research, including improved CAR T cell formulations (now in their second and third generations), has shown hopeful outcomes. In the NCI trial, anti-CD19 CAR T cells were used to treat twenty-two patients with low-dose conditioning chemotherapy-resistant diffuse large B cell lymphoma (DLBCL). One patient had a stable illness, while 17 had a full response (SD).[73–76]

Researchers at the University of Pennsylvania used CD19-specific CAR T cells to effectively cure the patient's multiple myeloma after myeloablative chemotherapy had failed to do so. There was no trace of the tumor left in the patient's cells after the drug was administered. Although just 0.5% of the patient's plasma cells displayed the CD19 antigen, CAR T cell therapy was effective because this population likely represented the least developed cancer reservoir that was resistant to conventional therapies.[77]

Over the last 10 years, more than 10 separate research organizations have been conducting clinical trials involving people with acute myeloid leukemia. To date, the researchers have conducted the first clinical trial specifically addressing the Lewis Y antigen. While the research did find a reduction in malignant blast counts in the short- and medium-term, patients did relapse over the study's intermediate and long-term follow-up periods.[78] It has also been shown that AML-specific CAR T cells that target CD33 and CD123 are effective. They are seen on the surface of normal and AML blast hematopoietic stem and progenitor cells (HSPCs).[79,80] In 2015, Wang et al. reported that a 41-year-old patient had temporary improvement after receiving CD33-specific CART cells. Acute myeloid leukemia may be diagnosed with the help of CD123, as proven in many studies.[81–83] In 2015, researchers at California's City of Hope National Medical Center began the first CART-123 experiment to test the effectiveness of a second-generation CAR. One patient who got a low dosage of CAR T cells achieved a morphologically leukemia-free status that persisted for 70 days,[84] whereas patients who received large doses of CAR T cells went into full remission. It was claimed that just three people in this group had gone into remission. No dose-limiting toxicities were seen, and individuals who had had cytopenias as a consequence of their treatment reported improvement by week 12.

Three recent years have been pivotal in the development of immune-cell-based oncotherapy, which has been in development for 31 years (since the first functional CAR T cells were generated) and 15 years (since their first clinical study).[85] In 2017, the FDA approved tisagenlecleucel[86] for the treatment of children with refractory B-ALL, and in 2018, axicabtagene ciloleucel[87] was approved for the treatment of adults with refractory diffuse large B cell lymphoma. For the treatment of refractory B-ALL in patients aged 0–25 and adult DLBCL, the European Medicines Agency (EMA) authorized tisagenlecleucel in 2018.[88]

4.6.3 UTILIZING CHIMERIC ANTIGEN RECEPTOR T CELLS TO FIGHT SOLID TUMORS

Only around 20% of clinical trials have looked at CAR T cells' effectiveness against solid tumors, and even fewer have looked at their effectiveness against leukemia. Multiple problems have arisen since the introduction of designer T cells for the treatment of solid tumors a decade ago.[89] Unfortunately, the vast majority of cancer types do not have a tumor-associated antigen that CAR T cells can target with high specificity. This results in a potentially lethal "on target, off tumor side effect" in healthy tissues. Second, CTLA-4 and PD-1, two checkpoint molecules associated with tumors, may inhibit T cell effector activities.[90–93] Finally, T cells that have been genetically altered in a solid tumor have a high rate of rapid and untimely demise. Still, several CAR T cell clinical studies targeting solid tumors in different organs have begun across the globe (lungs, esophagus, pancreas, stomach, breast, colon, brain, and ovary).

The human epidermal growth factor receptor 2 has been the subject of most clinical trials (ErbB2; EGFR2; and HER2). Tumor cells from the breast, lungs, stomach, glia, and pancreas all express HER2 strongly. Animal studies have shown the effectiveness of HER2-specific CAR T cells in the treatment of tumors expressing the target antigen.[94] Clinical adoption has been delayed despite promising findings.[95] Patients with HER2-positive osteosarcoma, glioma, rhabdomyosarcoma, and breast cancer who participated in clinical trials of CAR T cell therapy did not show a response to treatment in the vast majority of cases.[96] A cytokine storm fatally injured a 39-year-old woman who was taking part in a research study for HER2-positive metastatic colon cancer. This regulatory bottleneck slowed the development and commercialization of a treatment for solid tumors.[97]

Mesothelin, a glycoprotein, is the second most often targeted antigen because it is highly expressed in tumor cells (mesothelioma, lungs, stomach, pancreas, and esophageal cancer). Its role in healthy tissues is unclear, but its expression in malignant tumors is a poor prognostic indicator.[98] Animal studies have shown that mesothelin-specific chimeric antigen receptor T cells (meso-CAR T cells) may effectively cure mesothelin-positive tumors. Two patients in a clinical investigation of meso-CAR T cells electroporated with mRNA had little success.[99]

Disialoganglioside (GD2) is a glycolipid that is overexpressed on the surfaces of neuroblastoma, melanoma, and osteosarcoma cells, indicating that it might be a valuable therapeutic target. Only 3/19 participants in a therapeutic study employing GD2-specific CAR T cells had a full disappearance of their BCM.[100]

Despite a lack of clinical evidence, the EGFRvIII variant oncogene merits investigation since it is a member of the EGFR receptor family.[101] Overexpression of this molecule on cell membranes is a hallmark of cancer cells like those responsible for glioblastoma and head and neck cancer, but it is absent in normal somatic cells.[102] Results from early experiments with CAR T cells that target EGFRvIII were promising. Several studies[103] are now looking at the efficacy of treatments centered on this target molecule.

The remarkable success of CD19-specific CAR T cells in treating leukemia patients has propelled this therapy to the forefront of cancer immunotherapy. However, because of the highly immunosuppressive environment and the absence of unique cancer antigens, no therapeutic advantage has been identified in solid tumors.[104] The advent of new methods, however, gives rise to optimism that these restrictions could eventually be overcome. Developing safety mechanisms, T cell activation-dependent constructions, combinatorial antigen recognition strategies, multiple-targeting CARs, and CARs that target unhealthy (stressed) tissues, in general, are all steps toward the goal of creating finely tuned CAR T cells that avoid undesirable toxicities while efficiently eliminating tumors. Targeting intracellular proteins may be a function of T cell receptors (TCRs) that are substituted for or integrated into CARs.[105] The MHC molecules are the ones responsible for delivering the intracellular proteins to the TCRs for recognition. Since CARs cannot target intracellular targets, they can only attack around 1% of the likely neoantigens in tumors.

4.6.4 PREVENTING AND TREATING CAR T CELL THERAPY-RELATED TOXICITIES

CAR T cell therapy does not eliminate quickly dividing cells like conventional adjuvant therapies like chemotherapy and radiation. The resulting treatment has a very different side-effect profile compared to traditional medication. The researcher has classified the probable CAR T cell treatment adverse effects into two groups.[106]

As the name implies, "on target, off tumor" toxicity occurs when a CAR T cell locates a tumor-specific antigen on normal tissue and destroys the cells expressing it. It is widely known that CD19-specific CAR T cells have this undesirable impact because the target antigen is expressed on the surface of both normal B lymphocytes and malignant cells. They are healthy because they do not have B cell aplasia. Hypo- or agammaglobulinemia due to B cell depletion responds well to therapy with human intravenous immunoglobulin.[107]

Another CAR T cell treatment adverse effect that is receiving a lot of attention at the moment is cytokine release syndrome (CRS). Abnormally high levels of activating cytokines may cause serious symptoms such as fever, low blood pressure, systemic hypoxia, and, in extreme cases, multiorgan failure (interferon, granulocyte-macrophage colony-stimulating factor, tumor necrosis factor, interleukin-6, and interleukin-10). The role of IL-6 as a causative agent in the development of CRS has been the subject of many recent clinical investigations. The monoclonal antibody tocilizumab, which targets the IL-6 receptor, has greatly improved patient survival rates for treating this once-fatal condition. Indicating the efficacy of CAR T cell therapy, complete suppression of CRS is not desired.

Neurotoxicity, or CAR T cell-derived encephalopathy, is one of the most dangerous side effects of CAR T cell treatment, second only to infection. Dementia, aphasia, disorientation, somnolence, ataxia, tremors, and even cerebral edema are all possible side effects of a stroke. However, nothing is known about the biochemical basis of high-grade CRS-associated neurotoxicity, and it will take years to assess new treatment schemes.[108]

4.6.5 DESIGNER T CELLS AND THEIR POTENTIAL FOR USE IN OTHER DISORDERS

Immune system cells modified for therapeutic reasons must be able to selectively target and destroy cancerous or infectious cells without harming healthy ones. Inspired by the positive results of CAR T cell-based cancer therapy, several scientists are investigating the use of CAR-modified immune cells to treat autoimmune and infectious disorders.[109]

Autoimmune illnesses are characterized by chronic tissue damage caused by autoreactive B lymphocytes and plasma cells attacking healthy tissue under the false assumption that it is foreign. A possible solution to removing these B cells is the widespread use of monoclonal antibodies like rituximab, which recognizes its target through the CD20 protein displayed on its surface and produces an antibody-mediated cytotoxic response (ADCC). The CD20 molecule is not only expressed on autoreactive B cell clones but also acts as a sign for eliminating the beneficial population, resulting in total and chronic B cell aplasia. The septic complications of this illness might be fatal.

Using genetically engineered immune cells as a therapy for autoimmune illnesses is a contentious topic with two opposing viewpoints. Antibody-mediated disorders like pemphigus vulgaris (PV) might benefit from the use of chimeric antigen receptor T cells (CAAR T cells), which seek out and destroy autoantibody-expressing B cell clones. It has been shown that autoantigen-based chimeric immunoreceptors may excite T cells, leading to the targeted destruction of autoreactive B lymphocytes.[110]

Using a target of carcinoembryonic antigen, chimeric antigen receptor-modified T cells (CAR Tregs) have been demonstrated to reduce the autoimmune response in murine colitis.[111–113]

The public's interest in this area has been piqued by the notion that certain infectious illnesses are incurable. The use of immune cells that have been genetically engineered might significantly alter the therapy of many diseases. Here, the author discusses some of the most promising ways in which CAR T cells are being used to combat infectious illnesses.[114–116]

4.7 CONCLUSIONS

In its place, immunosuppressive therapy has given way to a broad range of oncoimmunologic medications that may be used to either directly target tumors or change immunological checkpoints as a result of the development of therapeutic antibodies. The topic of clinical cancer research using chimeric antigen receptors (CARs) created on designer T cells is cutting-edge and expanding rapidly. Current clinical studies including CAR T cell regimens are focused on their potential to treat multiresistant lymphoid malignancies. To successfully treat CD19-positive lymphomas and leukemias, CAR T cells have been used in the last decade. While there has been some promising research, the clinical outcomes for patients with solid tumors have been disappointing. A deadly cytokine release syndrome occurred at the same time as the conclusion of a clinical investigation on the HER2 tumor-specific antigen. This tragedy has shown the importance of evaluating the various parts of the CAR in terms of its activity, effectiveness, durability, and lifespan. Also, boosting the sensitivity and long-term potential of tumor detection technologies is an urgent need.

In the clinic, CAR T cells that target antibodies are being utilized to treat autoimmune diseases and viral infections. The author needs conclusive solutions to two important scientific questions in this field within the next decade. The first is whether modified T cells are effective in the treatment of solid tumors on their own or in combination with checkpoint inhibitors.

KEYWORDS

- **phenotypic plasticity**
- **immunotherapy**
- **monoclonal antibodies**
- **tumor**

REFERENCES

1. Ahmed, N. et al. HER2-Specific Chimeric Antigen Receptor Modifed Virus-Specific T Cells for progressive glioblastoma: a Phase 1 Dose-Escalation Trial. *JAMA Oncol.* **2017,** *3,* 1094–1101.
2. Ahmed, N. et al. Human Epidermal Growth Factor Receptor 2 (HER2) –Specific Chimeric Antigen Receptor-Modifed, T Cells for the Immunotherapy of HER2-Positive Sarcoma. *J. Clin. Oncol.* **2015,** *33,* 1688–1696.
3. Ali, S. et al. The European Medicines Agency Review of Kymriah (Tisagenlecleucel) for the Treatment of Acute Lymphoblastic Leukemia and Difuse Large B-Cell Lymphoma. *Oncologist* **2020,** *25,* e321–e327.
4. Anthony-Gonda, K. et al. Multi Specific Anti-HIV duoCAR-T Cells Display Broad In Vitro Antiviral Activity and Potent In Vivo Elimination of HIV-Infected Cells in a Humanized Mouse Model. *Sci. Transl. Med.* **2019**.
5. Barok, M. et al. Trastuzumab Decreases the Number of Circulating and Disseminated Tumor Cells Despite Trastuzumab Resistance of the Primary Tumor. *Cancer Lett.* **2008,** *260,* 198–208.

6. Barok, M. et al. Trastuzumab Causes Antibody-Dependent Cellular Cytotoxicity-Mediated Growth Inhibition of Submacroscopic JIMT-1 Breast Cancer Xenografts Despite Intrinsic Drug Resistance. *Mol. Cancer Ther.* **2007,** *6*, 2065–2072.
7. Barok, M.; Puhka, M.; Vereb, G.; Szollosi, J.; Isola, J.; Joensuu, H. Cancer-Derived Exosomes from HER2-Positive Cancer Cells Carry Trastuzumab-Emtansine into Cancer Cells Leading to Growth Inhibition and Caspase Activation. *BMC Cancer* **2018,** *18*, 504.
8. Baselga, J.; Norton, L.; Albanell, J.; Kim, Y. M.; Mendelsohn, J. Recombinant Humanized Anti-HER2 Antibody (Herceptin) Enhances the Antitumor Activity of Paclitaxel and Doxorubicin Against HER2/neu Overexpressing Human Breast Cancer Xenografts. *Cancer Res.* **1998,** *58*, 2825–2831.
9. Batra, S. A. et al. Glypican-3-Specific CAR T cells Coexpressing IL15 and IL21 Have Superior Expansion and Antitumor Activity Against Hepatocellular Carcinoma Cancer. *Immunol Res.* **2020,** *8*, 309– 320.
10. Beatty, G. L. et al. Mesothelin-Specific Chimeric Antigen Receptor mRNA-engineered T Cells Induce Anti-Tumor Activity in Solid Malignancies Cancer. *Immunol Res.* **2014,** *2*, 112–120.
11. Beatty, G. L.; O'Hara, M. Chimeric Antigen Receptor-Modified T Cells for the Treatment of Solid Tumors: Defining the Challenges and Next Steps. *Pharmacol. Ther.* **2016,** *166*, 30–39.
12. Bergsland, E.; Dickler, M. N. Maximizing the Potential of Bevacizumab in Cancer Treatment. *Oncologist* **2004,** *9* (Suppl 1), 36–42.
13. Blat, D.; Zigmond, E.; Alteber, Z.; Waks, T.; Eshhar, Z. Suppression of Murine Colitis and Its Associated Cancer by Carcinoembryonic Antigen-Specific Regulatory T Cells. *Mol. Ther.* **2014,** *22*, 1018–1028.
14. Bonifant, C. L.; Jackson, H. J.; Brentjens, R. J.; Curran, K. J. Toxicity and Management in CAR T-Cell Therapy. *Mol. Ther.* Oncolytics *3*, 16011.
15. Bonifant, C. L. et al. CD123-Engager T Cells as a Novel Immunotherapeutic for Acute Myeloid Leukemia. *Mol. Ther.* **2016,** *24*, 1615–1626.
16. Breman, E. et al. Overcoming Target Driven Fratricide for T Cell Therapy. *Front. Immunol.* **2018,** *9*, 2940.
17. Brentjens, R. J. et al. Safety and Persistence of Adoptively Transferred Autologous CD19-Targeted T Cells in Patients with Relapsed or Chemotherapy Refractory B-Cell Leukemias. *Blood* **2011,** *118*, 4817–4828.
18. Bridgeman, J. S.; Hawkins, R. E.; Bagley, S.; Blaylock, M.; Holland, M.; Gilham, D. E. The Optimal Antigen Response of Chimeric Antigen Receptors Harboring the CD3zeta Transmembrane Domain Is Dependent Upon Incorporation of the Receptor into the Endogenous TCR/CD3 Complex. *J. Immunol.* **2010,** *184*, 6938–6949.
19. Bridgeman, J. S.; Ladell, K.; Sheard, V. E.; Miners, K.; Hawkins, R. E.; Price, D. A.; Gilham, D. E. CD3zeta-Based Chimeric Antigen Receptors Mediate T Cell Activation via cis- and Trans-Signalling Mechanisms: Implications for Optimization of Receptor Structure for Adoptive Cell Therapy. Clin. Exp. Immunol. **2014,** *175*, 258–267.
20. Brudno, J. N.; Kochenderfer, J. N. Toxicities of Chimeric Antigen Receptor T Cells: Recognition and Management. *Blood* **2016,** *127*, 3321– 3330.
21. Brudno, J. N.; Kochenderfer, J. N. Recent Advances in CAR T-cell Toxicity: Mechanisms, Manifestations and Management. *Blood Rev.* **2019,** *34*, 45–55.

22. Carpenito, C. et al. Control of Large, Established Tumor Xenografts with Genetically Retargeted Human T Cells Containing CD28 and CD137 domains. *Proc. Natl. Acad. Sci. USA* **2011,** *106*, 3360–3365. https ://doi.org/10.1073/pnas.0813101106
23. Carraway, K. L.; Price-Schiavi, S. A.; Komatsu, M.; Jepson, S.; Perez, A.; Carraway, C. A. Muc4/Sialomucin Complex in the Mammary Gland and Breast Cancer. *J. Mammary Gland Biol. Neoplasia* **2001,** *6*, 323–337. https://doi.org/10.1023/a:1011327708973
24. Davies, A. J.; Rohatiner, A. Z.; Howell, S. et al. . Tositumomab and Iodine I 131 Tositumomab for Recurrent Indolent and Transformed B-Cell Non-Hodgkin's Lymphoma. *J. Clin. Oncol.* **2004,** *22*, 1469–1479.
25. Davis, T. A.; Grillo-Lopez, A. J.; White, C. A. et al. . Rituximab anti-CD20 Monoclonal Antibody Therapy in Non-Hodgkin's Lymphoma: Safety and Efficacy of Re-treatment. *J. Clin. Oncol.* **2000,** *18*, 3135–3143.
26. Davis, T. A.; Kaminski, M. S.; Leonard, J. P. et al. . The Radioisotope Contributes Significantly to the Activity of Radioimmunotherapy. *Clin. Cancer Res.* **2004,** *10*, 7792–7798.
27. (a) Denny, W. A. . Tumor-Activated Prodrugs—A New Approach to Cancer Therapy. *Cancer Invest.* **2004,** *22*, 604–619. (b) De Santes, K.; Slamon, D.; Anderson, S. K. et al. . Radiolabeled Antibody Targeting of the HER-2/neu Oncoprotein. *Cancer Res.* **1992,** *52*, 1916–1923.
28. Di Gaetano, N.; Cittera, E.; Nota, R. et al. . Complement Activation Determines the Therapeutic Activity of Rituximab In Vivo. *J. Immunol.* **2003,** *171*, 1581–1587.
29. Dinndorf, P. A.; Andrews, R. G.; Benjamin, D. et al. . Expression of Normal Myeloid-Associated Antigens by Acute Leukemia Cells. *Blood* **1986,** *67*, 1048–1053.
30. Dong, H.; Strome, S. E.; Salomao, D. R. et al. . Tumor-Associated B7-H1 Promotes T-Cell Apoptosis: A Potential Mechanism of Immune Evasion. *Nat. Med.* **2002,** *8*, 793–800.
31. Dowell, J. A.; Korth-Bradley, J.; Liu, H. et al. . Pharmacokinetics of Gemtuzumab Ozogamicin, an Antibody-Targeted Chemotherapy Agent for the Treatment of Patients with Acute Myeloid Leukemia in First Relapse. *J. Clin. Pharmacol.* **2001,** *41*, 1206–1214.
32. Dowsett, M.; Bartlett, J.; Ellis, I. O. et al. . Correlation Between Immunohistochemistry (HercepTest) and Fluorescence In Situ Hybridization (FISH) for HER-2 in 426 Breast Carcinomas from 37 Centres. *J. Pathol.* **2003,** *199*, 418–423.
33. Dranoff, G. . Cytokines in Cancer Pathogenesis and Cancer Therapy. *Nat. Rev. Cancer* **2004,** *4*, 11–22.
34. Dranoff, G. . CTLA-4 Blockade: Unveiling Immune Regulation. *J. Clin. Oncol.* **2005,** *23*, 662–664. Dupont, J.; Rothenberg, M.; Springs, D. et al. . Safety and Pharmacokinetics of Intravenous VEGF Trap in a Phase, I.; Clinical Trial of Patients with Advanced Solid Tumors. *J. Clin. Oncol.* **2005,** *23* (Suppl. 16S), 3029.
35. Ehrlich, P. . On Immunity with Special Reference to Cell Life: Croonian Lecture. In *The Collected Papers of Paul Ehrlich*; Himmelweir, B.; Ed.; Vol. II; Immunology and Cancer Research, Pergammon: London, 1956; pp 148–192, 195–196.
36. Essand, M.; Gronvik, C.; Hartman, T. et al. . Radioimmunotherapy of Prostatic Adenocarcinomas: Effects of 131I-labelled E4 Antibodies on Cells at Different Depth in DU 145 Spheroids. *Int. J. Cancer* **1995,** *63*, 387–394.

37. Fan, Z.; Masui, H.; Altas, I.; Mendelsohn, J. . Blockade of epidermal growth factor receptor function by bivalent and monovalent fragments of 225 anti-epidermal growth factor receptor monoclonal antibodies. *Cancer Res.* 53, 4322–4328.
38. Flavell, R. A.; Gorelik, L. . Transforming Growth Factor-B in T Cell Biology. *Nat. Rev. Immunol.* **2002,** *2*, 46–53.
39. Freeman, G. J.; Long, A. J.; Iwai, Y. et al. . Engagement of the PD-1 Immunoinhibitory Receptor by a Novel B7 Family Member Leads to Negative Regulation of Lymphocyte Activation. *J. Exp. Med.* **2000,** *192*, 1027–1034.
40. Freeman, S. D.; Kelm, S.; Barber, E. K. et al. . Characterization of CD33 as a New Member of the Sialoadhesin Family of Cellular Interaction Molecules. *Blood* **1995,** *85*, 2005–2012.
41. Gabrilovich, D.; Ishida, T.; Oyama, T. et al. . Vascular Endothelial Growth Factor Inhibits the Development of Dendritic Cells and Dramatically Affects the Differentiation of Multiple Hematopoietic Lineages In Vivo. *Blood* **1998,** *92*, 4150–4166.
42. (a) Gerber, H. P.; Ferrara, N. . Pharmacology and Pharmacodynamics of Bevacizumab as Monotherapy or in Combination with Cytotoxic Therapy in Preclinical Studies. *Cancer Res.* **2005,** *65*, 671–680. (b) Goldstein, G. . An Overview of Orthoclone OKT3 Transplantation Proceedings. In: "*Proceedings of an International Symposium on Monoclonal Antibody Therapy with Orthoclone OKT3 in Renal Transplantation.*" *Transplant. Proc.* **1986,** *18*, 927–930.
43. Goldstein, N. I.; Prewett, M.; Zuklys, K. et al. . Biological Efficacy of a Chimeric Antibody to the Epidermal Growth Factor Receptor in a Human Tumor Xenograft Model. *Clin. Cancer Res.* **1995,** *1*, 1311–1318.
44. Gordon, M. S.; Margolin, K.; Talpaz, M. et al. . Phase I Safety and Pharmacokinetic Study of Recombinant Human Anti-Vascular Endothelial Growth Factor in Patients with Advanced Cancer. *J. Clin. Oncol.* **2001,** *19*, 843–850.
45. Gorelik, L.; Flavell, R. A. . Immune-Mediated Eradication of Tumors Through the Blockade of Transforming Growth Factor-Beta Signaling in T Cells. *Nat. Med.* **2002,** *7*, 1118–1122.
46. Grazette, L. P.; Boecker, W.; Matsui, T. et al. . Inhibition of ErbB2 Causes Mitochondrial Dysfunction in Cardiomyocytes: Implications for Herceptin-Induced Cardiomyopathy. *J. Am. Coll. Cardiol.* **2004,** *44*, 2231–2238.
47. Gura, T. . Therapeutic Antibodies: Magic Bullets Hit the Target. *Nature* **2002,** *417*, 584–586. Hale, G.; Bright, S.; Chumbley, G. et al. . Removal of T Cells from Bone Marrow for Transplantation: A Monoclonal Antilymphocyte Antibody That Fixes Human Complement. *Blood* **1983,** *62*, 873–882.
48. Hale, G.; Dyer, M. J.; Clark, M. R. et al. Remission Induction in Non-Hodgkin Lymphoma with Reshaped Human Monoclonal Antibody CAMPATH-1H. *Lancet* **1998,** *2*, 1394–1399.
49. Hamann, P. R.; Hinman, L. M.; Hollander, I. et al. Gemtuzumab Ozogamicin, a Potent and Selective Anti-CD33 Antibody-Calicheamicin Conjugate for Treatment of Acute Myeloid Leukemia. *Bioconjug. Chem.* **2002,** *13*, 47–58.
50. Herbst, R. S.; Arquette, M.; Shin, D. M. et al. Phase II Multicenter Study of the Epidermal Growth Factor Receptor Antibody Cetuximab and Cisplatin for Recurrent and Refractory Squamous Cell Carcinoma of the Head and Neck. *J. Clin. Oncol.* **2005,** *23*, 5578–5587.

51. Jain, P. et al. Outcomes and Management of Patients with Mantle Cell Lymphoma After Progression on Brexucabtagene Autoleucel Therapy. *Br. J. Haematol.* **2020**. https://doi.org/10.1111/bjh.17197
52. James, S. E. et al. Antigen Sensitivity of CD22-Specific Chimeric TCR is Modulated by Target Epitope Distance from the Cell Membrane. *J. Immunol.* **2008,** *180*, 7028–7038.
53. Jensen, M. C. et al. Human T Lymphocyte Genetic Modification with Naked DNA. *Mol. Ther.* **2000,** *1*, 49–55.
54. Kawalekar, O. U. et al. Distinct Signaling of Coreceptors Regulates Specific Metabolism Pathways and Impacts Memory Development in CAR T Cells. *Immunity* **2016,** *44*, 712.
55. Johnson, L. A. et al. Gene Therapy with Human and Mouse T-Cell Receptors Mediates Cancer Regression and Targets Normal Tissues Expressing Cognate Antigen. *Blood* **2009,** *114*, 535–546.
56. Johnson, L. A. et al. Rational Development and Characterization of Humanized Anti-EGFR Variant III Chimeric Antigen Receptor T Cells for Glioblastoma. *Sci. Transl. Med.* **2015,** *7*, 275–222.
57. Jensen, M. C. et al. Human T Lymphocyte Genetic Modification with Naked DNA. *Mol. Ther.* **2000,** *1*, 49–55.
58. Kochenderfer, J. N. et al. Donor-Derived CD19-Targeted T Cells Cause Regression of Malignancy Persisting After Allogeneic Hematopoietic Stem Cell Transplantation. *Blood* **2013,** *122*, 4129–4139.
59. Kochenderfer, J. N. et al. Lymphoma Remissions Caused by AntiCD19 Chimeric Antigen Receptor T Cells Are Associated with High Serum Interleukin-15 Levels. *J. Clin. Oncol.* **2017,** *35*, 1803–1813.
60. Kruse, R. L. et al. HBsAg-Redirected T Cells Exhibit Antiviral Activity in HBV-Infected Human Liver Chimeric Mice. *Cytotherapy* **2018,** *20*, 697–705.
61. Le Joncour, V. et al. A Novel Anti-HER2 Antibody-Drug Conjugate XMT-1522 for HER2-Positive Breast and Gastric Cancers Resistant to Trastuzumab emTansine. *Mol. Cancer Ther.* **2019,** *18*, 1721–1730.
62. Lee, D. W. et al. T Cells Expressing CD19 Chimeric Antigen Receptors for Acute Lymphoblastic Leukaemia in Children and Young Adults: A Phase 1 Dose-Escalation Trial. *Lancet* **2015,** *385*, 517–528.
63. Li, W. et al. Redirecting T Cells to Glypican-3 with 4–1bb Zeta Chimeric Antigen Receptors Results in Th1 Polarization and Potent Antitumor Activity. *Hum Gene Ther.* **2016**. https://doi.org/10.1089/ hum.2016.025
64. Lofer, A. et al. A Recombinant Bispecific Single-Chain Antibody, CD19 x CD3, Induces Rapid And High Lymphoma-Directed Cytotoxicity by Unstimulated T Lymphocytes. *Blood* **2000,** *95*, 2098–2103.
65. Long, A. H. et al. 4-1BB Costimulation Ameliorates T Cell Exhaustion Induced by Tonic Signaling of Chimeric Antigen Receptors. *Nature Med.* **2015,** *21*, 581–590.
66. Louis, C. U. et al. Antitumor Activity and Long-Term Fate of Chimeric Antigen Receptor-Positive T Cells in Patients with Neuroblastoma. *Blood* **2011,** *118*, 6050–6056.
67. Mata, M. et al. Toward Immunotherapy with Redirected T Cells in a Large Animal Model: Ex Vivo Activation, Expansion, and Genetic Modification of Canine T Cells. *J. Immunother*. **2014,** *37*, 407–415.
68. Mathew, J.; Perez, E. A. Trastuzumab Emtansine in Human Epidermal Growth Factor Receptor 2-Positive Breast Cancer: A Review. .*Curr. Opin. Oncol.* **2011,** *23*, 594–600.

69. Maude, S. L. et al. Chimeric Antigen Receptor T Cells for Sustained Remissions in Leukemia. *N Engl. J. Med.* **2014,** *371*, 1507–1517.
70. Meyer, S.; Leusen, J. H.; Boross, P. Regulation of Complement and Modulation of Its Activity in Monoclonal Antibody Therapy of Cancer. *MAbs* **2014,** *6*, 1133–1144.
71. Mitsuyasu, R. T. et al. Prolonged Survival and Tissue Trafcking Following Adoptive Transfer of CD4zeta Gene-Modified Autologous CD4(+) and CD8(+) T Cells in Human Immunodefciency Virus-Infected Subjects. *Blood* **2000,** *96*, 785–793
72. Monjezi, R. et al. Enhanced CAR T-cell Engineering Using Nonviral Sleeping Beauty Transposition from Mini Circle Vectors. *Leukemia* **2017,** *31*, 186–194.
73. Moorkens, E.; Vulto, A. G.; Huys, I. An Overview of Patents on Therapeutic Monoclonal Antibodies in Europe: Are They a Hurdle to Biosimilar Market Entry? *MAbs* **2020,** *12*, 1743517.
74. Morgan, R. A.; Yang, J. C.; Kitano, M.; Dudley, M. E.; Laurencot, C. M.; Rosenberg, S. A. Case Report of a Serious Adverse Event Following the Administration of T Cells Transduced with a Chimeric Antigen Receptor Recognizing ERBB2. *Mol. Ther.* **2010,** *18*, 843–851.
75. Nadler, L. M. et al. Serotherapy of a Patient with a Monoclonal Antibody Directed Against a Human Lymphoma-Associated Antigen. *Cancer Res.* **1980,** *40*, 3147–3154.
76. Nagy, P.; Friedlander, E.; Tanner, M.; Kapanen, A. I.; Carraway, K. L.; Isola, J.; Jovin, T. M. Decreased Accessibility and Lack of Activation of ErbB2 in JIMT-1, a Herceptin-Resistant, MUC4-Expressing Breast Cancer Cell Line. *Cancer Res.* **2005,** *65*, 473–482.
77. Nakajima, M.; Sakoda, Y.; Adachi, K.; Nagano, H.; Tamada, K. Improved Survival of Chimeric Antigen Receptor-Engineered, T. (CAR-T) and Tumor-Specific T Cells Caused by Anti-Programmed Cell Death Protein 1 Single-Chain Variable Fragment Producing CAR-T Cells. *Cancer Sci.* **2019,** *110*, 3079–3088.
78. Neelapu, S. S. et al. Axicabtagene Ciloleucel CAR T-Cell Therapy in Refractory Large B-Cell Lymphoma. *N. Engl. J. Med.* **2017,** *377*, 2531–2544.
79. Neelapu, S. S. et al. Chimeric Antigen Receptor T-Cell Therapy—Assessment and Management of Toxicities. *Nat. Rev. Clin. Oncol.* **2018,** *15*, 47–62.
80. Nesic, D. et al. Cryo-Electron Microscopy Structure of the Alpha IIb Beta3-Abciximab Complex Arterioscler Thromb. *Vasc. Biol.* **2020,** *40*, 624–637.
81. O'Leary, M. C. et al. FDA Approval Summary: Tisagenlecleucel for Treatment of Patients with Relapsed or Refractory B-Cell Precursor Acute Lymphoblastic Leukemia. *Clin Cancer Res.* **2019,** *25*, 1142–1146.
82. Wang, Y.; Liu, Y.; Du, Y.; Yin, W.; Lu J. The Predictive Role of Phosphatase and Tensin Homolog (PTEN) Loss, Phosphoinositol-3 (PI3) Kinase (PIK3CA) Mutation, and PI3K Pathway Activation in Sensitivity to Trastuzumab in HER2-Positive Breast Cancer: A Meta-Analysis. *Curr. Med. Res. Opin.* **2013,** *29*, 633–642.
83. Oppermans, N.; Kueberuwa, G.; Hawkins, R. E.; Bridgeman, J. S.; (Transgenic T-Cell Receptor Immunotherapy for Cancer: Building on Clinical Success. *Ther. Adv. Vaccines Immunother.* **2020,** *8*, 2515135520933509.
84. Pegram, M.; Hsu, S.; Lewis, G. et al. Inhibitory Effects of Combinations of HER-2/neu Antibody and Chemotherapeutic Agents Used for Treatment of Human Breast Cancers. *Oncogene* **1999,** *18*, 2241–2251.
85. Piro, L. D.; White, C. A.; Grillo-Lopez, A. J. et al. Extended Rituximab (Anti-CD20 Monoclonal Antibody) Therapy for Relapsed or Refractory Low-Grade or Follicular Non-Hodgkin's Lymphoma. *Ann. Oncol.* **1999,** *10*, 655–661.

86. Phan, G. Q.; Yang, J. C.; Sherry, R. M. et al. Cancer Regression and Autoimmunity Induced by Cytotoxic T Lymphocyte-Associated Antigen 4 Blockade in Patients with Metastatic Melanoma. *Proc. Natl. Acad. Sci. USA* **2003,** *100*, 8372–8377.
87. Press, O. W.; Corcoran, M.; Subbiah, K. et al. A Comparative Evaluation of Conventional and Pretargeted Radioimmunotherapy of CD20-Expressing Lymphoma Xenografts. *Blood* **2001,** *98*, 2535–2543.
88. Press, O. W.; Eary, J. F.; Appelbaum, F. R. et al. Phase II trial of 131I-B1 (anti-CD20) Antibody Therapy with Autologous Stem Cell Transplantation for Relapsed B Cell Lymphomas. *Lancet* **1995,** *346*, 336–340.
89. Press, O. W.; Shan, D.; Howell-Clark, J. et al. Comparative Metabolism and Retention of Iodine-125, Yttrium-90, and Indium-111 Radioimmunoconjugates by Cancer Cells. *Cancer Res.* **1996,** *56*, 2123–2129.
90. Presta, L. G.; Chen, H.; O'Connor, S. J. et al. Humanization of an Anti-Vascular Endothelial Growth Factor Monoclonal Antibody for the Therapy of Solid Tumors and Other Disorders. *Cancer Res.* **1997,** *57*, 4593–4599.
91. Queen, C.; Schneider, W. P.; Selick, H. E. et al. . A Humanized Antibody That Binds to the Interleukin 2 Receptor. *Proc. Natl. Acad. Sci. USA* **1989,** *86*, 10029–10033.
92. Rai, K. R.; Freter, C. E.; Mercier, R. J. et al. Alemtuzumab in Previously Treated Chronic Lymphocytic Leukemia Patients Who Also Had Received Fludarabine. *J. Clin. Oncol.* **2002,** *20*, 3891–3897.
93. Ravetch, J. V.; Lanier, L. L. Immune Inhibitory Receptors. *Science* **2000,** *290*, 84–89.
94. Rebello, P.; Cwynarski, K.; Varughese, M. et al. Pharmacokinetics of CAMPATH-1H in BMT Patients. *Cytotherapy* **2001,** *3*, 261–267.
95. Reff, M. E.; Carner, K.; Chambers, K. S. et al. Depletion of B Cells In Vivo by a Chimeric Mouse Human Monoclonal Antibody to CD20. *Blood* **1994,** *83*, 435–445.
96. Riechmann, L.; Clark, M.; Waldmann, H. et al. Reshaping Human Antibodies for Therapy. *Nature* **1998,** *332*, 323–327.
97. Riethmuller, G.; Holz, E.; Schlimok, G. et al. Monoclonal Antibody Therapy for Resected Dukes' C Colorectal Cancer: Seven-Year Outcome of a Multicenter Randomized Trial. *J. Clin. Oncol.* **1998,** *16*, 1788–1794.
98. Scaltriti, M. et al. Lapatinib, a HER2 Tyrosine Kinase Inhibitor, Induces Stabilization and Accumulation of HER2 and Potentiates Trastuzumab-Dependent Cell Cytotoxicity. *Oncogene* **2009,** *28*, 803–814.
99. Schmidt, E.; Hunzelmann, N.; Zillikens, D.; Brocker, E. B.; Goebeler, M. Rituximab in Refractory Autoimmune Bullous Diseases. *Clin. Exp. Dermatol.* **2006,** *31*, 503–508.
100. Schubert, M. L. et al. Feasibility and Safety of CD19 Chimeric Antigen Receptor T Cell Treatment for B Cell Lymphoma Relapse After Allogeneic Hematopoietic Stem Cell Transplantation. *Biol Blood Marrow Transplant* **2020,** *26*, 1575–1580.
101. Servais, E. L. et al. Mesothelin Overexpression Promotes Mesothelioma Cell Invasion and MMP-9 Secretion in an Orthotopic Mouse Model and in Epithelioid Pleural Mesothelioma Patients. *Clin Cancer Res.* **2012,** *18*, 2478–2489.
102. Smith, T. T. et al. In Situ Programming of Leukaemia-Specific T Cells Using Synthetic DNA Nanocarriers. *Nat. Nanotechnol.* **2017,** *12*, 813–820.
103. Steinman, L. The Use of Monoclonal Antibodies for Treatment of Autoimmune Disease. *J. Clin. Immunol.* **1990,** *10*, 30S–38S; discussion 38S–39S.
104. Szoor, A.; Toth, G.; Zsebik, B.; Szabo, V.; Eshhar, Z.; Abken, H.; Vereb, G. Trastuzumab Derived HER2-Specific CARs for the Treatment of Trastuzumab-Resistant Breast

Cancer: CAR T Cells Penetrate and Eradicate Tumors That Are Not Accessible to Antibodies. *Cancer Lett.* **2020,** *484*, 1–8.
105. Tang, J.; Yu, J. X.; Hubbard-Lucey, V. M.; Neftelinov, S. T.; Hodge, J. P.; Lin, Y. Trial Watch: The Clinical Trial Landscape for PD1/PDL1 Immune Checkpoint Inhibitors. *Nat. Rev. Drug Discov.* **2018,** *17*, 854–855.
106. Tang, X. Y. et al. Third-Generation CD28/4–1BB Chimeric Antigen Receptor T Cells for Chemotherapy Relapsed or Refractory Acute Lymphoblastic Leukaemia: A Non-Randomised, Open-Label Phase I Trial Protocol. *BMJ Open* **2016,** *6*, e013904.
107. Tawara, I. et al. Safety and Persistence of WT1-Specific T-Cell Receptor Gene-Transduced Lymphocytes in Patients with AML and MDS. *Blood* **2017,** *130*, 1985–1994.
108. Tendeiro Rego, R.; Morris, E. C.; Lowdell, M. W. T-Cell Receptor Gene-Modified Cells: Past Promises, Present Methodologies and Future Challenges. *Cytotherapy* **2019,** *21*, 341–357.
109. Toth, G.; Szollosi, J.; Abken, H.; Vereb, G.; Szoor, A. A Small Number of HER2 Redirected CAR T Cells Significantly Improves Immune Response of Adoptively Transferred Mouse Lymphocytes Against Human Breast Cancer Xenografts. *Int. J. Mol. Sci.* **2020**. https://doi.org/10.3390/ijms21031039
110. Toth, G.; Szoor, A.; Simon, L.; Yarden, Y.; Szollosi, J.; Vereb G. The Combination of Trastuzumab and Pertuzumab Administered at Approved Doses May Delay. Development of trastuzumab Resistance by Additively Enhancing Antibody-Dependent Cell-Mediated Cytotoxicity. *MAbs* **2016,** *8*, 1361–1370.
111. Tsumoto, K.; Isozaki, Y.; Yagami, H.; Tomita, M.; Future Perspectives of Therapeutic Monoclonal Antibodies. *Immunotherapy* **2019,** *11*, 119–127.
112. van Vollenhoven, R. F. et al. Long-Term Safety of Rituximab in Rheumatoid Arthritis: 9.5-Year Follow-Up of the Global Clinical Trial Programme with a Focus on Adverse Events of Interest in RA Patients. *Ann. Rheum. Dis.* **2013,** *72*, 1496–1502.
113. Varadi, T. et al. Binding of Trastuzumab to ErbB2 Is Inhibited by a High Pericellular Density of Hyaluronan. *J. Histochem. Cytochem.* **2012,** *60*, 567–575.
114. Walseng, E. et al. A TCR-Based Chimeric Antigen Receptor. *Sci. Rep.* **2017,** *7*, 10713.
115. Wang, Q. S. et al.. Treatment of CD33-Directed Chimeric Antigen Receptor-Modified T Cells in One Patient with Relapsed and Refractory Acute Myeloid Leukemia. *Mol. Ther.* **2015,** *23*, 184–191.
116. O'Leary, M. C. et al. FDA Approval Summary: Tisagenlecleucel for Treatment of Patients with Relapsed or Refractory B-Cell Precursor Acute Lymphoblastic Leukemia. *Clin Cancer Res.* **2019,** *25*, 1142–1146.
117. Xu, J. Y. et al. Mesothelin-Targeting Chimeric Antigen Receptor-Modified T Cells by Piggy Bac Transposon System Suppress the Growth of Bile Duct Carcinoma. *Tumour Biol.* **2017,** *39*, 1010428317695949.

CHAPTER 5

Regulatory T Cells in Tumor Immunity: Role of Toll-Like Receptors

ABSTRACT

In this chapter, the author explores how toll-like receptor (TLR) signaling regulates Treg activity and how re-establishing Treg-mediated suppression of antigen-specific antitumor immunity may considerably improve cancer therapy. In this review, the author sums up the current state of knowledge on regulatory T cells (Tregs) and the regulation of Tregs in cancer by tumor-infiltrating innate immune cells such as dendritic cells (DCs). TLRs are a kind of PRR produced in the germ line that plays a critical role in controlling inflammation and immunological responses. Macrophages and DCs use TLRs to identify foreign invaders and aberrant cells, including cancer. Adaptive immunity helps the body fight off infections and tumor cells by generating more specific immunity and immunological memory when the body's natural defenses are no longer enough to do the job. Among the immune cells that infiltrate tumors, the presence of CD8+ T cells is associated with improved patient survival, whereas the presence of other immune cells, such as Tregs, is associated with a worse prognosis.

5.1 INTRODUCTION

Anti-infective immune system modifications may cause autoimmune and pathologic inflammation. Managing the immune response such that it has a balance of beneficial and harmless effects on the host is essential. Proteins known as pattern recognition receptors (PRRs) play a crucial role in maintaining order. About two decades after Charles Janeway Jr. initially proposed the existence of PRRs,[1] their existence has been scientifically proven.[2] By recognizing certain pathogen molecular structures, pathogen recognition receptors (PRRs)

Advances in Immunological Treatment of Cancer: Cellular Mechanism and Regulation.
Rishabha Malviya, Arun Kumar Singh, and Deepika Yadav (Authors)

play an important role in the initiation of the innate immune response and the enhancement of subsequent phases of the immune response (such as adaptive immunity). There is a rapidly expanding family of pattern recognition receptors (PRRs)[3,4] that includes, but is not limited to, toll-like receptors (TLRs), nucleotide-binding region and leucine-rich repeat-containing molecules (NLRs), RIG-I-like receptors (RLRs), and C-type lectin receptors (CLRs). One kind of key sensor for bacterial, fungal, viral, and protozoal products is the toll-like receptor (TLR). It is now known that toll-like receptors (TLRs) may activate both the innate and adaptive immune systems.

To avoid autoimmunity, self-reactive T and B cells are eliminated by the body at different phases of development (central tolerance), whereas autoreactive cells in the periphery are similarly eliminated, inhibited, or regulated (peripheral tolerance). CD4+CD25+Foxp3 Tregs play an essential role in peripheral T cell tolerance[5] by reducing resting and infection-induced activation of self-reactive cells in host organs. There is growing evidence that Tregs produce TLRs in addition to serving as a vital immune suppressive cellular mechanism for controlling TLR hyperactivity and avoiding sepsis and autoimmune diseases. Direct TLR activation on Treg, however, has been shown to decrease Treg activity and boost the immune response.[6] The effects of TLR expression and function on Treg will be analyzed in detail. To improve host defenses and lessen infection-related damage (thus, TLR activation), the author proposes that the suppressive function of Treg may be increased or lowered depending on the kind of infection.

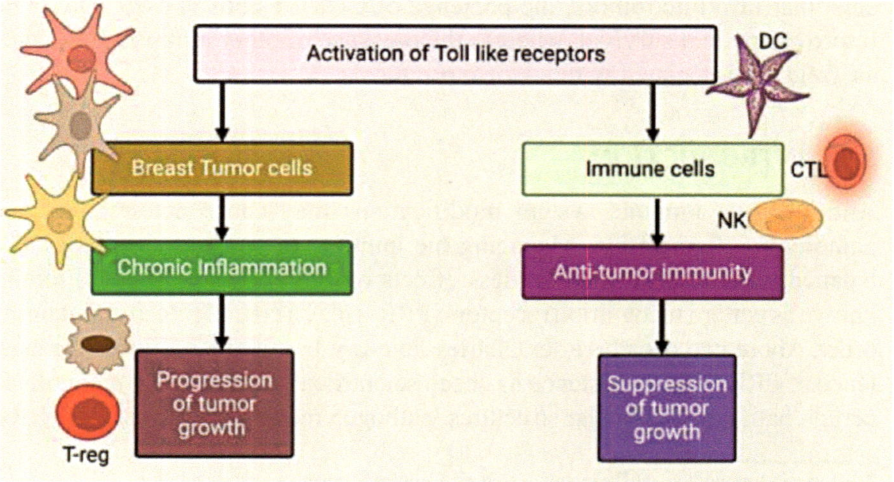

FIGURE 5.1 Schematic diagram of the mechanism of action of T cells in tumor growth.

Cancer patients may benefit from both curative and palliative chemotherapy.[7] As their name implies, cytotoxic drugs are designed to kill cancer cells. Typically, medicine induces cell death just in the desired ones (also known as "programmed cell death"). Although the death of healthy cells is never the goal of a chemotherapeutic procedure, it is a possibility.[8] Because most drugs only affect certain stages of the cell cycle, they might leave healthy cells exposed at otherwise crucial times. Iatrogenic side effects of chemotherapy include myelosuppression, hair loss, and mucositis. Due to the dose-limiting nature of these side effects, enhancing specificity is a primary objective of rational medicine design. It is well-known that myelosuppression, a common side effect of chemotherapy, reduces the effectiveness of immune responses in fighting cancer.[9] In this chapter, the researcher argues that lymph depletion is not always incompatible with immunotherapy and may potentially contribute to the induction of anticancer immune responses. However, the immunosuppressive properties of chemotherapeutic drugs are far-reaching.[10,11] The presentation of tumor antigens to the immune system relies on T cell activation and dendritic cell (DC) synthesis, both of which may be suppressed by these therapies. Some cancer treatments, including chemotherapy, might compromise the immune response by killing off healthy cells along with the cancerous ones. Concerns have been raised that the loss of lymphocytes (Fig. 5.1) due to chemotherapy may lead to a decrease in the number of regulatory CD4+ CD25+ T cells (Tregs), whose job it is to inhibit the immune system from reacting to tumor-associated antigens.[12] Here, the author takes a look at two situations that show how the immune system may interact with chemotherapy in a complicated way. It is only recently that researchers have begun to tease out the details of this connection, particularly the role that immunotherapeutic intervention plays. Even though chemotherapy is still the mainstay of cancer treatment, it is becoming more apparent that the two are not incompatible and may even have a synergistic effect.

5.2 IMMUNE CELLS IN IMMUNOSURVEILLANCE AND TUMOR DESTRUCTION

Cancer immunosurveillance is a notion proposed by Burnet and Thomas almost 50 years ago, and since then it has been backed by some studies. The innate immune system is the first line of defense against pathogens and cancer cells that have been genetically changed. The innate immune cells natural killer (NK), natural killer T (NKT), and -T cells have been shown to have a significant role in protecting the host against cancerous cells.[13]

Toll-like receptors (TLRs) are pattern recognition receptors (PRRs) that are present in the germ line and are responsible for recognizing invading pathogens and transformed tumor cells.[14] TLRs are very important regulators of inflammation and immune responses. The host's final line of defense, adaptive immunity, generates specific antibodies and immunological memory to destroy invading pathogens and cancer cells that have undergone mutation.[15]

Soluble chemicals, such as interferon (IFN)-, TRAIL (tumor necrosis factor-related apoptosis-inducing ligand), and TNFR1 (tumor necrosis factor), help the host in its fight against cancer (TNF-). Researchers have shown that CD8+ T cells are associated with improved survival in cancer patients, but the presence of regulatory T cells (Tregs) is associated with a poorer prognosis. Collectively, our results emphasize the importance of both the innate and adaptive immune systems in the tumor microenvironment and cancer immunosurveillance.[16]

From the very beginning of a disease, tumor cells and immune cells engage in a dynamic and continuing conflict, with the latter sometimes becoming triumphant. Chronic inflammation occurs when the immune system is unable to fully eliminate infection or repair damaged tissue.[17] Although the processes by which chronic inflammation directly leads to cancer are poorly understood, it is known that TLR-mediated recognition of invading pathogens or wounded tissues enhances the innate immune response, and this contributes to cancer (particularly through mechanisms that involve the nuclear factor [NF]-B pathway). Evidence suggests that bacterial and viral infections, together with the tissue damage they induce, are important sources of chronic inflammation that attracts even more immune cells. The immune system's ability to eliminate pathogens quickly has a curative effect on inflammation. Chronic inflammation occurs when the body's initial line of defense against pathogens, damaged tissues, or changed cells is ineffective.[18] When T cells are constantly activated by tumor cells and tumor-infiltrating macrophages/DCs, tumors may create or store Tregs. Cytokines and chemokines are secreted by immune cells in large quantities, which encourage blood vessel formation and tumor growth. Because of this, the inflammatory state of the tumor microenvironment may determine whether immune cells repress or drive tumor development.[19]

5.2.1 TLRs AND THEIR SIGNALING PATHWAYS

Innate immunity (immunity that does not rely on the adaptive immune system) is mostly facilitated by toll-like receptors (TLRs), which have been demonstrated to play an important role in a variety of innate immunological functions.[20] Type 1 integral membrane glycoproteins, of which TLRs are members, are characterized by the presence of a cytoplasmic toll/IL-1R homology (TIR) region and an external site with a different number of leucine-rich repeat (LRR) motifs. Both humans and mice have TLRs. These TLRs can recognize a limited but highly conserved collection of molecular structures called pathogen-associated molecular patterns (PAMPs).[21] TLR2 recognizes peptidoglycan, which is limited to gram-positive bacteria, whereas TLR4 recognizes lipopolysaccharide (LPS), which is found solely in gram-negative bacteria.[22] Receptors belonging to the toll-like family (TLR3-9) may recognize certain sequences of DNA and RNA. For example, during viral infection, toll-like receptor 9 (TLR9) may recognize single-stranded viral RNAs or guanosine-related analogues (loxoribine and imidazoquinoline).[23] In contrast, toll-like receptor 3 (TLR3) can recognize only double-stranded RNA. Toll-like receptors (TLRs) are expressed by a wide variety of cell types, including dendritic cells (DCs), macrophages, T cells, neutrophils, eosinophils, mast cells, monocytes, epithelial cells, and tumor cells.[24]

Only TLR 2, 4, 5, and 6 are surface receptors, whereas TLRs are located in endosomal storage compartments.[25] Each kind of toll-like receptor (TLR) normally only has one adaptor protein, MyD88, that binds to it. MyD88, a signal transducer protein, is required for efficient signal transduction from the overwhelming majority of toll-like receptors (TLRs) to downstream signaling molecules. Recent years have seen a lot of studies devoted to downstream effector pathways such as TLR-activated epigenetic modifications like NF-B, IRF-5, and IRF-7.[26] MyD88 creates a complex with TLR ligands, which in turn activates the kinase IRAK after TLR ligand binding. It has been shown that IRAK/MyD88 complexes promote TRAF6 activation.[27] Protein TRAF6 helps link the TNFR to its receptor. Both TRAF6 and NEMO, a subunit of the NF-kappa B activating inhibitor NF-kappa B kinase (IKK) complex, are ubiquitin-conjugating enzyme targets of TRAF6 (UBC13 and UEV1A). TGF-activated kinase 1 (TAK1), together with its binding proteins, may initiate the NF-kappa B (NF-B) and mitogen-activated protein kinase (MAPK) signaling pathways (TAB1, TAB2, and TAB3). Essential to the production of inflammatory cytokines is the action of these systems. Interactions between MyD88 and the IFN-regulated transcription factors IRF-5 and IRF-7[28] suggest that this molecule may have a role in enhancing proinflammatory cytokine or type I IFN responses. TLR3 does not need MyD88 but rather the adaptor protein TRIF to induce IFN- in response to pathogen recognition. TLR3 binds RIP1, TRAF3, and TRAF6 to induce the

TRAF family-member-associated NF-B activator (TANK) binding kinase 1 (TBK1) and/or inducible IkB kinase (IKK-i),[29] which in turn induces the production of type I IFN cytokines.

5.2.2 EXPRESSION OF TLRs BY REGULATORY T CELLS

There are 13 different kinds of TLRs in mammals.[30] Depending on their location inside the cell, TLRs are classified as either intracellular TLRs or cell surface TLRs. Different types of RNA and unmethylated CpG DNA are recognized by TLRs inside cells (TLR3, TLR7, and TLR8).[31] Toll-like receptors 2 and 6 recognize triacylated and diacylated bacterial lipopeptides; toll-like receptors 4 and 5 recognize lipoproteins; and toll-like receptor 11 recognizes flagellin, a profilin-like protein from apicomplexan protozoa, and an unknown cellular component(s) from human pathogenic organisms.[32,33] In contrast to TLR1, TLR2, and TLR4, TLR10, TLR12, and TLR13 remain unidentified.[34]

While TLRs are commonly generated by DCs, B cells, and M/monocytes, the coding sequences for TLR12 and TLR13 have not yet been discovered.[35] Evidence indicated that both classical T cells[36] and T cells expressed TLRs. By using RT-PCR,[37] researchers found that freshly isolated T cells from human peripheral blood or tonsils (>95% pure) had varied quantities of TLR1-10 mRNA.[38] Further analysis revealed that CD4+ and CD8+ T cells produced similar levels of mRNA for several TLRs. Even though it is difficult to make a direct comparison of TLR expression levels across studies due to the use of different internal controls,[39,40] it is interesting to note that CD4+ T cells express TLRs at a higher level than CD8+ T cells do. CD4+ T lymphocytes in rats mostly express TLR3, 5, 6, and 9; in mice, this ranges from TLR1 to TLR9.[41] The theory that non-T cell contamination led to the detection of TLR mRNA in these tests was debunked. TLR1 is expressed not only in the Jurkat CD4+ T cell line[42] but also in pure human PBMC to more than 99.9%. Intracellular labeling of TLR2 as well as surface staining of TLR2 offer evidence[43] for the presence of a subset of TLRs in human and animal CD4+ T cells.

The expression of TLRs in CD4+CD25 effector T cells (Teff) and CD4+CD25+Foxp3+ regulatory T cells (Treg) has been the subject of substantial research.[44] CD45RBhi naive CD4+ T cells, activated/memory CD45RBlowCD4+CD25 Teff, and CD45RBlowCD4+CD25+ Treg all extensively expressed TLR1, 2, and 6 mRNA, which is consistent with the functional heterodimeric coupling of TLR2/1 and TLR2/6.[45] Notwithstanding their lack of treatment, naïve CD4+ T cells isolated from human cord blood maintained substantial amounts of TLR2 and TLR4 mRNA or cytoplasmic protein expression.[46,47] Surface TLR2 and TLR4 were elevated after this treatment.[48] Moreover, CD45RO+ memory CD4+ T cells in adult peripheral blood exhibit substantial quantities of surface TLR2, in contrast to CD45RA+ naïve CD4+ T cells. High surface TLR2 levels were seen on CD45RA+ cells after anti-CD3 activation, and these levels were further elevated by cotreatment with the synthetic TLR2 ligand Pam3CSK4. Memory CD4+ T cells that express CD45RO did not expand on the other hand. Similar to humans, the author found that TLR2 surface expression is restricted to anti-CD3-activated CD4+CD25 Teff and CD4+CD25+ Treg but not to resting T cells. Quiescent T cells were shown to have TLR2 located in their cytoplasm.[49]

Newly obtained CD45RBlowCD4+CD25 Teff and CD45RBlowCD4+ CD25+ Treg expressed TLR4 mRNA, in contrast to CD45RBhi CD4+ T cells, which became CD45RBlow after 3 days of in vitro activation with plate-bound anti-CD3 and anti-CD28.[50] Three to four times as much mRNA was found in Treg than in Teff.[51] Intriguingly, it is well-known that LPS causes fast mRNA and protein-level downregulation of TLR4 in macrophages following contact. This may account for LPS tolerance. This occurred when LPS was administered to newly isolated human Teff and Treg.[52] These findings suggest that TLR4 expression is dynamic within individual cells and that TLR4 has different functions depending on its level of expression.

TLR5 mRNA or protein has been detected in the peripheral blood of humans and animals with CD44low CD4+ T cells, CD45RBlow Teff, and CD45RBhigh Treg.[53] TLR5 expression was confirmed in vitro-generated Th cells and Treg clones.[54] TLR5 was detected on naïve and memory CD4+ CD45RChi and CD4+ CD45RClow T cells, as well as two Ag-specific CD4+ T cell clones in rats.[55] TLR5 expression was up to four times higher in murine Treg than in Teff, just as TLR4 expression was in the former group. Recent research shows, however, that activation of anti-CD3/anti-CD8 antibodies significantly downregulates TLR5 expression.[56]

TLR3 expression is not present in either memory CD4+ T cells or non-neoplastic CD4+ T cells. TLR3 has been detected in CD4+CD25 T cells from mice without anti-CD3 activation by many research groups. It was shown that TLR3 was expressed by two Ag-specific CD4+ T cell clones, as well as naïve and memory CD4+ CD45RChi and CD4+ CD45RClow T cells from rats.[57] More evidence suggests that TLR3 protein is localized in the cytoplasm of mouse CD4+CD45RBhi T cells.

CD45RBlow Teff and Treg populations were positive for TLR7 and TLR8, but CD45RBhi naïve CD4+ T cells were negative for both of these receptors. TLR7 expression was also reduced in CD45RBhi naive CD4+ T cells. It was shown that CD4+ T cells from the lamina propria of mice exhibited large quantities of TLR7, joining CD4+ T cells, CD62Lhi T cells, and CD45RBhi T cells as TLR7-expressing cells. TLR7/8 expression in CD4(+) CD25(+) regulatory T cells (Tregs) and CD4(+)CD45RA(+) naïve T cells was detected in human peripheral blood.[58]

Scientists have been trying to figure out how TLRs contribute to immunity since it was discovered that TLR9 transcripts, like TLR3 transcripts, were missing from mouse CD4+ T cells.[59] According to our findings, certain CD4+ T lymphocytes express TLR9. TLR9 expression was found on CD4+CD45+CD62+ naive T cells and CD4+CD44+CD25+CD25- memory T cells.[60] TLR9 protein expression was verified both by external and intracellular labeling. TLR9 is expressed by a variety of rat immune cell subsets, including naive CD4+ CD45RChi T cells, memory CD4+ CD45RChi T cells, and regulatory CD4+ CD45RChi Tregs.[61]

Despite having a nearly full sequence, the TLR10 gene in mice has not been shown to have an ortholog. In humans, TLR10 was discovered in the same genomic region as TLR1, TLR6, and TLR2, all of which are known to heterodimerize with TLR2. TLR10 is capable of both homodimer and heterodimer formation with TLR1 and TLR2. The MyD88 gene is expressed by B cells and plasmacytoid DCs. Recent research has shown that in contrast to CD4+CD25+ Teff, TLR10 is constitutively expressed at the mRNA and protein levels in human CD4+CD25+ Treg.[62]

The vast majority of TLRs are expressed by CD4+CD25+ T regulatory T cells (Treg) and CD4+CD25+ T effector T cells (Teff), with TLR3 in Treg and TLR10 in Teff being the sole exceptions. Additionally, Treg expresses a broader repertoire of TLRs than Teff does, and they do so in a more discriminatory manner. Some examples of TLRs are TLR2, TLR4, TLR5, TLR7/8, and TLR10. The expression levels of the toll-like receptors TLR2, TLR4, TLR5, and TLR7 may shift as a result of changes in the activation/functional status of a subset of CD4+ T cells. The maintenance of Ag-specific effector and memory T cells may rely on TLRs acting as costimulatory receptors.[63]

5.2.3 FUNCTION OF TLR STIMULATION ON REGULATORY T CELLS

The expression of TLRs by human and mouse Treg demonstrates the physiological need and importance of direct TLR signaling for modulating Treg activity. Without TLR3, the MyD88-IRAK-4-NF-kB-dependent

TRIF-TBK-1-IRF-3 signaling pathway is used by all other TLRs. MyD88 mutant mice have been used in in vitro as well as in vivo studies of TLR signaling in CD4+ T cells.[64] It has been shown that the TLR9 ligand CpG DNA directly stimulates proliferation, prevents energy, improves survival, and increases IgG production in murine CD4+ T cells in response to a T cell-dependent Ag by activating MyD88 and phosphatidylinositol 3-kinase (PI-3 kinase). Deletion of IRAK-4 reduced the proliferation and effector activity of CD4+ and CD8+ T cells in response to TCR ligation in vivo and in vitro,[65] showing the importance of innate immunological signaling in T cell activation. Transplantation of MyD88/CD45RBhi CD4+ T cells into mice with a T cell adoptive transfer model of colitis significantly reduced disease severity and etiology compared to WT CD4+ T cells. The decline in cell proliferation and the production of effector cytokines like IL-17 contributed to this pattern.[66] The protective effect was also significantly diminished when MyD88/Treg was cotransplanted with WT Treg.[67] In vitro treatment with anti-CD3 and TLR ligands (Pam3CSK4, polyI: C, LPS, and CpG) resulted in increased proliferation of WT Teff, which was consistent with in vivo observations (except for polyI: C). MyD88-deficient Treg, on the other hand, showed less suppressive activity in vitro[68] compared to WT Treg. All of our results point to a functional role for TLRs in CD4+CD25 Teff and CD4+CD25+ Treg.[69]

Mutant Candida albicans cause some resistance to systemic infection in TLR2 null mice. The number of CD4+CD25+ Treg was considerably lower in TLR/ and MyD88/ animals compared to TLR4/mice but was comparable to that in WT mice.[70] In wild-type (WT) mice, the TLR2 agonist Pam3CSK4 boosted the number of CD4+Foxp3+ Treg. Although the TLR4 agonist LPS did not stimulate Treg growth in vitro, anti-CD3 and IL-2 did.[71] Peptidoglycan, a TLR2 ligand, has been shown to increase Treg survival in vitro when used alone. Pam3SCK4 therapy briefly reduced Foxp3 expression in vitro and resulted in a loss of suppressive efficacy in vitro and in vivo using adoptive transfer models of colitis, leishmaniasis, and C. albicans infections. Because Pam3SCK4 inhibited IL-2 production by Teff, Treg suppressor activity was reduced. Human heat shock protein 60 (HSP60) or its p277 peptide alone may enhance a TLR2-dependent expansion of human Treg activity in vitro.[72] This is because HSP60 and its p277 peptide interact with target cells and release soluble factors that inhibit CD4+ Teff and CD8+ T cell proliferation and cytokine production. This enhancement of Treg activity required the activation of phosphatidylinositol 3-kinase (PI3K) and p38 mitogen-activated protein kinase (MAPK) (p38). Inhibition of extracellular signal-regulated kinases (ERKs) enhanced HSP60-mediated reduction of Tregs. More research into the distinctions between exogenous and endogenous TLR ligands, as well as those between mouse and human TLR ligands, is required. Although it has been shown that the stress-inducible proteins HSP60, HSP70, and gp96 are endogenous ligands for toll-like receptors 2 and 4,[73] several studies have demonstrated that these proteins are unable to activate TLR2 and TLR4 in the absence of LPS. LPS may promote cytokine production on its own, but LPS chaperoned by HSP60 or gp96 seems to be far more potent.[74]

Treg proliferation and survival may be enhanced by LPS-induced TLR4 activation independently of TCR ligation, as was recently found. After LPS treatment, Treg suppressive activity increased both in vivo and in vitro.[75] The researcher recently showed that gp96 (grp94, HSP90b1), a luminal HSP, is required for the folding and assembly of almost all TLRs. This was shown, for example, by gene knockdown[76] and gain-of-function experiments. In transgenic mice, TLR4 hyper-responsiveness was shown when gp96 was generated ectopically on the cell surface; however, TLR responses were completely eliminated when gp96 was absent.[77] Notably, commensal bacteria, DCs, and IL-12 are needed for the autoimmune disease to develop in these gp96-transgenic mice.[78] Since gp96 transgenic mice have been shown to have increased in vivo TLR4 activity,[79] it may be worthwhile to investigate whether or not TLR4 has direct effects on Tregs. The author hypothesized that either a drop in Treg, which has been linked to autoimmunity, or an increase in Treg, which may prevent inflammation, would occur, given that both of these phenomena have been documented in the literature. Treg numbers and percentages in the spleen and thymus were not significantly different between WT and gp96 transgenic mice, as determined by intracellular labeling for the Foxp3 protein. According to our findings, transgenic Treg had a much higher per-cell inhibitory impact than WT Treg. In gp96 transgenic mice, the author found that TLR4 was always activated, leading us to conclude that this was the cause of the increased Treg activity.[80] The truth is that Treg expresses gp96 at an offtarget site. It is clear that TLR4 is necessary for the improved suppression by transgenic Treg since Treg from TLR4-null gp96 transgenic mice lost their enhanced potential to block CD4+CD25 Teff growth. It was hypothesized that increasing TLR4 activity in vivo would boost Treg function and decrease TLR4-dependent autoimmunity.[81] The effects of LPS-stimulated neutrophils on Treg have also been studied in humans. Preincubation of Treg with LPS increased their ability to inhibit IL-8, IL-6, and TNF- production by human neutrophils.

Human Treg and CD14+ monocytes were shown to have similar levels of TLR5 mRNA and protein expression.[82] Flagellin, a TLR5 agonist, has intriguing effects on Teff and Treg cells, such as the transient downregulation of TLR5 surface expression through direct ligation. In contrast, Treg responded well to flagellin treatment with increased Foxp3 expression. To stimulate proliferation and IL-2 production in Teff, flagellin and anti-CD3 were as potent as anti-CD28. Contrary to expectations, the addition of anti-CD3 and flagellin did not restore Treg responsiveness in the absence of antigen-presenting cells (APCs) but rather increased Treg suppressive activity in vitro.

In an experiment designed to see whether DCs had any influence on Treg function, CpG-A, but not LPS, generated a reduced suppressor function of human Treg, regardless of whether APCs were present. This study provided conclusive evidence that TLR is essential for regulating Treg activity in humans. Pretreatment of human CD4+ Treg cell lines in culture with CpG-A/B did not inhibit the proliferation of naïve CD4+ T cells in response to anti-CD3 stimulation (thus, without contamination of any APCs). Notably, the CpG motif in CpG-A has no influence on the loss of suppressive activity in Treg cell lines or freshly isolated Treg, but the poly G tail of CpG-A is responsible for this. The TLR8-MyD88-IRAK-4 pathway in Treg was discovered to be essential for the control of suppressive activity in both Ag-specific and natural Treg by the use of siRNA interference. Adopting TLR8 ligand-pretreated Treg into tumor-bearing mice resulted in a more robust antitumor immune response. Understanding that direct TLR8 ligation on Treg may control human Treg is a key finding for the immunotherapy of cancer, infection, and autoimmune disease.[83]

Direct stimulation by CpG increased the proliferation of both Teff and Treg in a dose-dependent manner in the absence of APCs without further augmenting cytokine production in both subsets of CD4+ T cells, despite the fact that TLR9 expression was lower in Teff and Treg from rats than from other TLRs.[84] In a traditional in vitro suppression investigation, CpG successfully eliminated Treg suppressiveness without changing Treg Foxp3 expression. The CpG preincubation process rendered Teff immune to Treg-mediated inhibition. Although CpG administration had no effect on Foxp3 expression in murine Treg, IL-2 production was much greater in WT than in MyD88/Teff cells. Teff and Treg showed resistance to Treg suppression in the presence of CpG only when sufficient MyD88 signaling occurred in either population. Similar to TLR2 on the Treg issue, it is unclear whether interleukin-2 (IL-2) plays a significant role in this process.[85]

However, TLR10's importance has been generally disregarded. Protein levels of TLR10 were clearly detectable and could be further increased by TCR activation; their expression was 200-fold higher in human Treg cells than in non-Treg cells. It has been hypothesized that Foxp3 binds to DNA in a cis-regulatory region near the TLR10 transcription start point. The expression of TLR10 was stimulated by Foxp3 in human Treg-depleted cells, but this effect seemed to require the formation of a cooperative complex between Foxp3 and NF-AT.[86]

Although TLR signaling is beyond the scope of this paper, it is critical for the in vitro and in vivo effector activities of CD4+CD25 Teff.[87] The surface expression of TLR2 is increased in human Teff that has been stimulated. Anti-CD3/IFN activation triggered Teff cell proliferation in response to direct stimulation of TLR2 and resulted in elevated IFN- and IL-2 production.[88] The TLR2 ligand Pam3CSK4 was discovered to stimulate the proliferation and survival of pre-existing Th1 cells in the mouse immune system without triggering a TCR response in a mouse model. Specifically, the MyD88-IRAK-4-MAPK pathway is switched on. Pam3CSK4 has the potential to convert human naive and memory CD4+ T cells into more potent effector cells. Those with high levels of the chemokine receptor CCR5 are more likely to contract HIV-1 subtypes X4 and R5. Ligation of toll-like receptors 3, 4, 5, 7, 8, and 9 has a direct effect on teff proliferation/survival and cytokine production. Toll-like receptor 4 (TLR4)-MyD88 pathway activation may promote CD3+ T cell adhesion to fibronectin in the presence of LPS in both humans and mice. The suppressor of cytokine 3 (SOC3), an inhibitor of cytokine signaling, was shown to decrease CXCL12-induced chemotactic responses in T cells when its levels were elevated.[89,90] Another kind of defective T cell that CpG may fix is called PKC-/CD4+ T cells. As a result of TCR and anti-CD28 ligation, PKC- is localized on T cells and attracted to immunological synapses.[91] Lack of adequate IL-2 production prevents PKC-/T cell replication in vitro. It has been revealed via in vivo research that PKC- plays several functions in immunological responses mediated by T helper cells (Th), T helper type 2 (Th2), and Th17.[92] In wild-type mice, Th17 cells and the generation of myosin-specific autoantibodies are required for the development of experimental autoimmune myocarditis (EAM). However, having a mutation in the PKC gene prevents these processes from occurring.[93] In PKC-/mice, EAM-like symptoms are induced by CpG infusion. In vitro, CpG restores the proliferation and survival of PKC-/CD4+ T cells by increasing the production of Bcl-XL and IL-17 in addition to IL-6 and TGF-.[94,95]

5.3 TLRs IN INNATE IMMUNITY, INFLAMMATION, AND CANCER DEVELOPMENT

Although TLRs are crucial in pathogen recognition and antibacterial or viral defense, the precise method through which they link inflammation and cancer is yet unknown.[96] There is strong evidence that TLR4 sequence variation is a critical risk factor for prostate cancer[97] since it has been demonstrated to dramatically enhance the prevalence of the illness. In immune cells, including macrophages, DCs, and T cells, TLRs play a role as sensors.[98] Cancer has been linked to chronic inflammation, which includes NF-B activation.[99] When bacteria, viruses, or injured tissues activate TLR, it triggers the NF-B pathway, which in turn produces inflammatory cytokines and chemokines.[100] The connection between cancer and the body's innate immune system is gaining more and more support.[101] Colon cancer risk is higher in those with inflammatory bowel diseases such as ulcerative colitis, Crohn's disease, and others.[102] Infection with Helicobacter pylori causes chronic inflammation in the stomach, which is a risk factor for the development of gastric cancer. Infection with the hepatitis B or hepatitis C virus (HBV or HCV, respectively) is a substantial risk factor for developing liver cancer.[103]

Thus, innate immunity generated through the TLR and NF-B pathways may construct inflammatory environments that attract immune cells, perpetuating chronic inflammatory states that result in the formation of immune suppressive cells, which may promote cancer initiation and proliferation.[104]

5.4 TUMOR-INFILTRATING IMMUNE CELLS IN THE TUMOR MICROENVIRONMENT

At the time of rapid growth, tumors often include several types of inflammatory and immune cells. Macrophages, dendritic cells (DCs), neutrophils, eosinophils, mast cells, gamma and delta effector and regulatory T cells, and many more fall into this category.[105] Many different types of cytokines, chemokines, and cytotoxic mediators may be produced by these cells according to conjecture. There are several soluble context-dependent mediators of cell death, such as tumor necrosis factor-alpha (TNF-), TRAIL, and interferons. Proteases, reactive oxygen and nitrogen species, and matrix metalloproteinases all play a role (IFNs).[106] The author will next look at the processes by which these factors improve tumor immunity or tolerance.[107,108]

5.4.1 TUMOR-INFILTRATING MACROPHAGES AND DCs

Macrophages that have found their way into tumors are called tumor-associated macrophages (TAMs), and they play a double role there. TAMs may change into macrophages, DCs, and Langerhans DCs after maturing from their hematopoietic progenitor cell or monocyte origins in the bone marrow.[109] Activation of IL-2 and IL-12 in TAMs may cause cancer cell death, but it also has the ability to promote tumor growth. Myeloid cell and epithelium-specific deletion of a floxed IKK allele (IKK-F/F) influence tumor initiation and progression.[110,111] This is because the NF-B pathway serves dual roles in tumorigenesis, first by protecting hematological malignancy cells from apoptosis and second by increasing the generation of proinflammatory cytokines in immune cells.[112] Tumor-infiltrating DCs with an immature phenotype that promotes tumor development may move to the tumor vasculature, where they may independently contribute to neovascularization and the progression of tumors expressing high amounts of the endothelium vascular endothelial growth factor (VEGF)-A. Tolerance is supported by immature DCs, whereas immunity is bolstered by mature DCs.[113] Myeloid suppressor cells inside tumors have been hypothesized to stimulate the expansion of Tregs, which in turn dampens antitumor immunity.[114,115] The tumor-infiltrating effector T cell during tumor control and progression of tumor is shown in Figure 5.2

5.5 CONCLUSIONS

When it comes to controlling the immune response, innate immunity and CD4+ Tregs are two of the most crucial players. Potentially effective in the fight against cancer are novel approaches that boost CD4+ effector T cells while decreasing CD4+ Tregs. However, the downstream signaling pathways connecting TLR8 or other players to the functional regulation of Tregs are not yet completely understood, despite the fact that Tregs can be downregulated and their suppressive function can be restored (e.g., in human cells by changing TLR8 signaling). The molecules required for the immunological suppression mediated by various subgroups of Tregs are still mostly unknown. The development of novel molecularly targeted drugs for the treatment of cancer depends on the successful identification and characterization of crucial pathways, such as those controlled by TLR ligands. It has been hypothesized that removing the suppressive features

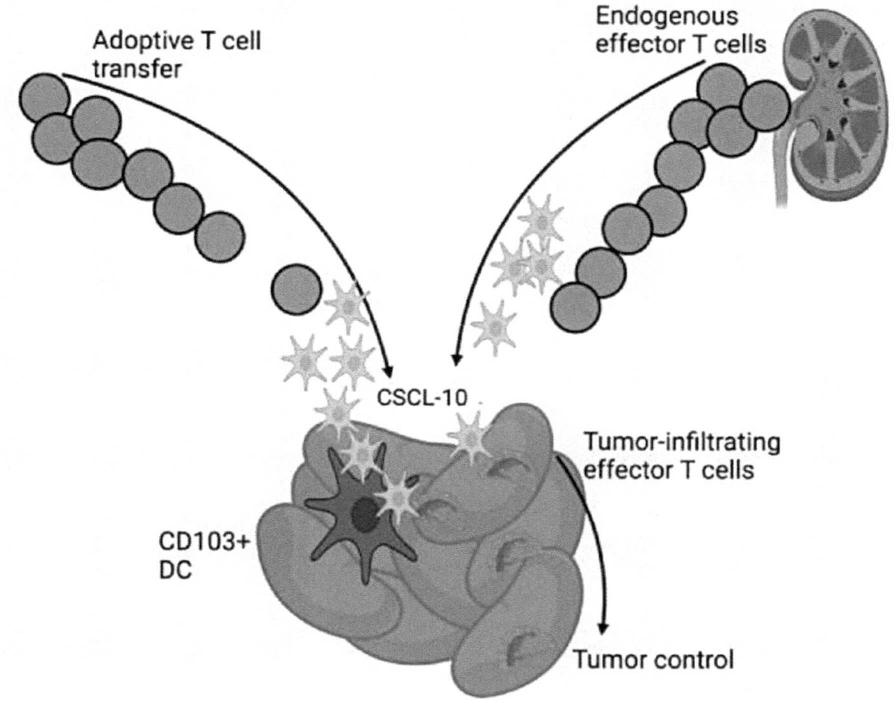

FIGURE 5.2 Schematic representation of tumor-infiltrating effector T cells during tumor control andprogression of the tumor.

of Treg cells would improve the therapeutic efficacy of vaccination techniques over the long run.

T cells express the vast majority of TLRs. In general, it seems that Treg has higher TLR expression than Teff does. Pathogen specificity, TLR expression/density/subcellular location, TLR accessory molecule (unc93b, gp96, PRAT4A, proteases, etc.) involvement, and downstream signaling cascades all play roles in determining whether TLR contact on Tregs increases or decreases Treg activity. The secrets of TLR biology in Tregs will be revealed by the deletion of each TLR in Tregs individually and the functional analysis of immune responses in the context of infection and autoimmunity.

KEYWORDS

- **tumor immunity**
- **receptors**
- **Treg activity**
- **tumor-infiltrating**
- **chemotherapy**

REFERENCES

1. Rosenberg, S. A. Progress in Human Tumour Immunology and Immunotherapy. *Nature* **2001,** *411*, 380–384.
2. Dunn, G. P.; Old, L. J.; Schreiber, R. D. The Three Es of Cancer Immunoediting. *Annu. Rev. Immunol.* **2004,** *22*, 329–260.
3. Old, L. J.; Chen, Y. T. New Paths in Human Cancer Serology. *J. Exp. Med.* **1998,** *187*, 1163–1167.
4. Wang, R. F.; Rosenberg, S. A. Human Tumor Antigens for Cancer Vaccine Development. *Immunol. Rev.* **1999,** *170*, 85–100.
5. Houghton, A. N.; Gold, J. S.; Blachere, N. E. Immunity Against Cancer: Lessons Learned from Melanoma. *Curr. Opin. Immunol.* **2001,** *13*, 134–140.
6. Wang, R. F. Enhancing Antitumor Immune Responses: Intracellular Peptide Delivery and Identification of MHC Class II-Restricted Tumor Antigens. *Immunol. Rev.* **2002,** *188*, 65–80.
7. Berzofsky, J. A.; Ahlers, J. D.; Janik, J.; Morris, J.; Oh, S.; Terabe, M. et al. Progress on New Vaccine Strategies Against Chronic Viral Infections. *J. Clin. Invest.* **2004,** *114*, 450–462.
8. Rosenberg, S. A.; Yang, J. C.; Restifo, N. P. Cancer Immunotherapy: Moving Beyond Current Vaccines. *Nat. Med.* **2004,** *10*, 909–915.
9. Dudley, M. E.; Wunderlich, J. R.; Robbins, P. F.; Yang, J. C.; Hwu, P.; Schwartzentruber, D. J. et al. Cancer Regression and Autoimmunity in Patients After Clonal Repopulation with Antitumor Lymphocytes. *Science* **2002,** *298*, 850–854.
10. Rosenberg, S. A.; Dudley, M. E. Cancer Regression in Patients with Metastatic Melanoma After the Transfer of Autologous Antitumor Lymphocytes. *Proc. Natl. Acad. Sci. USA* **2004,** *101* (Suppl. 2):14639–14645.
11. Wrzesinski, C.; Restifo, N. P. Less Is More: Lymphodepletion Followed by Hematopoietic Stem Cell Transplant Augments Adoptive T-Cell-Based Antitumor Immunotherapy. *Curr. Opin. Immunol.* **2005,** *17*, 195–201.
12. Wang, H. Y.; Deen, A. L.; Guangyong, P.; Zhong, G.; Yanchun Li, Yukiko, K. et al. Tumor-Specific Human CD4+ Regulatory T Cells and Their Ligands: Implication for Immunotherapy. *Immunity* **2004,** *20*, 107–118.

13. 13.Wang, H. Y.; Peng, G.; Guo, Z.; Shevach, E. M.; Wang R-F. Recognition of a New ARTC1 Peptide Ligand Uniquely Expressed in Tumor Cells by Antigenspecific CD4+ Regulatory T Cells. *J. Immunol.* **2005,** *174*, 2661–2670.
14. Sakaguchi, S.; Sakaguchi, N.; Shimizu, J.; Yamazaki, S.; Sakihama, T.; Itoh, M. et al. Immunologic Tolerance Maintained by CD25+ CD4+ Regulatory T Cells: Their Common Role in Controlling Autoimmunity, Tumor Immunity, and Transplantation Tolerance. *Immunol. Rev.* **2001,** *182*, 18–32.
15. Wood, K. J.; Sakaguchi, S. Regulatory T Cells in Transplantation Tolerance. *Nat. Rev. Immunol.* **2003,** *3*, 199–210.
16. Hori, S.; Nomura, T.; Sakaguchi, S. Control of Regulatory T Cell Development by the Transcription Factor foxp3. *Science* **2003,** *299*, 1057–1061.
17. Khattri, R.; Cox, T.; Yasayko, S. A.; Ramsdell, F. An essential role for Scurfin in CD4(+) CD25(+) T regulatory Cells. *Nat. Immunol.* **2003,** *4*, 337–342.
18. Fontenot, J. D.; Gavin, M. A.; Rudensky, A. Y. Foxp3 Programs the Development and Function of CD4(+)CD25(+) Regulatory T Cells. *Nat. Immunol.* **2003,** *4*, 330–336.
19. Sakaguchi, S. Naturally Arising CD4+ Regulatory T cells for Immunologic Self-Tolerance and Negative Control of Immune Responses. *Annu. Rev. Immunol.* **2004,** *22*, 531–562.
20. Shevach, E. M. CD4+ CD25+ Suppressor T Cells: More Questions Than Answers. *Nat. Rev. Immunol.* **2002,** *2*, 389–400.
21. Roncarolo, M. G.; Levings, M. K. The Role of Different Subsets of T Regulatory Cells in Controlling Autoimmunity. *Curr. Opin. Immunol.* **2000,** *12*, 676–683.
22. Francois Bach, J. Regulatory T Cells Under Scrutiny. *Nat. Rev. Immunol.* **2003,** *3*, 189–198.
23. Weiner, H. L. Induction and Mechanism of Action of Transforming Growth Factor-Beta-Secreting Th3 Regulatory Cells. *Immunol. Rev.* **2001,** *182*, 207–214.
24. Berendt, M. J.; North, R. J. T-Cell-Mediated Suppression of Anti-Tumor Immunity: An Explanation for Progressive Growth of an Immunogenic Tumor. *J. Exp. Med.* **1980,** *151*, 69–80.
25. North, R. J. Cyclophosphamide-Facilitated Adoptive Immunotherapy of an Established Tumor Depends on Elimination of Tumor-Induced Suppressor T Cells. *J. Exp. Med.* **1982,** *155*, 1063–1074.
26. North, R. J. Gamma-Irradiation Facilitates the Expression of Adoptive Immunity Against Established Tumors by Eliminating Suppressor T Cells. *Cancer Immunol. Immunother.* **1984,** *16*, 175–181.
27. North, R. J.; Awwad, M. T Cell Suppression as an Obstacle to Immunologically-Mediated Tumor Regression: Elimination of Suppression Results in Regression. *Prog. Clin. Biol. Res.* **1987,** *244*, 345–358.
28. Awwad, M.; North, R. J. Immunologically Mediated Regression of a Murine Lymphoma After Treatment with Anti-L3T4 Antibody: A Consequence of Removing L3T4+ Suppressor T Cells from a Host Generating Predominantly Lyt-2+ T Cell-Mediated Immunity. *J. Exp. Med.* **1988,** *168*, 2193–2206.
29. 28.North, R. J.; Awwad, M. Elimination of Cycling CD4+ Suppressor T Cells with an Anti-Mitotic Drug Releases Non-Cycling CD8+ T Cells to Cause Regression of an Advanced Lymphoma. *Immunology* **1990,** *71*, 90–95.

30. 29.Turk, M. J.; Guevara-Patino, J. A.; Rizzuto, G. A.; Engelhorn, M. E.; Houghton, A. N. Concomitant Tumor Immunity to a Poorly Immunogenic Melanoma Is Prevented by Regulatory T Cells. *J. Exp. Med.* **2004,** *200*, 771–782.
31. Sutmuller, R. P.; van Duivenvoorde, L. M.; van Elsas, A.; Schumacher, T. N.; Wildenberg, M. E.; Allison, J. P. et al. Synergism of Cytotoxic T Lymphocyte Associated Antigen 4 Blockade and Depletion of CD25(+) Regulatory T Cells in Antitumor Therapy Reveals Alternative Pathways for Suppression of Autoreactive Cytotoxic T Lymphocyte Responses. *J. Exp. Med.* **2001,** *194*, 823–832.
32. Tanaka, H.; Tanaka, J.; Kjaergaard, J.; Shu, S. Depletion of CD4+ CD25+ Regulatory Cells Augments the Generation of Specific Immune T Cells in Tumor-Draining Lymph Nodes. *J. Immunother.* **2002,** *25*, 207–217.
33. 32.Mukherji, B.; Guha, A.; Chakraborty, N. G.; Sivanandham, M.; Nashed, A. L.; Sporn, J. R. et al. Clonal Analysis of Cytotoxic and Regulatory T Cell Responses Against Human Melanoma. *J. Exp. Med.* **1989,** *169*, 1961–1976.
34. Chakraborty, N. G.; Twardzik, D. R.; Sivanandham, M.; Ergin, M. T.; Hellstrom, K. E.; Mukherji, B. Autologous Melanoma-Induced Activation of Regulatory T Cells That Suppress Cytotoxic Response. *J. Immunol.* **1990,** *145*, 2359–2364.
35. 34.Woo, E. Y.; Chu, C. S.; Goletz, T. J.; Schlienger, K.; Yeh, H.; Coukos, G. et al. Regulatory CD4(+)CD25(+) T Cells in Tumors from Patients with Early Stage Non-Small Cell Lung Cancer and Late-Stage Ovarian Cancer. *Cancer Res.* **2001,** *61*, 4766–4772.
36. Liyanage, U. K.; Moore, T. T.; Joo, H. G.; Tanaka, Y.; Herrmann, V.; Doherty, G. et al. Prevalence of Regulatory T Cells Is Increased in Peripheral Blood and Tumor Microenvironment of Patients with Pancreas or Breast Adenocarcinoma. *J. Immunol.* **2002,** *169*, 2756–2761.
37. Curiel, T. J.; Coukos, G.; Zou, L.; Alvarez, X.; Cheng, P.; Mottram, P. et al. Specific Recruitment of Regulatory T Cells in Ovarian Carcinoma Fosters Immune Privilege and Predicts Reduced Survival. *Nat. Med.* **2004,** *10*, 942–949.
38. Viguier, M.; Lemaitre, F.; Verola, O.; Cho, M. S.; Gorochov, G.; Dubertret, L. et al. Foxp3 Expressing CD4(+)CD25(High) Regulatory T Cells Are Over Represented in Human Metastatic Melanoma Lymph Nodes and Inhibit the Function of Infiltrating T Cells. *J. Immunol.* **2004,** *173*, 1444–1453.
39. Belkaid, Y.; Piccirillo, C. A.; Mendez, S.; Shevach, E. M.; Sacks, D. L. CD4+CD25+ Regulatory T Cells Control Leishmania Major Persistence and Immunity. *Nature* **2002,** *420*, 502–507.
40. Peng, G.; Guo, Z.; Kiniwa, Y.; Voo, K. S.; Peng, W.; Fu, T. et al. Toll-like Receptor 8 Mediated-Reversal of CD4+ Regulatory T Cell Function. *Science* **2005,** *309*, 1380–1384.
41. (a) Taguchi, O.; Kojima, A.; Nishizuka, Y. Experimental Autoimmune Prostatitis After Neonatal Thymectomy in the Mouse. *Clin. Exp. Immunol.* **1985,** *60*, 123–129. (b) McCullagh, P. Interception of the Development of Self Tolerance in Fetal Lambs. Eur *J. Immunol.* **1989,** *19*, 1387–1392.
42. Bausinger, H.; Lipsker, D.; Ziylan, U.; Manie, S.; Briand, J. P.; Cazenave, J. P.; Muller, S.; Haeuw, J. F.; Ravanat, C.; de la Salle, H.; Hanau, D. Endotoxin-Free Heat-Shock Protein 70 Fails to Induce APC Activation. *Eur. J. Immunol.* **2002,** *32*, 3708–3713.
43. Gao, B.; Tsan, M. F. Recombinant Human Heat Shock Protein 60 Does Not Induce the Release of Tumor Necrosis Factor Alpha from Murine Macrophages. *J. Biol. Chem.* **2003,** *278*, 22523–22529.

44. Osterloh, A.; Kalinke, U.; Weiss, S.; Fleischer, B.; Breloer, M. Synergistic and Differential Modulation of Immune Responses by Hsp60 and lipopolysaccharide. *J. Biol. Chem.* **2007,** *282*, 4669–4680.
45. Liu, B.; Li, Z. Endoplasmic Reticulum HSP90b1 (gp96, grp94) Optimizes B-Cell Function via Chaperoning Integrin and TLR But Not Immunoglobulin. *Blood* **2008,** *112*, 1223–1230.
46. Yang, Y.; Liu, B.; Dai, J.; Srivastava, P. K.; Zammit, D. J.; Lefrancois, L.; Li, Z. Heat Shock Protein gp96 Is a Master Chaperone for Toll-Like Receptors and Is Important in the Innate Function of Macrophages. *Immunity* **2007,** *26*, 215–226.
47. Liu, B.; Yang, Y.; Dai, J.; Medzhitov, R.; Freudenberg, M. A.; Zhang, P. L.; Li, Z. TLR4 Up-Regulation at Protein or Gene Level Is Pathogenic for Lupus-like Autoimmune Disease. *J. Immunol.* **2006,** *177*, 6880–6888.
48. Dai, J.; Liu, B.; Cua, D. J.; Li, Z. Essential Roles of IL-12 and Dendritic Cells But Not IL-23 and Macrophages in Lupus-Like Diseases Initiated by Cell Surface HSP gp96. *Eur. J. Immunol.* **2007,** *37*, 706–715.
49. Dai, J.; Liu, B.; Ngoi, S. M.; Sun, S.; Vella, A. T.; Li, Z. TLR4 Hyper Responsiveness via Cell Surface Expression of Heat Shock Protein gp96 Potentiates Suppressive Function of Regulatory T Cells. *J. Immunol.* **2007,** *178*, 3219–3225.
50. LaRosa, D. F.; Gelman, A. E.; Rahman, A. H.; Zhang, J.; Turka, L. A.; Walsh, P. T. CpG DNA Inhibits CD4+CD25+ Treg Suppression Through Direct MyD88-Dependent Costimulation of Effector CD4+ T cells. *Immunol. Lett.* **2007,** *108*, 183–188.
51. Thibault, S.; Tardif, M. R.; Barat, C.; Tremblay, M. J. TLR2 Signaling Renders Quiescent Naive and Memory CD4+ T Cells More Susceptible to Productive Infection with X4 and R5 HIV-Type 1. *J. Immunol.* **2007,** *179*, 4357–4366.
52. Zanin-Zhorov, A.; Tal-Lapidot, G.; Cahalon, L.; Cohen-Sfady, M.; Pevsner-Fischer, M.; Lider, O.; Cohen, I. R. Cutting Edge: T Cells Respond to Lipopolysaccharide Innately via TLR4 Signaling. *J. Immunol.* **2007,** *179*, 41–44.
53. Monks, C. R.; Kupfer, H.; Tamir, I.; Barlow, A.; Kupfer, A. Selective Modulation of Protein Kinase C-Theta During T-Cell Activation. *Nature* **1997,** *385*, 83–86.
54. Bi, K.; Tanaka, Y.; Coudronniere, N.; Sugie, K.; Hong, S.; van Stipdonk, M. J.; Altman, A. Antigen-Induced Translocation of PKC-Theta to Membrane Rafts Is Required for T Cell Activation. *Nat. Immunol.* **2001,** *2*, 556–563.
55. Sun, Z.; Arendt, C. W.; Ellmeier, W.; Schaeffer, E. M.; Sunshine, M. J.; Gandhi, L.; Annes, J.; Petrzilka, D.; Kupfer, A.; Schwartzberg, P. L.; Littman, D. R. PKC-Theta Is Required for TCR-Induced NF-kappaB Activation in Mature But Not Immature T Lymphocytes. *Nature* **2000,** *404*, 402–407.
56. Marsland, B. J.; Nembrini, C.; Schmitz, N.; Abel, B.; Krautwald, S.; Bachmann, M. F.; Kopf, M. Innate Signals Compensate for the Absence of PKC-{Theta} During In Vivo CD8(+) T Cell Effector and Memory Responses. *Proc. Natl. Acad. Sci. USA* **2005,** *102*, 14374–14379.
57. Marsland, B. J.; Soos, T. J.; Spath, G.; Littman, D. R.; Kopf, M. Protein Kinase C Theta Is Critical for the Development of In Vivo T Helper (Th)2 Cell But Not Th1 Cell Responses. *J. Exp. Med.* **2004,** *200*, 181–189.
58. Salek-Ardakani, S.; So, T.; Halteman, B. S.; Altman, A.; Croft, M. Differential Regulation of Th2 and Th1 Lung Inflammatory Responses by Protein Kinase C Theta. *J. Immunol.* **2004,** *173*, 6440–6447.

59. Tan, S. L.; Zhao, J.; Bi, C.; Chen, X. C.; Hepburn, D. L.; Wang, J.; Sedgwick, J. D.; Chintalacharuvu, S. R.; Na, S. Resistance to Experimental Autoimmune Encephalomyelitis and Impaired IL-17 Production in Protein Kinase C Theta-Deficient Mice. *J Immunol.* **2006,** *176*, 2872–2879.
60. Fontenot, J. D.; Gavin, M. A.; Rudensky, A. Y. Foxp3 Programs the Development and Function of CD4+CD25+ Regulatory T Cells. *Nat. Immunol.* **2003,** *4*, 330–336.
61. Khattri, R.; Cox, T.; Yasayko, S. A.; Ramsdell, F. An Essential Role for Scurfin in CD4+CD25+ T Regulatory Cells. *Nat. Immunol.* **2003,** *4*, 337–342.
62. Miyake, K.; Yamashita, Y.; Hitoshi, Y.; Takatsu, K.; Kimoto, M. Murine B Cell Proliferation and Protection from Apoptosis with an Antibody Against a 105-kD Molecule: Unresponsiveness of X-Linked Immunodeficient B Cells. *J. Exp. Med.* **1994,** *180*, 1217–1224.
63. Nagai, Y.; Shimazu, R.; Ogata, H.; Akashi, S.; Sudo, K.; Yamasaki, H.; Hayashi, S.; Iwakura, Y.; Kimoto, M.; Miyake, K. Requirement for MD-1 in Cell Surface Expression of RP105/CD180 and B-Cell Responsiveness to Lipopolysaccharide. *Blood* **2002,** *99*, 1699–1705.
64. Divanovic, S.; Trompette, A.; Atabani, S. F.; Madan, R.; Golenbock, D. T.; Visintin, A.; Finberg, R. W.; Tarakhovsky, A.; Vogel, S. N.; Belkaid, Y.; Kurt-Jones, E. A.; Karp, C. L. Negative Regulation of Toll-Like Receptor 4 Signaling by the Toll-Like Receptor Homolog RP105. *Nat. Immunol.* **2005,** *6*, 571–578.
65. Gorczynski, R. M.; Kai, Y.; Miyake, K. MD1 Expression Regulates Development of Regulatory T Cells. *J. Immunol.* **2006,** *177*, 1078–1084.
66. Tabeta, K.; Hoebe, K.; Janssen, E. M.; Du, X.; Georgel, P.; Crozat, K.; Mudd, S.; Mann, N.; Sovath, S.; Goode, J.; Shamel, L.; Herskovits, A. A.; Portnoy, D. A.; Cooke, M.; Tarantino, L. M.; Wiltshire, T.; Steinberg, B. E.; Grinstein, S.; Beutler, B. The Unc93b1 Mutation 3d Disrupts Exogenous Antigen Presentation and Signaling via Toll-Like Receptors 3, 7 and 9. *Nat. Immunol.* **2006,** *7*, 156–164.
67. Takahashi, K.; Shibata, T.; Akashi-Takamura, S.; Kiyokawa, T.; Wakabayashi, Y.; Tanimura, N.; Kobayashi, T.; Matsumoto, F.; Fukui, R.; Kouro, T.; Nagai, Y.; Takatsu, K.; Saitoh, S.; Miyake, K. A Protein Associated with Toll-Like Receptor (TLR) 4 (PRAT4A) Is Required for TLR-Dependent Immune Responses. *J. Exp. Med.* **2007,** *204*, 2963–2976.
68. Wakabayashi, Y.; Kobayashi, M.; Akashi-Takamura, S.; Tanimura, N.; Konno, K.; Takahashi, K.; Ishii, T.; Mizutani, T.; Iba, H.; Kouro, T.; Takaki, S.; Takatsu, K.; Oda, Y.; Ishihama, Y.; Saitoh, S.; Miyake, K. A Protein Associated with Toll-Like Receptor 4 (PRAT4A) Regulates Cell Surface Expression of TLR4. *J. Immunol.* **2006,** *177*, 1772–1779.
69. Randow, F.; Seed, B. Endoplasmic Reticulum Chaperone gp96 Is Required for Innate Immunity But Not Cell Viability. *Nat. Cell Biol.* **2001,** *3*, 891–896.
70. Liu, G.; Zhao, Y. Toll-Like Receptors and Immune Regulation: Their Direct and Indirect Modulation on Regulatory CD4+ CD25+ T Cells. *Immunology* **2007,** *122*, 149–156.
71. Hemmi, H.; Kaisho, T.; Takeuchi, O.; Sato, S.; Sanjo, H.; Hoshino, K. et al. Small Anti-Viral Compounds Activate Immune Cells via the TLR7 MyD88- Dependent Signaling Pathway. *Nat. Immunol.* **2002,** *3*, 196–200.
72. Akira, S.; Takeda, K. Toll-Like Receptor Signalling. *Nat. Rev. Immunol.* **2004,** *4*, 499–511.

73. Crozat, K.; Beutler, B. TLR7: A New Sensor of Viral Infection. *Proc. Natl. Acad. Sci. USA* **2004,** *101*, 6835–6836.
74. Takaoka, A.; Yanai, H.; Kondo, S.; Duncan, G.; Negishi, H.; Mizutani, T. et al. Integral Role of IRF-5 in the Gene Induction Programme Activated by Toll-Like Receptors. *Nature* **2005,** *434*, 243–249.
75. Honda, K.; Yanai, H.; Negishi, H.; Asagiri, M.; Sato, M.; Mizutani, T. et al. IRF-7 Is the Master Regulator of type-I Interferon-Dependent Immune Responses. *Nature* **2005,** *434*, 772–777.
76. Honda, K.; Ohba, Y.; Yanai, H.; Negishi, H.; Mizutani, T.; Takaoka, A. et al. Spatiotemporal Regulation of MyD88-IRF-7 Signalling for Robust Type-I Interferon Induction. *Nature* **2005,** *434*, 1035–1040.
77. Serra, P.; Amrani, A.; Yamanouchi, J.; Han, B.; Thiessen, S.; Utsugi, T. et al. CD40 Ligation Releases Immature Dendritic Cells from the Control of Regulatory CD4+CD25+ T Cells. *Immunity* **2003,** *19*, 877–889.
78. Munn, D. H.; Sharma, M. D.; Lee, J. R.; Jhaver, K. G.; Johnson, T. S.; Keskin, D. B. et al. Potential Regulatory Function of Human Dendritic Cells Expressing Indoleamine 2,3-Dioxygenase. *Science* **2002,** *297*, 1867–1870.
79. Fallarino, F.; Grohmann, U.; Hwang, K. W.; Orabona, C.; Vacca, C.; Bianchi, R. et al. Modulation of Tryptophan Catabolism by Regulatory T Cells. *Nat. Immunol.* **2003,** *4*, 1206–1212.
80. Vicari, A. P.; Chiodoni, C.; Vaure, C.; Ait-Yahia, S.; Dercamp, C.; Matsos, F. et al. Reversal of Tumor-Induced Dendritic Cell Paralysis by CpG Immunostimulatory Oligonucleotide and Anti-Interleukin 10 Receptor Antibody. *J. Exp. Med.* **2002,** *196*, 541–549.
81. Pasare, C.; Medzhitov, R. Toll Pathway-Dependent Blockade of CD4+CD25+ T Cell-Mediated Suppression by Dendritic Cells. *Science* **2003,** *299*, 1033–1036.
82. Fehervari, Z.; Sakaguchi, S. Control of Foxp3+ CD25+CD4+ Regulatory Cell Activation and Function by Dendritic Cells. *Int. Immunol.* **2004,** *16*, 1769–1780.
83. Kubo, T.; Hatton, R. D.; Oliver, J.; Liu, X.; Elson, C. O.; Weaver, C. T. Regulatory T Cell Suppression and Energy are Differentially Regulated by Proinflammatory Cytokines Produced by TLR-Activated Dendritic Cells. *J. Immunol.* **2004,** *173*, 7249–7258.
84. Yamazaki, S.; Iyoda, T.; Tarbell, K.; Olson, K.; Velinzon, K.; Inaba, K. et al. Direct Expansion of Functional CD25+ CD4+ Regulatory T Cells by Antigen-Processing Dendritic Cells. *J. Exp. Med.* **2003,** *198*, 235–247.
85. Klein, L.; Khazaie, K.; von Boehmer, H. In Vivo Dynamics of Antigenspecific Regulatory T Cells Not Predicted from Behavior In Vitro. *Proc. Natl. Acad. Sci. USA* **2003,** *100*, 8886–8891.
86. Fisson, S.; Darrasse-Jeze, G.; Litvinova, E.; Septier, F.; Klatzmann, D.; Liblau, R. et al. Continuous Activation of Autoreactive CD4+ CD25+ Regulatory T Cells in the Steady State. *J. Exp. Med.* **2003,** *198*, 737–746.
87. Obermajer, N.; Urban, J.; Wieckowski, E.; Muthuswamy, R.; Ravindranathan, R.; Bartlett, D. L. et al. Promoting the Accumulation of Tumor-Specific T Cells in Tumor Tissues by Dendritic Cell Vaccines and Chemokine-Modulating Agents. *Nat. Protoc.* **2018,** *13*, 335–357.
88. O'Neill, L. A.; Bryant, C. E.; Doyle, S. L. Therapeutic Targeting of Toll-Like Receptors for Infectious and Inflammatory Diseases and Cancer. *Pharmacol. Rev.* **2009,** *61*, 177–197.

89. Ou, L.; Lin, S.; Song, B.; Liu, J.; Lai, R.; Shao, L. The Mechanisms of Graphene-Based Materials-Induced Programmed Cell Death: A Review of Apoptosis, Autophagy, and Programmed Necrosis. *Int. J. Nanomed.* **2017,** *12*, 6633–6646.
90. Ouyang, L.; Shi, Z.; Zhao, S.; Wang, F. T.; Zhou, T. T.; Liu, B. et al. Programmed Cell Death Pathways In Cancer: A Review of Apoptosis, Autophagy, and Programmed Necrosis. *Cell Prolif.* 2012, *45*, 487–498.
91. Pasare, C.; Medzhitov, R. Toll Pathway-Dependent Blockade of CD4+ CD25+ T Cell-Mediated Suppression by Dendritic Cells. *Science* **2003,** *299*, 1033–1036.
92. Pradere, J. P.; Dapito, D. H.; Schwabe, R. F. The Yin and Yang of Toll-Like Receptors in Cancer. *Oncogene* **2014,** *33*, 3485–3495.
93. Pulendran, B.; Smith, J. L.; Caspary, G.; Brasel, K.; Pettit, D.; Maraskovsky, E. et al. Distinct Dendritic Cell Subsets Differentially Regulate the Class of Immune Response In Vivo. *Proc. Natl. Acad. Sci. USA*1999, *96*, 1036–1041.
94. Reshma, S. C.; Syama, S.; Mohanan, P. V. Nano-Biointeractions of PEGylated and Bare Reduced Graphene Oxide on Lung Alveolar Epithelial Cells: A Comparative In Vitro Study. *Colloids Surf. B Biointerfaces* **2016,** *140*, 104–116.
95. Ryffel, B.; Fremond, C.; Jacobs, M.; Parida, S.; Botha, T.; Schnyder, B. et al. Innate Immunity to Mycobacterial Infection in Mice: Critical Role for Toll-Like Receptors. *Tuberculosis* **2005,** *85*, 395–405.
96. Sakaguchi, S. (2005). Naturally Arising Foxp3-Expressing CD25+CD4+ Regulatory T Cells in Immunological Tolerance to Self and Non-Self. Nat. Immunol. **2005,** *6*, 345–352.
97. Sato, Y.; Goto, Y.; Narita, N.; Hoon, D. S. Cancer Cells Expressing Toll-Like Receptors and the Tumor Microenvironment. *Cancer Microenviron.* **2009,** *2*, 205–214.
98. Schölch, S.; Rauber, C.; Tietz, A.; Rahbari, N. N.; Bork, U.; Schmidt, T. et al. Radiotherapy Combined with TLR7/8 Activation Induces Strong Immune Responses Against Gastrointestinal Tumors. *Oncotarget* **2015,** *6*, 4663–4676.
99. Sharma, P.; Allison, J. P. The Future of Immune Checkpoint Therapy. *Science* **2015,** *348*, 56–61.
100. Sharpe, A. H.; Freeman, G. J. The B7-CD28 Superfamily. *Nat. Rev. Immunol.* **2002,** *2*, 116–126.
101. Sheyhidin, I.; Nabi, G.; Hasim, A.; Zhang, R. P.; Ainiwaer, J.; Ma, H. et al. Overexpression of TLR 3, TLR 4, TLR7 and TLR9 in Esophageal Squamous Cell Carcinoma. *World, J. Gastroenterol.* **2011,** *17*, 3745–3751.
102. Shi, C. S.; Kehrl, J. H. TRAF6 and A20 Regulate Lysine 63-Linked Ubiquitination of Beclin-1 to Control TLR4-Induced Autophagy. *Sci. Signal.* **2010,** *3*, ra42.
103. Singh, V. K.; Ducey, E. J.; Fatanmi, O. O.; Singh, P. K.; Brown, D. S.; Purmal, A. et al. CBLB613: A TLR 2/6 Agonist, Natural Lipopeptide of Mycoplasma Arginini, as a Novel Radiation Countermeasure. *Radiat. Res.* 2012, *177*, 628–642.
104. Singh, V. K.; Romaine, P. L.; Seed, T. M. Medical Counter Measures for Radiation Exposure and Related Injuries: Characterization of Medicines, FDA-Approval Status and Inclusion into the Strategic National Stockpile. *Health Phys.* **2015,** *108*, 607–630.
105. Takaki, H.; Shime, H.; Matsumoto, M.; Seya, T. Tumor Cell Death by Pattern-Sensing of Exogenous RNA: Tumor Cell TLR3 Directly Induces Necroptosis by Poly(I:C) In Vivo, Independent of Immune Effector-Mediated Tumor Shrinkage. *Oncoimmunology* **2017,** *6*, e1078968.

106. Tan, M. L.; Ooi, J. P.; Ismail, N.; Moad, A. I.; Muhammad, T. S. Programmed Cell Death Pathways and Current Antitumor Targets. *Pharm. Res.* **2009,** *26*, 1547–1560.
107. Vignali, D. A.; Collison, L. W.; Workman, C. J. How Regulatory T Cells Work. *Nat. Rev. Immunol.* **2008,** *8*, 523–532.
108. Vigneron, N. Human Tumor Antigens and Cancer Immunotherapy. *Biomed. Res. Int.* **2015**, 948501.
109. Yoneda, K.; Sugimoto, K.; Shiraki, K.; Tanaka, J.; Beppu, T.; Fuke, H. et al. Dual Topology of Functional Toll-Like Receptor 3 Expression in Human Hepatocellular Carcinoma: Differential Signaling Mechanisms of TLR3-Induced NF-kappaB Activation and Apoptosis. *Int. J. Oncol.* **2018,** *33*, 929–936.
110. Zhan, Z.; Xie, X.; Cao, H.; Zhou, X.; Zhang, X. D.; Fan, H. et al. Autophagy Facilitates TLR4- and TLR3-Triggered Migration and Invasion of Lung Cancer Cells Through the Promotion of TRAF6 Ubiquitination. *Autophagy* **2014,** *10*, 257–268.
111. Zhang, D.; Li, L.; Jiang, H.; Knolhoff, B. L.; Lockhart, A. C.; Wang-Gillam, A. et al. Constitutive IRAK4 Activation Underlies Poor Prognosis and Chemoresistance in Pancreatic Ductal Adenocarcinoma. *Clin. Cancer Res.* 2017, *23*, 1748–1759.
112. Zhang, Y.; Luo, F.; Cai, Y.; Liu, N.; Wang, L.; Xu, D. et al. TLR1/TLR2 Agonist Induces Tumor Regression by Reciprocal Modulation of Effector and Regulatory T Cells. *J. Immunol.* **2011,** *186*, 1963–1969.
113. Zhao, X. L.; Lin, Y.; Jiang, J.; Tang, Z.; Yang, S.; Lu, L. et al. High-Mobility Group Box 1 Released by Autophagic Cancer-Associated Fibroblasts Maintains the Stemness of Luminal Breast Cancer Cells. *J. Pathol.* **2017,** *243*, 376–389.
114. Zhivotovsky, B. Systems Biology Analysis of Cell Death Pathways in Cancer: How Collaborative and Interdisciplinary Research Helps. In *Cancer Systems Biology, Bioinformatics and Medicine*; Cesario, A., Marcus, F., Eds.; Springer: Dordrecht.
115. Zhou, M.; Mcfarland-Mancini, M. M.; Funk, H. M.; Husseinzadeh, N.; Mounajjed, T.; Drew, A. F. Toll-Like Receptor Expression in Normal Ovary and Ovarian Tumors. *Cancer Immunol. Immunother.* **2009,** *58*, 1375–1385.
116. Mahoney, K. M.; Rennert, P. D.; Freeman, G. J. Combination Cancer Immunotherapy and New Immunomodulatory Targets. *Nat. Rev. Drug Discov.* **2015,** *14*, 561–584.

CHAPTER 6

Cancer Immunotherapy: Understanding Immune tolerance

ABSTRACT

The search for effective immunotherapy against cancer is one of the most serious problems immunologists face. To combat melanoma, the author expanded our laboratory studies into human clinical trials based on immunological approaches. The major histocompatibility complex (MHC) molecules on melanoma cells exhibited a collection of peptide antigens that were broadly recognized by patient's T lymphocytes. They are now routinely included in peptide-based vaccinations, which elicit CD8+ and CD4+ T-cell responses in 80–100% of patients. Several individuals saw objective, clinical regressions of their tumors, and those who were vaccinated had a better stage-specific survival rate. New clinical studies will explore the feasibility of including melanoma helper peptides (MHP) in multi-peptide vaccines targeting CD8 cells. The effectiveness of combination therapy and the modulation of host-tumor interactions will be the focus of future studies. Animal studies that mirror human trials are providing crucial insights into the immunobiology of the melanoma-host interaction. Given that most melanoma antigens are derived from pigment-producing proteins, the researcher investigated self-tolerance to tyrosinase (Tyr) and the effects of vaccination on T cells in a self-tolerogenic environment. Dendritic cells primed with peptide immunogens improved anticancer immunity in a mouse model. This opens the door to new approaches for treating cancer by using vaccines.

6.1 INTRODUCTION

Melanoma-antigen-specific T cells have been found in tumors, regional lymph nodes (LN), and peripheral blood, indicating the establishment of an

Advances in Immunological Treatment of Cancer: Cellular Mechanism and Regulation.
Rishabha Malviya, Arun Kumar Singh, and Deepika Yadav (Authors)

innate immune response in the overwhelming majority of patients.[1] After being cultured in vitro and transferred adoptively, melanoma-reactive CD8 T lymphocytes may affect tumor regression by recognizing tumor-controlling Ags.[2,3] A therapeutic vaccine and better CD8 T cells for adoptive immunotherapy will be created using these Ags. Proteins involved in pigment production are the primary source of peptide Ags presented by class I MHC molecules and recognized by CD8 T lymphocytes in normal melanocytes.[4] There is a group of proteins called melanocyte differentiation proteins (MDP) that includes transient receptor potential (TRP)-1 and TRP-2.[5] Given their ability to display peptides from a wide variety of Ags, class I and II major histocompatibility complex (MHC) molecules are found almost everywhere.[6] The current and potential use of MDPs as immunological targets in the development of vaccines for tumor control across a wide range of patients is supported by the clinical effectiveness of MDP-directed immunotherapies.[7,8]

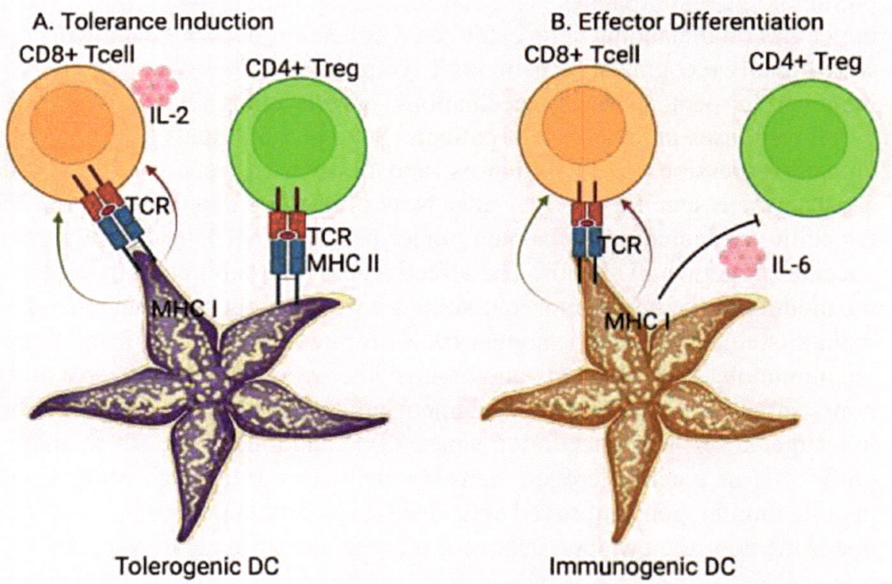

FIGURE 6.1 Diagrammatically representation of CD4+ during immunogenic treatment of cancer.

Importantly, vitiligo, a skin condition characterized by the autoimmune destruction of melanocytes, in humans and animals has been related to the spontaneous or immunotherapy-induced regression of melanoma.[9] Our research team has spent the last 15 years finding these antigens, incorporating them into melanoma vaccines, and studying their immunogenicity in animal models.

6.2 IMMUNE SYSTEM TUMOR DETECTION

It has been hypothesized that the immune system can prevent or eliminate most malignancies in their early stages because tumor development is infrequent in immunocompetent hosts.[10,11] Overexpression of cellular proteins or neoantigens resulting from mutations in tumor genes may serve as immune recognition targets.[12] Evidence from clinical trials seems to support this hypothesis since lymphocyte infiltration of tumors has been linked to improved survival in several different kinds of solid tumors.[13] The high incidence of multiple malignancies among organ-donation recipients undergoing lifelong immunosuppressive therapy has led to the concept that T-cell immune surveillance against tumors, which plays a role in humans. Recent research[14] has shown that people with HIV-1 infection and AIDS are at a higher risk of acquiring all forms of lymphomas, even those with no obvious viral relationship, including Kaposi's sarcomas.

Burnet's proposal of immune system surveillance has been met with skepticism ever since it was first proposed.[15] This is because nude animals with T-cell defects do not exhibit faster tumor development compared to normal mice.[16] While T-cell-deficient animals developed tumors at the same rate as control mice,[17,18] it was hypothesized that the innate immune systems of the mice would compensate. Tumor incidence is higher in Rag/ and STAT-1/ mice, which are severely immunocompromised and have abnormalities in both the innate and adaptive immune systems,[19] and this has brought the topic back into the limelight. Tumors do not appear in these mice until they have been alive for at least a year, suggesting that immune surveillance regulates the slow and irregular development of tumors in these animals. Tumor growth might be slowed by the body's natural immune system. To evade elimination by class I-restricted cytotoxic T lymphocytes,[20] tumors must downregulate the expression of class I MHC, which suggests that natural killer (NK) cells may play a particularly important role in regulating the growth of these tumors. Graft-versus leukemia (GVL) consequences in the context of MHC haploidentical hematopoietic stem-cell transplantation have been the subject of recent debates.[21,22]

Tumor-bearing hosts may respond to tumors in the following two ways—intrinsically and in response to stimulation.[23] It may be possible to utilize the expression profile of a tumor to identify tumor-specific antigens as opposed to just-associated ones (TAA).[24] Tumor-associated antigens (TAAs) are encoded by genes in normally functioning cells, but genuine tumor-specific antigens are encoded by mutations in these same genes.[25]

Proteins encoded by the Ras oncogene, chimeric fusion proteins like BCR-ABL, and alterations to the immunoglobulin heavy-chain locus are all possibilities of tumor-specific antigens (idiotype [Id] of B-cell neoplasias).[26] Differentiation antigens such as a prostate-specific antigen, alpha-fetoprotein, breast mucin, MART-1 (a melanoma antigen recognized by T lymphocytes), and tyrosinase are all connected with testis cancer (TAAs).[27] In contrast to somatic cells, spermatozoa tend to exhibit cancer testicular antigens. During cancer's numerous developmental phases, these genes are often activated. Members of the melanoma antigen-encoding family are included as well.[28] Murine double minute 2 (MDM-2) oncoprotein and Wilm's tumor gene are both examples of TAAs that are amplified or overexpressed in malignancies and may be targeted by immune responses.

To identify tumor antigens recognized by T lymphocytes, genomic DNA from tumors is often transfected into cells expressing a specific MHC. Transfecting human T cells with the gene encoding the cancer antigen gave them the ability to attack and destroy tumors with remarkable efficiency. They are presented to CD8+ T lymphocytes through MHC class I molecules. Although CD8+ tumor-reactive T lymphocytes that are MHC class I restricted have been identified, immunization trials using class I presented peptides have shown comparatively poor tumor responses.[29] Poor immune responses to these peptides have been linked to a lack of tumor-specific CD4 support.[30,31] Vaccinating a mouse model with a virus-producing an MHC class II T-helper epitope led to the development of CD4+ T cells that recognized and attacked tumor cells that lacked the MHC class II antigen. Crosspriming and CD4-mediated immune response orchestration were necessary for tumor rejection because CD8+ cytotoxic T lymphocyte (CTL) cells needed to recognize a distinct viral antigen.[32] Studies have shown that CD4+ T lymphocytes may recognize cancer antigens presented by MHC class II (such as mutant CDC27[33]). It may be critical to identify additional tumor antigens expressed by MHC class II if the author were to elicit efficient anti-cancer immune responses. CD4+ T cells may play a role in the initiation and maintenance of the immune response by inducing antigen-presenting cells (APC) to deliver antigens in a costimulatory and CTL-supportive manner.[34–36] Efforts to identify major histocompatibility complex (MHC) class II-restricted tumor antigens (TAAs) have increased in recent years.

Tumor antigens that elicit an immune response may also be discovered by screening a phage expression library using serum from cancer patients.[37] This method has led to the discovery of several antigens linked to testicular cancer. The importance of these antigens for immunization in vivo is still debatable, despite their shown function in T cell-dependent immunity in vitro. There should be a T-cell response to these serologically established antigens since T cells are essential for assisting with humoral immune responses.[38] The NY-ESO-1 antigen was discovered by a serologic screening of cDNA expression libraries, and further research confirmed that it could stimulate both CD8 and CD4 T-cell responses.[39] Characteristics of both cellular and humoral immune responses are present in hosts harboring tumors. It has been shown that tumors expressing NY-ESO-1 may elicit a humoral immune response in 40–50% of patients.[40] In a recent study employing tetramers and enzyme-linked immunosorbent spot testing, 10 out of 11 persons who acquired spontaneous antibodies also had a CD8+ T-cell response. Only those who additionally had an antibody response had their CD8+ T-cell responses limited to HLA-A2.[41] Four out of seven patients with NY-ESO-1-expressing tumors had CD8+ NY-ESO-1-specific T-cell responses after receiving an immunization through the cutaneous route. Improved T-cell immunity is linked to both remission and long-term health.[42] However, individuals who had an antibody response to NY-ESO-1 before vaccination did not show an increase in antigen-specific CD8+ T cells despite the treatment's success. Data suggest that this immune response may be overwhelmed by tumor growth (or an antigen loss mutation permits tumor growth) or may simply be an epiphenomenon associated with tumor progression, even though spontaneous T- and B-cell responses can be detected in tumor-bearing hosts and T-cell responses can be induced by vaccination strategies.[43] These beliefs are supported by data showing a correlation between the amount of cancer tissue and the frequency of antibody detection.

6.3 CD8 T-CELL RECEPTOR-RECOGNIZED TUMOR ANTIGENS: A STRUCTURAL DEFINITION

Using mass spectrometry and melanoma patient-derived cytotoxic T lymphocyte (CTL) lines, the author was able to positively identify some of the peptides included in the vaccinations the author evaluated.[44] According to our preliminary research, all MDP-derived HLA-restricted peptides were either gp100- or Tyr-based. All of these mutations were discovered in melanoma tissue samples from various people, suggesting their relevance to a wide variety of tumors. In addition, T cells from many patients recognized them, indicating a preexisting immune response that may be strengthened by vaccination. Two of these four peptides were found to have undergone posttranslational modifications that were not previously detectable but had a substantial effect on T-cell recognition when analyzed by mass spectrometry.[45] One such process was the conversion of asparagine in the Tyr369-377 peptide to aspartic acid. It has been observed that N-linked carbohydrates are eliminated from epitopes formed from a broad range of glycoproteins. Peptide N-glycanase plays a role in bringing about this transformation during the processing of antigens. Second, a disulfide bond[46] was formed between a free cysteine residue and one of the cysteines in the Tyr243-251 peptide. Tyr peptide contains two cysteine residues, which permits disulfide exchange to remove the free cysteine, but cysteinylation has been seen in other epitopes. Surprisingly, the author discovered that cytotoxic T-lymphocytes (CTL) primed to recognize the original melanoma peptide and could still recognize a variant of the peptide in which a serine residue substituted Cys244.[47] These results highlight the importance of using an indirect approach to detecting antigens. In recent years, it has become clear that melanoma cells express two additional peptides that are not derived from MDP. Nearly all melanomas (90%) express RLSNRLLLR, which has been traced back to a hitherto unidentified cancer-testis antigen (TAG).[48] The author has demonstrated that several class I MHC molecules exhibit a number of the antigens reported by other groups,[49] and researchers have also identified the naturally processed versions of many of the antigens. To create a vaccine against cancer, scientists have utilized mass spectrometry to study the differently expressed peptides of cancer cells. Transformed cells are characterized by aberrant phosphorylation of cell signaling molecules, which may serve as an antigen for the immune system to reject the cells.[50] Through the use of mass spectrometry, the author discovered that immunization may induce CD8 T cells that recognize phosphorylated peptides since these peptides are normally processed and presented by class I MHC molecules. Our most recent work[51] reveals a subset of phosphorylated peptides that are being investigated as potential cancer vaccines, and researchers know that the repertory of these peptides changes with tumor type.

The immune system's tolerance of self-antigens has been a key obstacle to cancer immunotherapy. As a result of research on immunological tolerance, cutting-edge methods of cancer immunotherapy have been developed.[52] Because immune effectors like anticancer monoclonal antibodies and chimeric antigen receptor T cells are developed in vitro rather than in vivo to create tumor-specific effectors, adoptive immunotherapy[53] offers the potential for safe and effective transmission to the recipient. Vaccinating against cancer with whole tumor cells, proteins, peptides, or dendritic cells has not been proven to be effective alone, but may work better when paired with other immunotherapeutic approaches. Agents that initiate and sustain T-cell activation are key components of immunomodulatory approaches to cancer immunotherapy.[54] New developments in checkpoint blocking offer much promise for countering inhibitory signals and keeping up an effective anticancer response. A growing body of knowledge on immunological tolerance and its countermeasures is driving the creation, testing, and, hopefully, approval of potent new combinations of treatments. Tolerance in the tumor microenvironment may be overcome using strategies like in situ immunization. As a consequence of encouraging outcomes seen with these treatments for a variety of cancer patients, research and development into cancer immunotherapies that aim to reduce immunological tolerance are increasing.[55]

Because of the exponential growth in our understanding of immunology over the last several decades, researchers now have better vaccinations against a wide range of infectious diseases. Due to this, many people are now hoping against the hope that immunotherapy for cancer may help them. Peripheral tolerance is characterized by the inhibition of autoreactive cells, whereas central tolerance is controlled by immunological editing in the thymus.[56] This introductory article will discuss immune tolerance and how it is being addressed in contemporary cancer immunotherapy.

Understanding the immune response and the function of immunological tolerance at each level is crucial before delving into the issue of breaking tolerance with immunotherapy. DCs process TAAs and deliver them to antigen-specific T cells, which then activate, proliferate, and maintain the T-cell response long enough to eradicate cancer (Fig. 6.1).[57] Tolerance to an antigen may develop after immunological suppression at any time. Defects in antigen absorption and presentation are a likely cause of an inadequate immune response.[58] The number of T cells that respond to particular tumor antigens may be greatly reduced by immune editing.[59] Tumor-specific lymphocyte activation and proliferation may be hampered by insufficient co-stimulatory signals.[60] Unfortunately, just though the immune system mounts a robust initial reaction, it doesn't mean it will keep going strong for long enough to cause the tumor to shrink. The accumulation of activated T cells at the tumor site is essential.[61] Fighting against fatigue and immunosuppression brought on by the cancer microenvironment is exhausting and difficult. Tumor cells may suppress the immune system in a variety of ways.[62] Tumor-specific immunity and immunological tolerance are shown in Figure 6.2.

Among these methods include tinkering with the antigen presentation mechanism and the release of immunosuppressive chemicals. Lymphocyte apoptosis might be hastened or negative regulatory pathways set in motion by these methods. Tumor cells, either directly or indirectly, that increase immunological tolerance are given a selective survival advantage and may multiply and ultimately form tumors.

Tolerance-busting strategies in cancer immunotherapy may be roughly classified according to their placement in the treatment's timeline. Examples of these therapies include vaccination, immunomodulatory drugs, and the adoption of healthy relatives.[63]

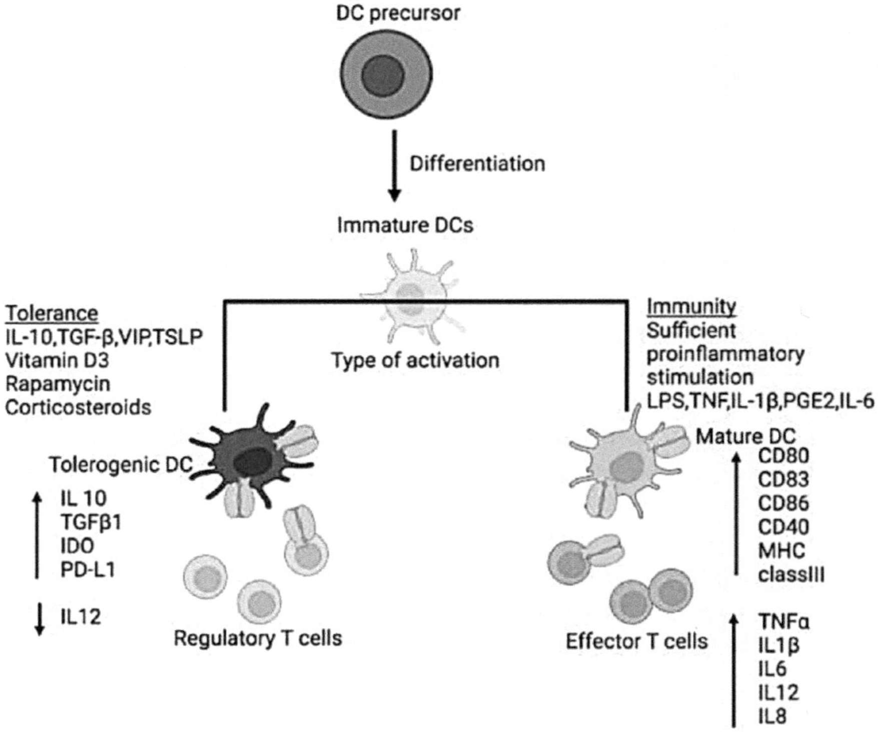

FIGURE 6.2 Tumor-specific immunity and immunological tolerance.

6.4 TRANSFERRING IMMUNOLOGICAL EFFECTORS VIA ADOPTION

The capacity to mass-produce immunological effectors (such antibodies and T cells) ex vivo, without in vivo antigen presentation and innate immune proliferation, is a major benefit of adoptive transfer as an immunotherapeutic method.

6.4.1 TUMOR-SPECIFIC MONOCLONAL ANTIBODIES

One of the most promising and effective methods of treating cancer is with the help of monoclonal antibodies (mAbs). The Food and Drug Administration (FDA) has approved a wide variety of monoclonal antibodies (mAbs) for the treatment of solid tumors and hematological malignancies. The anti-VEGF antibody bevacizumab, the anti-CD20 antibody rituximab, and the anti-HER2 antibody trastuzumab are only a few examples.[64] Antibody-dependent cell-mediated cytotoxicity, phagocytosis, and complement-dependent cytotoxicity are all possible methods by which mAbs inhibit tumor development.[65] More and more evidence shows that the lysis of cancer cells generated by monoclonal antibodies (mAbs) enhances DCs' capacity to take up and cross-present tumor antigens, hence facilitating the establishment of adaptive immune responses.[66] This demonstrates how influencing a single part of the immune response may have far-reaching consequences.

Research into mAbs of varying specificities and structural modifications to reduce immunogenicity were early steps in the development of mAb treatment.[67] To improve their interactions with immune effector cells like natural killer cells and DCs, scientists are attempting to alter their shape.[68]

Immunoconjugates may combine the selectivity of mAbs with the efficiency of cytotoxic moieties. Both hematologic malignancies and solid tumors may benefit from antibody-drug conjugates.[69] Two such examples are the radio conjugates 90Y-ibritumomab tiuxetan and 131I-tositumomab, both of which are used to treat lymphoma. Since immunoconjugates depend on immune identification of the target antigen, they are an effective method of overcoming immune tolerance; yet they are not often referred to as "immunotherapy" since cancer cell lysis may occur for reasons other than immunologically mediated lysis.

By redirecting the T cell's innate specificity toward malignant cells expressing the target antigen, bispecific mAbs may be used to take advantage of the T cell's cytolytic potential. For this purpose, an antigen on immune effector cells is fused with tumor antigens. Blinatumomab (anti-CD19 and anti-CD20) and catumaxomab (anti-EPCAM and anti-CD3) are also examples of bispecific antibodies.[70] Since bispecific antibodies are quickly removed, their clinical utility is limited to infusions that occur constantly in the clinic. Despite this, there is ongoing research and development into the efficacy of bispecific mAbs in the clinic against several cancer types.

6.4.2 ADOPTIVE T-CELL TRANSFER (ACT)

Lymphocytes are harvested from the patient, multiplied in vitro with the help of several growth factors, and then re-infused to treat cancer.[71,72] This treatment is known as "adoptive T-cell transfer." Interleukin 2 (IL-2) co-infusion was used to further boost T-cell activation. The idea behind transplanting tumor-infiltrating lymphocytes is that they include a disproportionately large number of antitumor T cells, thereby increasing the transplanted cells' tumor-specificity and effectiveness. Patients with metastatic melanoma treated with autologous tumor-infiltrating lymphocytes have obtained objective responses at 50% and extended remissions with these adjustments.[73]

The technique of adoptive transfer of T cells that have been genetically modified to produce chimeric antigen receptors (CARs) has gained a lot of attention in recent years. CARs are single-chain chimeric ligands made by fusing the complementarity-determining region (CDR) of an antibody with the signaling domain of a T-cell receptor (TCR).[74] Autologous T cells may be activated and proliferated in response to their antigen in vivo via the transfer of CAR genes. This has the potential to create immunologic memory against the precise target antigen,[75] which might be used to successfully eradicate a substantial tumor burden in a therapeutic context. The structures used to create CARs have been refined over time via preclinical and clinical testing. Results from trials utilizing CD19-based CARs to treat B-cell malignancies have been encouraging.[76] Research conducted in the last several years has indicated that the majority of patients who have CD19+ chronic lymphocytic leukemia react well to therapy, with 14 of 20 patients (82%) exhibiting a full response.[77]

It is not easy to put CARs into practice. Unfortunately, the present state of the art in CAR-T treatment is both time-consuming and financially prohibitive. A "cytokine storm," which may develop when many T cells get activated, has been linked to serious side effects. Resistance may emerge as a result of diminished target antigen expression.[78] One other problem with CARs is that they may unintentionally attack healthy cells that just faintly show signs of expressing the target antigen. However, owing to the formation of CAR-based immunologic memory, this autoimmunity may persist for a long time and be difficult to suppress. It is for this reason that the pool of potential antigen targets for CARs is so small. Despite these challenges, the author is moving into a period of rapid improvement in CAR constructs, the development of standard protocols that enable more automated generation of CARs, the careful identification of patient populations most likely to benefit from CAR therapy with acceptable toxicities, and an improved and standardized approach to managing toxicities.

6.4.3 VACCINATION

Therapeutic cancer vaccines expose the patient's immune system to cancer-related antigens, peptides, or even whole tumor cells. As the antigen-expressing tumor is already present, immune tolerance develops in such people. The use of immunological adjuvants, antigen selection and modification, antigen presentation cells such as DCs, and other strategies have all been investigated to help immunizations get around this resistance.[79]

The effectiveness of vaccinations against human papillomavirus (HPV) cancer is enhanced by the fact that immunity to the virus does not have to be developed against foreign antigens (HPV). In 2009, the FDA authorized Cervarix®, a vaccination against the human papillomavirus (HPV) that causes cervical cancer.[80] The creation of a therapeutic cancer vaccine has been far more challenging than originally anticipated. This is true regardless of whether the antigen is whole tumor cells, tumor antigens, or peptides. Immunological editing, faulty antigen processing and presentation, and trouble maintaining an effective immune response are all possible roadblocks to a cancer vaccine's complete efficacy. The author is, however, figuring out how to not just survive, but flourish despite these challenges. Less malignant cells and weaker immunological tolerance are common in those with a low illness load.[81]

6.4.4 WHOLE-CELL VACCINES

If immunological tolerance is more common for a given antigen, cell-based vaccinations may be advantageous because they allow the host to generate immune responses against various TAAs.[82] GVAX is one such vaccine, and it's created from two irradiated prostate cancer cell lines that secrete granulocyte-macrophage colony-stimulating factor (GM-CSF). Randomized clinical investigations failed to reveal an increase in survival with GVAX,[83] even though it was proven to produce an immunological response. For GVAX, immunization was performed using allogeneic cell lines; however, this may not be the optimal method for eliciting an immune response against TAAs. Using modified autologous cells in a clinical setting is more challenging. In any case, studies are being conducted to investigate this potential. Researchers are looking for ways to increase its capacity to break tolerance, such as by further genetic modification of the tumor cell, fusion of the tumor cell with antigen-presenting cells, and the use of different immune adjuvants.[84]

6.4.5 PROTEIN ANTIGEN OR PEPTIDE-BASED IMMUNIZATION

By skipping the antigen processing step, immunization using intact peptides produced from tumor-related antigens may be able to overcome immune resistance. T cells are activated when these peptides bind to MHC molecules (often class I) on DCs.[85] First clinical investigations of immunization using free peptides showed low effectiveness, even though peptides are safe and inexpensive to manufacture. In certain cases, the use of immunological stimulants to increase the activation of peptide-specific T cells might be useful. Clinical trials treating individuals with advanced melanoma have shown promising outcomes when large doses of interleukin-2 (IL-2) are combined with short peptides derived from glycoprotein 100 (gp100), a melanocyte differentiation antigen. The response rate and progression-free survival (PFS) were considerably improved when vaccination was combined with IL-2, compared to when either therapy was administered alone.[86] A phase-III clinical trial for individuals with follicular lymphoma found encouraging results when using subcutaneous GM-CSF in combination with an anti-idiotype fusion antibody produced from keyhole limpet hemocyanin (KLH). The median time before illness development was significantly prolonged by the patient-specific Id protein vaccine[87] as compared to the placebo immunization (KLH + GM-CSF).

Unfortunately, a lot of work has to be done before peptide-based immunizations can be called a complete success. The correct peptide must be selected, tumor cells must produce the peptide in a heterogeneous fashion, DC activation must be boosted at the site of vaccination to maximize T-cell activation, peripheral tolerance must be avoided once an immune response has been generated, and tumor-induced immunosuppressive mechanisms must be blocked. To be effective in therapy, peptide vaccines will likely need additional strategies for overcoming immune tolerance beyond the use of typical immunological adjuvants.

6.4.6 DCs LOADED WITH ANTIGENS AS A VACCINE

Because DCs are essential for the processing and presenting of tumor antigens, but because DCs may be defective or inadequate in an amount in cancer patients, this realization prompted the creation of DC-based vaccinations. This problem is circumvented by promoting DC proliferation and maturation in vitro using patient-derived precursors and a variety of cytokines and growth factors. To further stimulate the immune response against the tumor, a second dosage of DCs that have been preconditioned with tumor antigens, peptides, or genetically modified to create a TAA may be administered to the patient.[88] One dendritic cell (DC) vaccine was licensed by the FDA in 2010 for the treatment of metastatic hormone-refractory prostate cancer. (sipuleucel-T). A fusion protein of granulocyte-macrophage colony-stimulating factor (GM-CSF) and prostatic acid phosphatase is carried by DCs in this immunization (PAP). In clinical studies, sipuleucel-T increased median survival by 4.1 months.[89]

This variation in DC subsets' ability to digest and deliver antigens and activate T cells[90,91] exemplifies the complexity of DC biology. Tolerogenic DCs may exist as a distinct subset of DCs. To make DC-based vaccines more effective, researchers have tried a variety of methods to enhance antigen uptake, maturation, and presentation in an immunogenic rather than tolerogenic environment.[92] Methods for activating DCs include the use of interleukins, immunostimulatory mAbs, and agonists for the Toll-like receptors (TLR4, TLR7, and TLR9). Therapeutic DC-based vaccines for the treatment of cancer have shown promise, although it is yet unclear how successful they will be in clinical settings.[93]

6.4.7 COMBINATION CD4/CD8 T-CELL VACCINATION MODULATES RESPONSES AND REGULATORY FUNCTION

According to the aforementioned studies, peptide vaccines have the potential to be employed to induce targeted CD8 T-cell responses in melanoma patients. Only between 3% and 6% of individuals show clinical improvement across trials, and most patients' T-cell responses are rather small.[94]

So, it's crucial to develop strategies to increase clinical response rates and the magnitude of the responses.[95] Our current vaccination technique in an adjuvant setting[96] involves using a combination of multi-peptide vaccination to recruit melanoma-specific CD4 T cells, strategies to modulate regulatory mechanisms, and continuous multi-peptide vaccination to activate CD8 cells.

Once only considered a possible source of IL-2, our understanding of helper T cells has expanded to encompass the critical function CD4 cells play in initiating and sustaining CD8 T-cell responses against malignancies. Furthermore, CD4 T-cell adoptive therapy may induce tumor protection in several model systems. Therefore, it seems that both CD8 and CD4 T cells mediate the protective immunity generated by cancer cell vaccines.[97]

Although stimulating helper T cells is crucial for efficient immunotherapy, the effect of this on cytotoxic T-lymphocyte (CTL) responses is not well understood. In a single experiment using vaccines, keyhole limpet hemocyanin (KLH) was demonstrated to increase CD8 T-cell responses to Tyr peptides linked with class I MHC. In a prior study, the author employed a tetanus helper peptide to generate nonspecific helper T-cell activation at the vaccination site.[98] Increased efficacy of cytotoxic T-cell epitope vaccination may be achieved by including helper epitopes from the same protein(s) in the vaccine, as opposed to using helper epitopes that are not specific to the antigen being vaccinated against.[99] Only a few small-scale studies have been conducted so far, each using a single class I-restricted peptide and a single class II-restricted peptide to test the hypothesis. Human trials with tumor-associated helper peptides in cancer patients are few, and the research that has been conducted so far has had a narrow focus. In contrast to immunization with a generic helper peptide, vaccination with a peptide linked with melanoma may produce antigen-reactive CD4+ T-cell responses and may increase CD8+ responses; however, further study is needed to substantiate these theories. To test these hypotheses, the researcher has initiated three clinical investigations.[100]

6.5 THERAPEUTIC IMMUNOMODULATION

Through research, the author has identified several cell types in the tumor microenvironment that promote local immunosuppression. Myeloid-derived suppressor cells (MDSCs)[101–103] have a strong suppressive effect on T lymphocytes and may be abundant in the cancer microenvironment. As Tregs may be triggered by MDSCs, they may be able to inhibit effector T cells. The presence of MDSCs and Tregs in the peripheral blood or at the tumor site has been associated with a worse prognosis. Immature DCs with low expression of costimulatory markers (CD80/CD86) and T-cell anergy may be promoted by tumor-derived IL-10 and vascular endothelial growth factor[104] (VEGF).

Approaches for selectively attacking immunosuppressive cells in the tumor microenvironment have proven challenging to develop so far. Instead, several medications that may influence immune tolerance induced by the tumor microenvironment[105] have been used in an attempt to tip the scales in favor of immune reactivity and away from immunological tolerance.

6.5.1 T-CELL ACTIVATION ENHANCEMENT

Research on using interleukins as a vaccine has focused on both individual interleukins and interleukin mixtures. To name a few, the author have interleukins (IL)-2, IL-7, IL-12, and IL-15.[106] Although each has its unique effects, they all have the commonality of activating T cells. For T-cell activation and proliferation to have an immunogenic impact, it is necessary to acquire a co-stimulatory signal from DCs. mAbs having the ability to deliver costimulatory signals are now being tested in humans. Examples include CD40, OX40 (CD134), 4-1BB (CD137), and GITR (CD357).[107]

6.5.2 CHECKPOINT BLOCKADE

Checkpoints in the T cell response that drastically limit the T cell's ability to respond might be crucial in avoiding autoimmunity. Nonetheless, they shorten the length and effectiveness of immune responses that target cancer. There are several molecules on the surface of T cells, including cytotoxic T-lymphocyte antigen 4, programmed death 1, T cell immunoglobulin and mucin domain-containing protein 3, and lymphocyte activation gene.[108] The cancer microenvironment may become unresponsive and immune suppression may follow if intratumoral cells produce these indications. Tregs use many of these chemicals to inhibit the activity of effector T cells.[109] This suggests that they might be targeted therapeutically to counteract the effects of immune tolerance.

One of the most fascinating and intriguing areas of cancer immunology and clinical oncology today is the study of these negative regulators and the use of checkpoint inhibition to diminish repressive signals. Only a small number of recent studies[110] have shown that checkpoint-blocking mAbs are successful in the clinic. These mAbs boost the immune response against tumors by recognizing a receptor or ligand and inhibiting their interaction. The anti-CTLA-4 mAb ipilimumab (Bristol Myers-Squibb) was approved by the FDA in 2011 for the treatment of metastatic melanoma after a phase III study showed the first overall survival advantage in this group of patients. Autoimmune symptoms such as enteropathies and endocrinopathies have been documented in patients on anti-CTLA-4.[111] Using PD-1 and PD-L1 in early clinical and preclinical studies has shown encouraging findings. For the treatment of solid tumors and high-risk hematologic malignancies, two anti-PD-1 mAbs (MDX-1106 and CT-011) and one anti-PD-L1 mAb (PD-L1-C19) are currently being evaluated (MDX-1105[112]). Inhibition of TIM-3 and LAG-3 in clinical studies is also ongoing. Results from animal studies suggest that combining PD1 inhibition with the blocking of these other pathways might increase the effectiveness of both treatments.

Promoting tumor shrinking while limiting unacceptable harm from increased T-cell autoreactivity to benign cells is a difficult balancing act that must be maintained while administering these medications. The use of checkpoint blocking in clinical practice as an immunotherapeutic technique will continue notwithstanding this setback. Medical oncologists may be unprepared to handle the autoimmune toxicities that may follow from checkpoint blockade, despite their extensive background in managing the side effects of cytotoxic medicines. This will need to change over the coming decade as the researcher understands more about checkpoint blocking and can include it in cancer immunotherapy for a larger percentage of patients.

6.5.3 SELF-TOLERANCE TO MTYR369 AND CANCER CONTROL

Here, the author examined the differences in immune responses to mTyr369 between mice immunized with either bone marrow-derived DC (BMDC) grown in GM-CSF and IL4, matured in the presence of TNFa or CD40L, and pulsed with synthetic peptides[113] or with a recombinant vaccinia virus expressing full-length mTyr369 (mTyr vac). Intracellular IFN-g accumulation (IFN-gICS) and labeling with mTyr369-specific HLA-A*0201 tetramers were used to identify mTyr369-specific CD8 T cells in peak responder spleen and peripheral LN samples.[114] The author has shown that T-cell characteristics can be altered by in vitro culture, but our ex vivo approach circumvents this issue. AAD albinos, in contrast to AADTyr mice, have strong immunological responses to mTyr369, demonstrating that self-tolerance to this antigen is based at least in part on decreased T-cell reactivity.

High-cycle PCR tests show that Tyr is largely expressed in melanocytes in the basal layer of the epidermis, hair follicles, and retinal pigment epithelium, despite widespread low-level expression in many organs, including the thymus. Because thymic tissue from AADTyr donors could not suppress responses to mTyr369 in neonatally thymectomized AAD albino hosts, it is clear that tolerance to mTyr369 cannot be acquired by exposing developing T cells to this Ag on thymic epithelial cells. Furthermore, the number or differentiation status of mTyr369-specific FH TCR transgenic (tgTCR) cells in the thymus of AADTyr and AAD albino mice were unaffected. This further demonstrates[115] that mature T cells encounter Ag in the periphery, rather than in the thymus, leading to tolerance to mTyr369.

Tolerance induction in CD8 T cells in response to Ags generated specifically in a broad variety of nonlymphoid peripheral organs has been studied using mice expressing the tgTCR. CD8 T cells exposed in this way show a restricted collection of activation markers, undergo weak proliferation and deletion, and fail to develop effector activity, as suggested by the available data. By limiting their numbers, CD8 cells may be kept in a dormant, or resting, state. To assess peripheral self-tolerance to mTyr369. The author adopted CFSE-labeled Thy 1.2FH CD8 cells from AAD albino donors and implanted them into Tyr or albino Thy 1.1AAD recipients. Albino patients' peripheral LN cells remained quiescent three days after being reisolated, but Tyr people' had had several cell divisions.[116] After 14 days, the number of endogenously Tyr-activated T lymphocytes dropped below the detection threshold, suggesting that apoptosis was the probable cause of death.[117] This work is the first to show that T-cell activation-induced deletion is a critical component of self-tolerance for Tyr, and it involves tolerance to an Ag generated in melanocytes.

6.6 CONCLUSION

Because of recent advances in our understanding of immune tolerance and its elimination strategies, cancer immunotherapy is experiencing a golden age. Cancer immunotherapy was widely considered a "failed theory" by oncologists and cancer researchers at the turn of the century after decades of unproductive clinical investigations with cancer vaccines. You have successfully shown your case, thus further discussion is unnecessary. When used against cancer, monoclonal antibodies developed for this purpose are very successful. Checkpoint blockade monoclonal antibodies (mAbs), which don't kill tumor cells directly, have shown remarkable therapeutic success. Given the promising results shown with CAR T cell therapy in the treatment of many cancers, this approach is anticipated to play a role in the development of cancer immunotherapy in the future. There is mounting evidence that vaccination may be successful in the treatment of advanced malignancies, despite the many obstacles that still need to be overcome.

There are several strategies for overcoming immunotolerance to cancer, and these strategies all affect the disease and immune response in unique ways. Oncology treatments nowadays often work in one of two ways: either by avoiding immune editing altogether or by directly targeting the underlying processes that lead to peripheral tolerance. Over the next decade, researchers want to get a better understanding of both the effectiveness of individual medications and the best ways to implement these treatments into comprehensive cancer immunotherapy programs.

It is concerning because tumor-infiltrating T lymphocytes are often tolerated, inactivated, or negatively regulated. Dendritic cells, which aid in the fight against cancer, may be affected by this as well. This is a key obstacle to the development of efficient immune therapies, as are other kinds of immunological escape. Immune escape would be unnecessary if the immune system had no bearing on melanoma progression. The body's antitumor immune response selectively destroys melanoma cells that lack immune escape phenotypes, which is shown by the high number of immune escape morphologies in metastatic melanomas. Melanoma vaccine clinical trials' findings corroborate this fact of biology. It has been shown scientifically that some people benefit from cancer vaccinations in ways that continue for a long time. However, outcomes from clinical trials of melanoma vaccines have been inconsistent.

It would be easy to write off cancer vaccines as doomed to failure given the dismal clinical response rates seen with them. To put this in perspective, think of all the medical and surgical techniques that "failed" just as dramatically in their early stages of research yet are now considered standard practice. Examples of effective therapies include chemotherapy for Hodgkin's disease, liver transplants, heart surgeries, and monoclonal antibodies. Better scientific equipment has finally made this a reality. Improving our understanding of the host-tumor interaction and creating new ways to boost the efficacy of immunization are two important challenges that must be answered by the next generation of cancer vaccine research.

KEYWORDS

- **tolerance**
- **immunotherapy**
- **melanoma**
- **monoclonal antibodies**
- **vaccines**

REFERENCES

1. Vesely, M. D.; Kershaw, M. H.; Schreiber, R. D.; Smyth, M. J. Natural Innate and Adaptive Immunity to Cancer. *Annu. Rev. Immunol.* **2011,** *29*, 235–271.
2. Mellman, I.; Coukos, G.; Dranoff, G. Cancer Immunotherapy Comes of Age. *Nature* **2011,** *480* (7378), 480–489.
3. Hollander, N. Immunotherapy for B-Cell Lymphoma: Current Status and Prospective Advances. *Front. Immunol.* **2012,** *3*, 3.
4. Kim, R.; Emi, M.; Tanabe, K. Cancer Immunoediting from Immune Surveillance to Immune escape. *Immunology* **2007,** *121* (1), 1–14.
5. Driessens, G.; Kline, J.; Gajewski, T. F. Costimulatory and Coinhibitory Receptors in Anti-Tumor Immunity. *Immunol Rev.* **2009,** *229* (1), 126–144.
6. Bellone, M.; Calcinotto, A. Ways to Enhance Lymphocyte Trafficking into Tumors and Fitness of Tumor in Filtrating Lymphocytes. *Front Oncol.* **2013,** *3*, 231.
7. Rabinovich, G. A.; Gabrilovich, D.; Sotomayor, E. M. Immunosuppressive Strategies That Are Mediated by Tumor Cells. *Annu. Rev. Immunol.* **2007,** *25*, 267–296.
8. Scott, A. M.; Wolchok, J. D.; Old, L. J. Antibody Therapy of Cancer. *Nat. Rev. Cancer* **2012,** *12* (4), 278–287.
9. Galluzzi, L.; Vacchelli, E.; Fridman, W. H. et al. Trial Watch: Monoclonal Antibodies in Cancer Therapy. *Oncoimmunology* **2012,** *1* (1), 28–37.
10. Topalian, S. L.; Weiner, G. J.; Pardoll, D. M. Cancer Immunotherapy Comes of Age. *J. Clin. Oncol.* **2011,** *29* (36), 4828–4836.
11. May, C.; Sapra, P.; Gerber, H. Advances in Bispecific Biotherapeutics for the Treatment of Cancer. *Biochem. Pharmacol.* **2012,** *84* (9), 1105–1112.
12. Mathur, R.; Weiner, G. J. Picking the Optimal Target for Antibody-Drug Conjugates. *Am. Soc. Clin. Oncol. Educ. Book.* 2013.
13. Galluzzi, L.; Vacchelli, E.; Eggermont, A. et al. Trial Watch: Adoptive Cell Transfer Immunotherapy. *Oncoimmunology* **2012,** *1* (3), 306–315.
14. Overwijk, W. W. Breaking Tolerance in Cancer Immunotherapy: Time to ACT. *Curr. OpinImmunol.* **2005,** *17* (2), 187–194.
15. Rosenberg, S. A.; Yannelli, J. R.; Yang, J. C. et al. Treatment of Patients with Metastatic Melanoma with Autologous Tumor-Infiltrating Lymphocytes and Interleukin-2. *J. Natl. Cancer Inst.* **1994,** *86*, 1159–1166.

16. Kershaw, M. H.; Westwood, J. A.; Slaney, C. Y.; Darcy, P. K. Clinical Application of Genetically Modified T Cells in Cancer Therapy. *Clin. Trans. Immunol.* **2014,** *3*, e16.
17. Grupp, S. A.; Frey, N. V.; Aplenc, R.; Barrett, D. M.; Chew, A.; Kalos, M. et al. T Cells Engineered with a Chimeric Antigen Receptor (CAR) Targeting CD19 (CTL019) Produce Significant In Vivo Proliferation, Complete Responses and Long-Term Persistence Without Gvhd in Children and Adults with Relapsed, Refractory ALL. *Poster Pres. Am. Soc. Hematol.* **2013,** *122* (21), 67.
18. Bocchia, M.; Bronte, V.; Colombo, M. P. et al. Antitumor Vaccination: Where West and. *Haematologica* **2000,** *85* (11), 1172–1206.
19. Chiang, C. L.; Kandalaft, L. E.; Coukos, G. Adjuvants for Enhancing the Immunogenicity of Whole Tumor Cell Vaccines. *Int. Rev. Immunol.* **2011,** *30* (2–3), 150–182.
20. Browning, M. J. Antigen Presenting Cell/Tumor Cell Fusion Vaccines for Cancer Immunotherapy. *Hum. Vaccin. Immunother.* **2013,** *9* (7), 1545–1548.
21. Slingluff, C. L., Jr. The Present and Future of Peptide Vaccines for Cancer: Single or Multiple, Long or Short, Alone or in Combination? *Cancer J.* **2011,** *17* (5), 343–350.
22. Schwartzentruber, D. J.; Lawson, D. H.; Richards, J. M.; Conry, R. M.; Miller, D. M.; Treisman, J. et al. Gp100 Peptide Vaccine and Interleukin-2 in Patients with Advanced Melanoma. *N. Engl. J. Med.* **2011,** *364* (22), 2119–2127.
23. Schuster, S. J.; Neelapu, S. S.; Gause, B. L.; Janik, J. E.; Muggia, F. M.; Gockerman, J. P. et al. Vaccination with Patient-Specific Tumor-Derived Antigen in First Remission Improves Disease-Free Survival in Follicular Lymphoma. *J. Clin. Oncol.* **2011,** *29* (20), 2787–2794.
24. Palucka, K.; Banchereau, J. Cancer Immunotherapy via Dendritic Cells. *Nat. Rev. Cancer* **2012,** *12* (4), 265–277.
25. Kantoff, P. W.; Higano, C. S.; Shore, N. D.; Berger, E. R.; Small, E. J.; Penson, D. F. et al. Sipuleucel-Timmunotherapy for Castration-Resistant Prostate Cancer. *N. Engl. J. Med.* **2010,** *363* (5), 411–422.
26. Galluzzi, L.; Senovilla, L.; Vacchelli, E. et al. Trial Watch: Dendritic Cell-Based Interventions for Cancer Therapy. *Oncoimmunology* **2012,** *1* (7), 1111–1134.
27. Vacchelli, E.; Galluzzi, L.; Eggermont, A. et al. Trial Watch: FDA-Approved Toll-Like Receptoragonists for Cancer Therapy. *Oncoimmunology* **2012,** *1* (6), 894–907.
28. Vacchelli, E.; Galluzzi, L.; Eggermont, A. et al. Trial Watch: Immunostimulatory Cytokines. *Oncoimmunology* **2012,** *1* (4), 493–506.
29. Wolchok, J. D.; Yang, A. S.; Weber, J. S. Immune Regulatory Antibodies: Are They the Next Advance? *Cancer J.* **2010,** *16* (4), 311–317.
30. Swann, J. B.; Smyth, M. J. Immune Surveillance of Tumors. *J. Clin. Invest.* **2007,** *117* (5), 1137–1146.
31. Aguilar, L. K.; Guzik, B. W.; Aguilar-Cordova, E. Cytotoxic Immunotherapy Strategies for Cancer: Mechanisms and Clinical Development. *J. Cell Biochem.* **2011,** *112* (8), 1969–197.
32. Grosso, J. F.; Jure-Kunkel, M. N. CTLA-4 Blockade in Tumor Models: An Overview of Preclinical and Translational Research. *Cancer Immun.* **2013,** *13*, 5.
33. Harshman, L. C.; Drake, C. G.; Wargo, J. A.; Sharma, P.; Bhardwaj, N. Cancer Immunotherapy Highlights from the 2014 ASCO Meeting. *Cancer Immunol. Res.* **2014,** *2* (8), 714–719.

34. Hodi, F. S.; O and, S. J.; McDermott, D. F.; Weber, R. W.; Sosman, J. A.; Haanen, J. B. et al. Improved Survival with Ipilimumab in Patients with Metastatic Melanoma. *N. Engl. J. Med.* **2010,** *363* (8), 711–723.
35. Pardoll, D. M. The Blockade of Immune Checkpoints in Cancer Immunotherapy. *Nat. Rev. Cancer*. **2012,** *12* (4), 252–264.
36. Ludgate, C. M. Optimizing Cancer Treatments to induce an Acute Immune Response: Radiation Abs Copal Effects, PAMPs, and DAMPs. *Clin. Cancer Res*. **2012,** *18* (17), 4522–4525.
37. Brody, J. D.; Ai, W. Z.; Czerwinski, D. K. et al. In Situ Vaccination with a TLR9 Agonist Induces Systemic Lymphoma Regression: A Phase I/II Study. *J. Clin. Oncol.* **2010,** *28* (28), 4324–4332.
38. Tanchot, C.; Lemonnier, F. A.; Perarnau, B. et al.. Differential Requirements for Survival and Proliferation of CD8 Naive or Memory T Cells. *Science* **1997,** *276*, 2057–2062.
39. Brocker, T.; Riedinger, M.; Karjalainen, K. Targeted Expression of Major Histocompatibility Complex (MHC) Class II Molecules Demonstrates That Dendritic Cells can Induce Negative But Not Positive Selection of Thymocytes In Vivo. *J. Exp. Med.* **1997,** *185*, 541–550.
40. Kappler, J. W.; Roehm, N.; Marrack, P. T Cell Tolerance by Clonal Elimination in the Thymus. *Cell* **1987,** *49*, 273–280.
41. Surh, C. D.; Sprent, J. T-Cell Apoptosis Detected In Situ During Positive and Negative Selection in the Thymus. *Nature* **1994,** *372*, 100–103.
42. Ramsdell, F.; Fowlkes, BJ: Clonal Deletion Versus Clonal Energy: The Role of the Thymus in Inducing Self Tolerance. *Science* **1990,** *248*, 1342–1348.
43. Kyewski, B.; Derbinski, J.; Gotter, J. et al.. Promiscuous Gene Expression and Central T-Cell Tolerance: More Than Meets the Eye. *Trends Immunol.* **2002,** *23*, 364–371.
44. Klein, L.; Klein, T.; Ruther, U. et al. CD4 T Cell Tolerance to Human C-Reactive Protein, an Inducible Serum Protein, Is Mediated by Medullary Thymic Epithelium. *J. Exp. Med.* **1998,** *188*, 5–16.
45. Sykes, M.; Auchincloss, H. J.; Sachs, D. H. Transplantation Immunology. In *Fundamental Immunology*; Paul, W. E., Ed.; Lippincott Williams& Wilkins: Philadelphia, 2003; pp 1481–1555.
46. Sakaguchi, S.; Sakaguchi, N.; Shimizu, J. et al. Immunologic Tolerance Maintained by CD25CD4 Regulatory T Cells: Their Common Role in Controlling Auto Immunity, Tumor Immunity, and Transplantation Tolerance. *Immunol. Rev.* **2001,** *182*, 18–32.
47. Dieckmann, D.; Plottner, H.; Berchtold, S. et al. Ex Vivo Isolation and Characterization of CD4CD25 T Cells with Regulatory Properties from Human Blood. *J. Exp. Med.* **2001,** *193*, 1303–1310.
48. Steinman, R. M.; Nussenzweig, M. C. Avoiding Horror Auto Toxicus: The importance of Dendritic Cells in Peripheral T Cell Tolerance. *Proc. Natl. Acad. Sci. USA* **2002,** *99*, 351–358.
49. Hawiger, D.; Inaba, K.; Dorsett, Y. et al. Dendritic Cells Induce Peripheral T Cell Unresponsiveness Under Steady State Conditions In Vivo. *J. Exp. Med.* **2001,** *194*, 769–779.
50. Albert, M. L.; Jegathesan, M.; Darnell, R. B. Dendritic Cell Maturation Is Required for the Cross-Tolerization of CD8 T Cells. *Nat. Immunol.* **2001,** *2*, 1010–1017.
51. Theobald, M.; Biggs, J.; Hernandez, J. et al. Tolerance to p53 by A2.1-Restricted Cytotoxic T Lymphocytes. *J. Exp. Med.* **1997,** *185*, 833–841.

52. Lee, P. P.; Yee, C.; Savage, P. A. et al. Characterization of Circulating T Cells Specific for Tumor-Associated Antigens in Melanoma Patients. *Nat. Med.* **1999,** *5*, 677–685.
53. Benlalam, H.; Labarriere, N.; Linard, B. et al. Comprehensive Analysis of the Frequency of Recognition of Melanoma-Associated Antigen (MAA) by CD8 Melanoma in Filtrating Lymphocytes (TIL), Implications for Immunotherapy. *Eur. J. Immunol.* **2001,** *31*, 2007–2015.
54. Kawakami, Y.; Dang, N.; Wang, X. et al. Recognition of Shared Melanoma Antigens in Association with Major HLA-A Alleles by Tumor in Filtrating T Lymphocytes from 123 Patients with Melanoma. *J. Immunother.* **2000,** *23*, 17–27.
55. Gervois, N.; Labarriere, N.; Le Guiner, S. et al. High Avidity Melanoma-Reactive Cytotoxic T Lymphocytes Are Efficiently Induced from Peripheral Blood Lymphocytes on Stimulation by Peptide-Pulsed Melanoma Cells. *Clin. Cancer Res.* **2000,** *6*, 1459–1467.
56. Staveley, O.; Carroll, K.; Sotomayor, E.; Montgomery, J. et al. Induction of Antigen-Specific T Cell Energy: An Early Event in the Course of Tumor Progression. *Proc. Natl. Acad. Sci. USA* **1998,** *95*, 1178–1183.
57. Sotomayor, E. M.; Borrello, I.; Rattis, F. M. et al. Cross-Presentation of Tumor Antigens by Bone Marrow-Derived Antigen-Presenting Cells Is the Dominant Mechanism in the Induction of T-Cell Tolerance During B-Cell Lymphoma Progression. *Blood* **2001,** *98*, 1070–1077.
58. Sotomayor, E. M.; Borrello, I.; Tubb, E. et al. Conversion of Tumor-Specific CD4 T-Cell Tolerance to T-Cell Priming Through In Vivo Ligation of CD40. *Nat. Med.* **1999,** *5*, 780–787.
59. Chen, Q.; Daniel, V.; Maher, D. W. et al. Production of IL-10 by Melanoma Cells: Examinationof Its Role in Immunosuppression Mediated by Melanoma. *Int. J. Cancer* **1994,** *56*, 755–760.
60. Tada, T.; Ohzeki, S.; Utsumi, K. et al. Transforming Growth Factor-Beta-Induced Inhibition of T Cell Function: Susceptibility Difference in T Cells of Various Phenotypes and Functions and Its Relevance to Immunosuppression in the Tumor-Bearing State. *J. Immunol.* **1991,** *146*, 1077–1082.
61. Inge, T. H.; Hoover, S. K.; Susskind, B. M. et al. Inhibition of Tumor-Specific cytotoxic T-Lymphocyte Responses by Transforming Growth Factor Beta 1. *Cancer Res.* **1992,** *52*, 1386–1392.
62. Gorelik, L.; Flavell, R. A. Immune-Mediated Eradication of Tumors Through the Blockade of Transforming Growth Factor-Beta Signaling in T cells. *Nat. Med.* **2001,** *7*, 1118–1122.
63. Strand, S.; Hofmann, W. J.; Hug, H. et al. Lymphocyte Apoptosis Induced by CD95 (APO-1/Fas) Ligand-Expressing Tumor Cells: A Mechanism of Immune Evasion? *Nat. Med.* **1996,** *2*, 1361–1366.
64. Ohshima, K.; Muta, K.; Nakashima, M. et al. Expression of Human Tumor-Associated Antigen RCAS1 in Reed-Sternberg Cells in Association with Epstein-Barr Virus Infection: A Potential Mechanism of Immune Evasion. *Int. J. Cancer.* **2001,** *93*, 91–96.
65. Gastman, B. R.; Atarshi, Y.; Reichert, T. E. et al. Fas Ligand Is Expressed on Human Squamous Cell Carcinomas of the Head and Neck, and It Promotes Apoptosis of T Lymphocytes. *Cancer Res.* **1999,** *59*, 5356–5364.

66. Nakashima, M.; Sonoda, K.; Watanabe, T. Inhibition of Cell Growth and Induction of Apoptotic Cell Death by the Human Tumor-Associated Antigen RCAS1. *Nat. Med.* **1999,** *5*, 938–942.
67. Walker, P. R.; Saas, P.; Dietrich, P. Y. Tumor Expression of Fas Ligand (CD95L) and the Consequences. *Curr. Opin. Immunol.* **1998,** *10*, 564–572.
68. Dong, H.; Strome, S. E.; Salomao, D. R. et al. Tumor-Associated B7–H1 Promotes T-Cell Apoptosis: A Potential Mechanism of Immune Evasion. *Nat. Med.* **2002,** *8*, 793–800.
69. Pitti, R. M.; Marsters, S. A.; Lawrence, D. A. et al. Genomic Amplification of a Decoy Receptor for Fas Ligand in Lung and Colon Cancer. *Nature* **1998,** *396*, 699–703.
70. Mizoguchi, H.; O and Shea, J. J.; Longo, D. L. et al. Alterations in Signal Transduction Molecules in T Lymphocytes from Tumor-Bearing Mice. *Science* **1992,** *258*, 1795–1798.
71. Finke, J. H.; Zea, A. H.; Stanley, J. et al. Loss of T-Cell Receptor Zeta Chain and p56lck in T-Cells in Filtrating Human Renal Cell Carcinoma. *Cancer Res.* **1993,** *53*, 5613–5616.
72. Willerford, D. M.; Chen, J.; Ferry, J. A. et al. Interleukin-2 Receptor Alpha Chain Regulates the Size and Content of the Peripheral Lymphoid Compartment. *Immunity* **1995,** *3*, 521–530.
73. Liyanage, U. K.; Moore, T. T.; Joo, H. G. et al. Prevalence of Regulatory T Cells Is Increased in Peripheral Blood and Tumor Microenvironment of Patients with Pancreas or Breast Adenocarcinoma. *J. Immunol.* **2002,** *169*, 2756–2761.
74. Woo, E. Y.; Chu, C. S.; Goletz, T. J. et al. Regulatory CD4CD25 T Cells in Tumors from Patients with Early-Stage Non-Small Cell Lung Cancer and Late-Stage Ovarian Cancer. *Cancer Res.* **2001,** *61*, 4766–4772.
75. Woo, E. Y.; Yeh, H.; Chu, C. S. et al. Cutting Edge: Regulatory T Cells from Lung Cancer Patients Directly Inhibit Autologous T Cell Proliferation. *J. Immunol.* **2002,** *168*, 4272–4276.
76. Rohrer, J. W.; Barsoum, A. L.; Dyess, D. L. et al. Human Breast Carcinoma Patients Develop Clonable on Cofetal Antigen-Specific Effector and Regulatory T Lymphocytes. *J. Immunol.* **1999,** *162*, 6880–6892.
77. Dudley, M. E.; Wunderlich, J. R.; Robbins, P. F. et al. Cancer Regression and Autoimmunity Inpatients After Clonal Repopulation with Antitumor Lymphocytes. *Science* **2002,** *298*, 850–854.
78. Dudley, M. E.; Wunderlich, J. R.; Yang, J. C. et al. A Phase I Study of Non Myeloablative Chemotherapy and Adoptive Transfer of Autologous Tumor Antigen-Specific T Lymphocytes in Patients with Metastatic Melanoma. *J. Immunother.* **2002,** *25*, 243–251.
79. Dudley, M. E.; Wunderlich, J.; Nishimura, M. I. et al. Adoptive Transfer of Cloned Melanoma-Reactive T Lymphocytes for the Treatment of Patients with Metastatic Melanoma. *J. Immunother.* **2001,** *24*, 363–373.
80. Rosenberg, S. A.; Yannelli, J. R.; Yang, J. C. et al. Treatment of Patients with Metastatic Melanoma with Autologous Tumor-in Filtrating Lymphocytes and Interleukin 2. *J Natl Cancer Inst.* **1994,** *86*, 1159–1166.
81. Yee, C.; Thompson, J. A.; Roche, P. et al. Melanocyte Destruction After Antigen-Specific Immunotherapy of Melanoma: Direct Evidence of T Cell-Mediated Vitiligo. *J. Exp. Med.* **2000,** *192*, 1637–1644.
82. Ochsenbein, A. F.; Klenerman, P.; Karrer, U. et al. Immune Surveillance Against a Solid Tumor Fails Because of Immunological Ignorance. *Proc. Natl. Acad. Sci. USA* **1999,** *96*, 2233–2238.

83. Ochsenbein, A. F.; Sierro, S.; Odermatt, B. et al. Roles of Tumour Localization, Second Signals and Cross Priming in Cytotoxic T-Cell Induction. *Nature* **2001,** *411*, 1058–1064.
84. Timmerman, J. M.; Czerwinski, D. K.; Davis, T. A. et al. Idiotype-Pulsed Dendritic Cell Vaccination for B-Cell Lymphoma: Clinical and Immune Responses in 35 Patients. *Blood* **2002,** *99*, 1517–1526.
85. Rasmussen, T.; Hansson, L.; Osterborg, A. et al. Idiotype Vaccination in Multiple Myeloma Induced a Reduction of Circulating Clonal Tumor B Cells. *Blood* **2002,** *101*, 4607–4610.
86. Zhu, D.; Rice, J.; Savelyeva, N. et al. DNA Fusion Vaccines Against B-Cell Tumors. *Trends Mol. Med.* **2001,** *7*, 566–572.
87. Liso, A.; Stockerl-Goldstein, K. E.; Auffermann-Gretzinger, S. et al. Idiotype Vaccination Using Dendritic Cells After Autologous Peripheral Blood Progenitor Cell Transplantation for Multiple Myeloma. *Biol Blood Marrow Transplant* **2000,** *6*, 621–627.
88. Timmerman, J. M.; Singh, G.; Hermanson, G. et al. Immunogenicity of a Plasmid DNA Vaccine Encoding Chimeric Idiotype in Patients with B-Cell Lymphoma. *Cancer Res.* **2002,** *62*, 5845–5852.
89. Bendandi, M.; Gocke, C. D.; Kobrin, C. B. et al. Complete Molecular Remissions Induced by Patient-Specific Vaccination Plus Granulocyte-Monocyte Colony-Stimulating Factor Against Lymphoma. *Nat. Med.* **1999,** *5*, 1171–1177.
90. Kwak, L. W.; Campbell, M. J.; Czerwinski, D. K. et al. Induction of Immune Responses in Patients with B-Cell Lymphoma Against the Surface-Immunoglobulin Idiotype Expressed by their Tumors. *N. Engl. J. Med.* **1992,** *327*, 1209–1215.
91. Harris, J. E.; Ryan, L.; Hoover, H. C., Jr, et al: Adjuvant Active Specific Immunotherapy for Stage II and III Colon Cancer with an Autologous Tumor Cell Vaccine: Eastern Cooperative Oncology Group Study E5283. *J. Clin. Oncol.* **2000,** *18*, *148*–157.
92. McAneny, D.; Ryan, C. A.; Beazley, R. M. et al. Results of a Phase I Trial of a Recombinant Vaccinia Virus That Expresses Carcinoembryonic Antigen in Patients with Advanced Colorectal Cancer. *Ann. Surg. Oncol.* **1996,** *3*, 495–500.
93. Gulley, J.; Chen, A. P.; Dahut, W. et al. Phase I Study of a Vaccine Using Recombinant Vaccinia Virus Expressing PSA (rV-PSA) in Patients with Metastatic Androgen-Independent Prostate Cancer. *Prostate* **2002,** *53*, 109–117.
94. Berinstein NL: Carcinoembryonic Antigen as a Target for Therapeutic Anticancer Vaccines: A Review. *J. Clin. Oncol.* **2002,** *20*, 2197–2207.
95. Fong, L.; Hou, Y.; Rivas, A. et al. Altered Peptide Ligand Vaccination with FLT3 Ligand Expanded Dendritic Cells for Tumor Immunotherapy. *Proc. Natl. Acad. Sci. USA* **2001,** *98*, 8809–8814.
96. Mackensen, A.; Herbst, B.; Chen, J. L. et al. Phase I Study in Melanoma Patients of a Vaccine with Peptide-Pulsed Dendritic Cells Generated In Vitro from CD34 Hematopoietic Progenitor Cells. *Int. J. Cancer* **2000**, *86*, 385–392.
97. Vermorken, J. B.; Claessen, A. M.; van Tinteren, H. et al. Active Specific Immunotherapy for Stage II and Stage III Human Colon Cancer: A Randomised Trial. *Lancet* **1999,** *353*, 345–350.
98. Soiffer, R.; Lynch, T.; Mihm, M. et al. Vaccination with Irradiated Autologous Melanoma Cells Engineered to Secrete Human Granulocyte-Macrophage Colony-Stimulating Factor Generates Potent Antitumor Immunity in Patients with Metastatic Melanoma. *Proc. Natl. Acad. Sci. USA* **1998,** *95*, 13141–13146.

99. Gong, J.; Koido, S.; Chen, D. et al. Immunization Against Murine Multiple Myeloma with Fusions of Dendritic and Plasmacytoma Cells Is Potentiated by Interleukin 12. *Blood* **2002,** *99*, 2512–2517.
100. Gong, J.; Chen, D.; Kashiwaba, M. et al. Induction of Antitumor Activity by Immunization with Fusions of Dendritic and Carcinoma Cells. *Nat. Med.* **1997,** *3*, 558–561.
101. Nestle, F. O.; Alijagic, S.; Gilliet, M. et al. Vaccination of Melanoma Patients with Peptide- or Tumor Lysate-Pulsed Dendritic Cells. *Nat. Med.* **1998,** *4*, 328–332.
102. Tjoa, B. A.; Simmons, S. J.; Elgamal, A. et al. Follow-Up Evaluation of a Phase II Prostate Cancer Vaccine Trial. *Prostate* **1999,** *40*, 125–129.
103. Murphy, G. P.; Tjoa, B. A.; Simmons, S. J. et al. Phase II Prostate Cancer Vaccine Trial: Report of a Study Involving 37 Patients with Disease Recurrence Following Primary Treatment. *Prostate* **1999,** *39*, 54–59.
104. Hock, H.; Dorsch, M.; Kunzendorf, U. et al. Vaccinations with Tumor Cells Genetically Engineered to Produce Different Cytokines: Effectivity Not Superior to a Classical Adjuvant. *Cancer Res.* **1993,** *53*, 714–716.
105. Dranoff, G.; Jaffee, E.; Lazenby, A. et al. Vaccination with Irradiated Tumor Cells Engineered to Secrete Murine Granulocyte-Macrophage Colony-Stimulating Factor Stimulates Potent, Specific, and Long-Lasting Anti-Tumor Immunity. *Proc Natl Acad. Sci. USA* **1993,** *90*, 3539–3543.
106. Chang, E. Y.; Chen, C. H.; Ji, H. et al. Antigen-Specific Cancer Immunotherapy Using a GM-CSF Secreting Allogeneic Tumor Cell-Based Vaccine. *Int. J. Cancer* **2000,** *86*, 725–730.
107. Jaffee, E. M.; Hruban, R. H.; Biedrzycki, B. et al. Novel Allogeneic Granulocyte-Macrophage Colony-Stimulating Factor-Secreting Tumor Vaccine for Pancreatic Cancer: A Phase I Trial of Safety and Immune Activation. *J. Clin. Oncol.* **2001**, *19*, 145–156.
108. Salgia, R.; Lynch, T.; Skarin, A. et al. Vaccination with Irradiated Autologous Tumor Cells Engineered to Secrete Granulocyte-Macrophage Colony-Stimulating Factor Augment Santitumor Immunity in Some Patients with Metastatic Non–Small-Cell Lung Carcinoma. *J. Clin. Oncol.* **2003,** *21*, 624–630.
109. Townsend, S. E.; Allison, J. P. Tumor Rejection After Direct Costimulation of CD8 T Cells by B7-Transfected Melanoma Cells. *Science* **1993,** *259*, 368–370.
110. Chen, L.; Ashe, S.; Brady, W. A. et al. Costimulation of Antitumor Immunity by the B7 Counter Receptor for the T Lymphocyte Molecules CD28 and CTLA-4. *Cell* **1992,** *71*, 1093–1102.
111. Cayeux, S.; Beck, C.; Aicher, A. et al. Tumor Cells Cotransfected with Interleukin-7 and B7.1 Genes Induce CD25 and CD28 on Tumor-in Filtrating T Lymphocytes and are Strong Vaccines. *Eur. J. Immunol.* **1995,** *25*, 2325–2331.
112. van Mierlo, G. J.; den Boer, A. T.; Medema, J. P. et al. CD40 Stimulation Leads to Effective Therapy of CD40 Tumors Through Induction of Strong Systemic Cytotoxic T Lymphocyte Immunity. *Proc. Natl. Acad. Sci. USA* **2002,** *99*, 5561–5566.
113. Diehl, L.; den Boer, A. T.; Schoenberger, S. P. et al. CD40 Activation In Vivo Overcomes Peptide-Induced Peripheral Cytotoxic T-Lymphocyte Tolerance and Augments Anti-Tumor Vaccine Efficacy. *Nat. Med.* **1999,** *5*, 774–779.
114. Schultze, J. L.; Cardoso, A. A.; Freeman, G. J. et al. Follicular Lymphomas Can Be Induced to Present Alloantigen Efficiently: A Conceptual Model to Improve Their Tumor Immunogenicity. *Proc. Natl. Acad. Sci. USA* **1995,** *92*, 8200–8204.

115. Schultze, J. L.; Michalak, S.; Seamon, M. J. et al. CD40-Activated Human B Cells: An Alternative Source of Highly Efficient Antigen Presenting Cells to Generate Autologous Antigen-Specific T Cells for Adoptive Immunotherapy. *J. Clin. Invest.* **1997,** *100,* 2757–2765.
116. Von Bergwelt-Baildon, M. S.; Vonderheide, R. H.; Maecker, B. et al. Human Primary and Memory Cytotoxic T Lymphocyte Responses Are Efficiently Induced by Means of CD40-Activated B Cells as Antigen-Presenting Cells: Potential for Clinical Application. *Blood* **2002,** *99,* 3319–3325.
117. Schultze, J. L.; Anderson, K. C.; Gilleece, M. H. et al. A Pilot Study of Combined Immunotherapy with Autologous Adoptive Tumour-Specific T-Cell Transfer, Vaccination with CD40-Activated Malignant B Cells and Interleukin 2. *Br. J. Haematol.* **2001,** *113,* 455–460.

CHAPTER 7

Immunity's Role in Suppressing Tumor Growth and Shaping Tumor Immunogenicity

ABSTRACT

Mutations and epigenetic alterations that disrupt regular cell growth and survival pathways lead to cell transformation and tumor development. Cancer cells often express either abnormal proteins or unusual levels of normal cellular proteins that function as tumor antigens, and whether an unmodified immune system can detect and eliminate such altered host-derived cells has been a topic of heated debate for the better part of a century. The author reviews the current status of this topic and highlights multiple significant studies that indicate immunity not only protects the host against cancer development (i.e., provides a cancer immunosurveillance function) but also may promote tumor growth, sometimes by developing more aggressive tumors. This article provides a comprehensive analysis of the growing body of scientific and clinical evidence supporting the concept of "cancer immunoediting," which emphasizes the immune system's paradoxical role in protecting the host while also promoting tumor progression.

7.1 INTRODUCTION

The idea that the immune system controls cancer growth is making a comeback,[1,2] after being debated for a century. Almost 50 years after its introduction by Burnet and Thomas, the idea of cancer immunosurveillance is still hotly contested. Because it is difficult to scientifically demonstrate whether or not the host's natural immune defense mechanisms can prevent

Advances in Immunological Treatment of Cancer: Cellular Mechanism and Regulation.
Rishabha Malviya, Arun Kumar Singh, and Deepika Yadav (Authors)

the development of cancers that are not caused by viruses.[3,4] Cancer immunosurveillance is an extrinsic tumor suppressor mechanism that has been confirmed to exist in recent years,[5] thanks to a mountain of evidence from both experimental investigations in mice models and persuasive studies in human patients.[6] Researchers have also come to understand that the immune system has a multifaceted role to play in fighting cancer.[7] Immunosuppressed cancer occurs when the body's immune system is unable to eradicate a tumor.[8] There are three steps involved in immunoediting for cancer: detection, equilibrium, and evasion. In this case, the immune system is involved in the growth and spread of cancerous cells rather than preventing their formation.[9] The terms "escape" and "elimination" allude to the standard method of cancer immunosurveillance, whereas "equilibrium" and "escape" describe the time of immune-mediated latency after inadequate tumor death, respectively.[10] Since both authors believe that learning more about how the immune system interacts with tumors in the wild would help us develop more effective immunologically based cancer therapies, the author may have reached similar conclusions on cancer immunoediting recently.[11]

7.2 "INTRINSIC" AGAINST "EXTRINSIC" CANCER-BLOCKING AGENTS

When somatic cells acquire mutations that promote their unchecked, abnormal growth, a cancerous evolutionary process ensues. This allows cancer cells to multiply uncontrollably.[12]

In large, long-lived species, where cells in regenerative tissues have the potential for neoplastic illness throughout life, preventing the development of dysregulated autonomously developing cells is vital for evolutionary success. As a result, several biological mechanisms have developed to control cell growth (Fig. 7.1). In case of abnormal growth, many "intrinsic" tumor suppression mechanisms are on standby, ready to trigger apoptosis, repair, or senescence.[13] Mitochondria are essential for the execution of the primary cell death program, which controls the cell's response to survival factor signals, stress, and injury. Once released, the many proapoptotic effectors housed in mitochondria activate executioner caspases inside the cell, resulting in cell death.[14] Mitochondrial permeability is regulated by the balance between the pro-apoptotic Bax/Bak and anti-apoptotic Bcl2/BclXL proteins. Apoptosis is regulated at several levels, including transcription and posttranscriptional, by caspase inhibitors known as inhibitors of apoptosis (IAPs), which also play a role in death-effector component synthesis. There may be a separate pathway leading to cell death; however, that is triggered by the ligation of "death receptors" on the surface of cells.[15] Two such receptors are the tumor necrosis factor (TNF) receptor (Fas/CD95) and TRAIL receptor 2. That's why, when stimulated by members of the TNF superfamily, these receptors (TRAIL-R2, DR5) form the "death-inducing signaling complex" (DISC). This combination is responsible for activating apical caspase8.[16]

Several proteins in a cell could be able to detect mutagenic insults. The ability of these proteins to directly affect the apoptotic-effector machinery[17] makes them effective intrinsic tumor suppressors. DNA damage, hypoxia, and dietary limitation are only a few examples of cellular stressors that trigger p53's activation of cellular programs for apoptosis, senescence, and repair. Besides these, other stresses on cells, including oncogenes, may kill cells by directly affecting various parts of the cell death machinery,[18] and not only the p53 protein. The relative importance of p53-independent apoptotic mechanisms versus p53-dependent apoptotic mechanisms in the suppression of carcinogenesis depends on the tumor type and the number and sequence of oncogenic mutations present in every particular transformed cell.[19] Cellular senescence may result from oncogene activation. This state may be distinguished from quiescence, which is characterized by reversible cell-cycle arrest[20] due to the unique morphological and gene-expression modifications that occur during this time,[21] thereby, reversing the senescence that the oncogene causes is a necessary step before cells may be converted. To maintain itself, cancer needs joint lesions that decouple mitogenic Ras signaling from senescence.[22] Overall, senescence and apoptosis provide significant roadblocks to the progression of any preneoplastic cell.

However, the presence of malignant cells may be "felt" by cells and the surrounding tissues via at least three different types of "extrinsic" processes.[23] To avoid committing suicide, epithelial cells, for example, are dependent on the extracellular matrix (ECM) for social support from a variety of trophic cues provided by their surrounding environment.[24] Because of this, the cells can avoid committing suicide. Death of cells, known in science as "anoikis," happens when cells get detached from the ECM and the normal tissue architecture collapses as a consequence of unregulated development.[25] In the latter, it seems that genes controlling cell junctions and proliferation are intertwined. These genes' functions are associated with cell polarity.[26] Third, leukocytes aid in the detection, elimination, and/or cytostasis of aberrant cell populations. Several different types of cell-killing strategies are used by immune system effector cells in their drive to eliminate tumors. These mechanisms include both mitochondrial and cell death receptor pathways.[27] Multiple intrinsic and extrinsic systems work together with exceptional efficiency and selectivity to inhibit tumor growth.[28] Even though every human being contains trillions of potentially cancerous cells, and each of those cells has hundreds of cancer-causing genes, and even if every one of those cells is exposed to a large mutation rate, the average individual will get cancer only once or twice in their lifetime.[29]

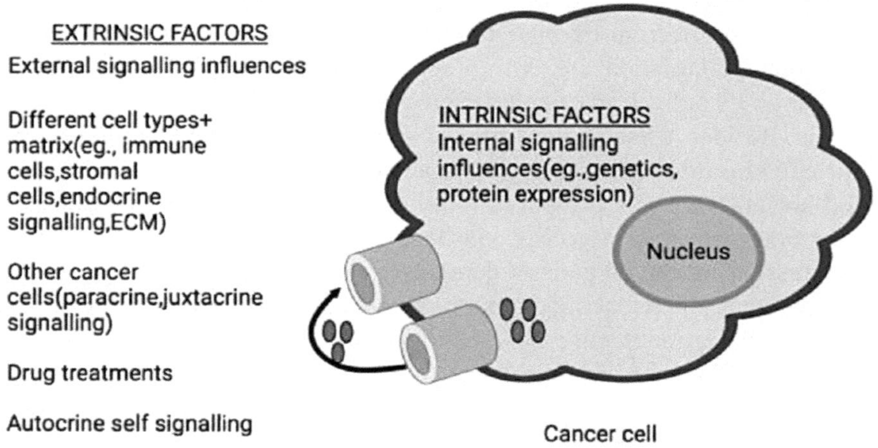

FIGURE 7.1 Schematic representation of extrinsic factors and intrinsic factors of during cancer development.

7.3 CANCER IMMUNOSURVEILLANCE THEORY: FROM CONFLICT TO RECOGNITION

One of the first people to propose that the host immune system actively combats neoplastic disease was Paul Ehrlich. It was a shocking finding made by William B. Coley that some patients with cancer who had bacterial infections also saw a decrease in their tumors in the same year (1891).[30]

Patients with advanced cancer who received injections of Coley's heat-inactivated bacterial cultures or surplus bacterial culture supernatants reported significant tumor shrinkage and improved survival. Recently, it has become clear that the compounds hitherto thought to be "Coley's poisons"[31] were produced by bacteria and have powerful immunomodulatory properties.

The notion of immunological regulation of cancer wasn't substantially explored, however, until the 1950s, when Medawar and colleagues elucidated the role of cellular components of the immune system in recognizing and controlling allograft rejection. Although this study did not refute tumor-specific immunity,[32] it did cast doubt on the concept since the experimental tumors were likely seen as foreign grafts.[33] However, not long later, inbred mouse strains were developed, making it possible to test the hypothesis that tumors developing in such animals may be immunologically distinguished from normal tissues of the same strain.[34] Tumor-specific immune responses were shown to exist when syngeneic mice that had been inoculated against a tumor were resistant to a second challenge with the same tumor cells.[35] Overall, the study's findings suggest that immune cells may recognize changed tissue by either reacting to soluble compounds generated by tumors or recognizing specific surface markers on cancer cells. The cancer immunosurveillance hypothesis[36] was presented by Burnet and Thomas in the 1980s, based on the idea that lymphocytes are in charge of eliminating newly emerging, nascent transformed cells.

Cancer incidence was shown to be higher in animals with compromised immune systems compared to those with healthy immune systems in subsequent tests.[37] Early attempts to test this hypothesis using mice with immunosuppression caused by neonatal thymectomy yielded conflicting results.[38] When compared to healthy mice, immunocompromised rodents had a significantly higher incidence of both virally induced cancers and spontaneously arising lymphomas. The majority of researchers, however, concluded that this finding only reflected the fact that immunocompromised hosts were more prone to acquiring oncogenic viruses.[39,40] Persistent infection, according to these researchers, is to blame for the high prevalence of this malignancy because it leads to unresolved antigenic stimulation, which leads to increased cell proliferation, which leads to somatic mutations, which leads to the development of lymphoma.[41] Nu/nu (nude athymic) mice of the CBA/H strain did not develop carcinogen [methylcholanthrene (MCA)]-induced tumors earlier or more frequently than their immunocompetent littermates, and the incidences of spontaneous, nonviral tumors in nude and wild-type mice were similar, casting doubt on the cancer immunosurveillance hypothesis.[42] Initial impressions were that these findings were very damaging to the immunosurveillance hypothesis of cancer.[43] The failures the researchers encountered may now be traced back to defects in the experiment's design.[44] For example, recent research has revealed that mice who forgo socks don't suffer from significantly diminished immune systems.[45] This resistance to sarcoma seems to have been overcome in the CBA/H breed of mice since these animals produce an isoform of the aryl hydroxylase enzyme.[46] Tumor monitoring lapses of three to seven months were shown to be inadequate for detecting cancer as a result of chance.[47] As a result of recent advances in gene targeting and transgenic animal technologies, as well as the availability of highly selective blocking monoclonal antibodies, the cancer immunosurveillance hypothesis may now be evaluated in molecularly described mouse models of immunodeficiency (mAb).[48] The past 15 years have seen a rebirth of interest in the cancer immunosurveillance theory,[49] which is based on the idea that better in vivo mouse cancer models may be translated to humans.

Immunocompetent people still develop malignancies that are resistant to many different treatments, even though there is substantial evidence showing that the immune system can detect and prevent cancer. It's likely that the host's first innate and adaptive immune responses didn't eliminate all of the altered cells, which might explain this clinical observation. The spread of low-immunogenicity tumors is facilitated by a weakened immune system. To highlight immunity's function in eradicating and changing neoplastic disease during homeostasis and evasion, the term "cancer immunoediting" was coined.[50] The elimination, equilibrium, and escape steps will be explained in details later. Here, the author examines the role of leukocytes, paying special attention to their ability to detect cellular change and their effector activities in the regulation, elimination, and shaping of malignancies.[51]

7.4 REMOVAL OF MALIGNANT GROWTHS

7.4.1 STRUCTURE OF CANCER IMMUNOEDITING'S TUMOR-KILLING PHASE

For the most comprehensive immunoediting experience, go forward to the elimination step.[52] This is when the idea of cancer immunosurveillance begins to take shape. Our working hypothesis is that transformed cells may evade their endogenous tumor-suppressor mechanisms before the immune system can exert its extrinsic tumor-suppressor functions. Oncogenic viral infection is only one example of a trigger that may simultaneously activate immunological and intrinsic tumor suppressor mechanisms.[53] Similar to host defense against microbial diseases, immunological rejection of a developing tumor likely requires cooperation between the innate and adaptive immune systems. The author postulates that the innate immune cells become aware of a tumor's presence and initiate an anticancer immune response[54] as a result of local tissue disruption caused by angiogenesis or tissue-invasive growth. A combination of chemokines generated by tumor cells and stromal proinflammatory compounds may encourage innate immune cells to concentrate on the site of the new "risk." As part of their role as sentinel cells in monitoring tissue stress, damage, and/or transformation, dendritic cells (DCs) may detect and respond to danger signals such as heat-shock proteins (HSPs) (released as a result of tumor-cell damage or necrosis), ECM-breakdown products (ECMBPs), and proinflammatory factors (including cytokines released by degenerating tumor cells and reactive host innate cells).[55] It has been proposed that cytokines such as interleukin-1 (IL-1), TNF-alpha (TNF-a), type I interferon (IFN), granulocyte-macrophage colony-stimulating factor (GM-CSF), and interleukin-15 (IL-15) enhance DC development and activity.[56] During the early stages of inflammation or cellular transformation, tumor cells release stress ligands for NKG2D,[57] which may rapidly attract NK cells, macrophages, gd T cells, and/or NKT cells to the "risk" zone.

By interacting with MHC/tumor-related peptide complexes or glycolipidCD1 complexes expressed on tumor cells, T cells and NKT cells may be able to recognize tumors in progress.[58]

IFNs are secreted by immune cells to restrict tumor progression and boost the immune response, and these cells may then activate cytotoxic effector pathways to destroy the changed cell (s). Attracting more innate immune cells to the tumor site may need initial IFNg release at the tumor site to stimulate local chemokine production.[59] A potential feedback loop between tumor-infiltrating macrophages (to generate IL-12) and tumor-infiltrating NK cells may be set in motion by-products of ECM remodeling (to produce IFN-g). IFNg-dependent pathways that may contribute to tumor necrosis include antiproliferative, proapoptotic, and angiostatic effects.[60] In contrast to activated macrophages, which may kill tumor cells by producing tumoricidal products including reactive oxygen and reactive nitrogen intermediates, activated NK cells may kill tumor cells via TRAIL-dependent or perforin-dependent mechanisms.

Establishing tumor-specific adaptive immune responses[61] is the next phase, and this is induced by tumor antigens produced by various cell death mechanisms. When immature DCs reach the tumor site, they are triggered either by interacting with tumor-infiltrating NK cells or by being exposed to cytokines released by innate immune cells fighting the tumor. Tumor antigens may be taken up by activated DCs either directly from tumor cell apoptotic bodies or indirectly through heat shock protein complexes carrying tumor antigens.[62] Mature dendritic cells that have become highly active in response to various chemokines and/or cytokines may move to lymph nodes to activate naive tumor-specific Th1 CD4+ T cells.[63] Antigenic tumor peptides are cross-presented on DC MHC class I molecules, activating and expanding tumor-specific CD8+ CTL with the help of Th1 cells. MHC class I and class II molecules may stimulate B-cell activation by delivering peptides derived from tumor-associated antigens to CD4+ and CD8+ T lymphocytes, respectively.[64] At this point, the host has likely established an adaptive immune response unique to the tumor and can eradicate it entirely. CD4+ and CD8+ T lymphocytes migrate to the tumor site and help kill antigen-positive tumor cells.[65] CD4+ T cells produce IL-2, and the host produces IL-15, both of which are necessary for the survival of tumor-specific CD8+ T cells.[66] Tumor-specific CD8+ T lymphocytes will trigger both direct and indirect mechanisms of cancer cell death as a result of the increased immunogenicity of tumor cells exposed to the IFNg. These CD8+ T cells certainly "skill" a significant number of tumor cells in vivo. However, tumor cells also produce large quantities of IFN-g when they contact macrophages; macrophages may trigger cell cycle arrest, apoptosis, angiostasis, and macrophage tumoricidal activity in tumor cells through these IFN-g-dependent pathways.[67] These two possibilities are not exclusive of one another and both occur in any particular tumor, albeit the relative contributions of each may differ. Therefore, to get rid of antigenically different neoplastic cells, the elimination phase of cancer immunoediting (also known as the cancer immunosurveillance phase) must be repeated.[68] The fact that cancer rates tend to be higher in ageing populations is especially remarkable given the loss in the immune system and, by extension, cancer immunosurveillance with age.

7.4.2 THE IMPORTANCE OF BOTH INNATE AND ADAPTIVE IMMUNITY

Innate and adaptive immune cells, among others, have emerged as key players in the evolution of innate immunity against tumors in recent years.

Studies that failed to find evidence of spontaneous cancer development in germ-free nude mice were re-examined in a more rigorous study that showed a lower frequency of B-cell lymphoma in these animals compared with their heterozygote littermates.[69] A similar increase in lymphoma incidence was seen in anytime-stimulated nude mice. Because they lack the ability to somatically alter their lymphocyte antigen receptors, RAG2-deficient mice (or those missing the essential cofactor RAG1) are unable to generate T, B, or NKT cells in the periphery. In contrast to other forms of inherited immunodeficiency, the DNA damage repair mechanisms of transforming nonimmune cells are unaffected by RAG2 loss (such as in SCID animals). At 18 months of age, RAG2 null 129/Sv animals had almost 50% more gastrointestinal tumors than wild-type mice.[70] Also, as compared to their wild-type counterparts, Helicobacter-negative RAG2/ 129/Sv mice were substantially more likely to develop gastrointestinal and lung epithelial cancers when raised in a pathogen-free mouse environment and given broad-spectrum antibiotics every other month. Consistent with these findings, it has been shown that RAG2deficient 129/Sv animals, who also lack STAT1, are predisposed to developing a broader spectrum of malignancies, such as colon and mammary adenocarcinomas, at a younger age. Subcutaneous injection of the carcinogen methylcholanthrene caused sarcomas to form more rapidly and often in 129/Sv RAG2/ mice than in strain-matched wild-type controls (MCA). Similar results were seen when MCA was given to RAG1/ C57BL/6 mice (Smyth, unpublished data). Finally, lymphocytes prevent tumor development in both chemically and naturally carcinogenic animals.[71]

7.4.3 OTHER LEUKOCYTES

It's possible that different types of leukocytes may considerably boost the immune response against tumors. Recent research has indicated a more systemic difficulty in the function of numerous granulocytes in beige mice, which may explain the higher risk of lymphoma in these animals.[72] Previously, it was considered that a lack of NK cell activity was to blame. Neutrophils have also been proven in several studies to have a role in tumor immunity. Research on breast tumor vaccination in mice revealed numerous mechanisms, including innate and adaptive immune components, that may play a role in cancer eradication through different pathways. The latest research by Cui et al. adds to the growing body of data supporting the centrality of innate immune cells in the immune system's role of surveillance (2003).[73] At least two autosomal dominant loci have been demonstrated to be critical for passing on a trait that confers resistance to age-related cancer in the mouse strain known as BALB/ c, which has been the subject of intensive study. Mice with this "spontaneous regression/complete remission" (SR/CR) phenotype are very effective in eliminating syngeneic and allogeneic tumor cells from their organs and tissues. Breeding for the SR/CR trait on a nude genetic background also resulted in resistance, demonstrating that innate immune cells are heavily involved in mediating the SR/CR phenotype . Tracing the genetic roots of the SR/CR trait[74] should help researchers better understand how the innate immune system regulates tumor growth. Creating mice on pure genetic backgrounds with deficiencies in leukocyte subsets (such as NK cells, DCs, and monocyte subsets) is necessary for studying the innate immune response to tumors.[75] For instance, eosinophil-deficient mice have recently been developed, allowing scientists to investigate the natural role of eosinophils in tumor immunosurveillance. As one example, it has been demonstrated that activated eosinophils may effectively eliminate tumor metastases.[76] To rapidly untangle the anticancer effect of innate immune components in wild-type mice, more selective function-modifying monoclonal antibodies are needed (for example, those that inhibit IFN production by particular DC subpopulations).[77] Surprisingly, there have been no reports of spontaneous tumor growth in animals missing one or more of the Toll-like receptor pathways, yet TLR4 and the MyD88 adaptor are needed for adjuvant-induced activation of antitumor CTL cytotoxicity. Immune surveillance is difficult to predict since it is affected by several variables, including the tumor's cell of origin, transformation process, anatomic location, stromal response, cytokine and chemokine microenvironment, and immunologic recognition mechanism.[78] To determine whether immunosurveillance of all cancer-vulnerable tissues of the body is universal or regionally variable,[79] it is still important to assess the influence of various immunologic components on tumor growth in different models (naturally occurring and those caused by chemicals).[80]

7.4.4 IMMUNE CELLS TARGET CANCER

Recent evidence suggests immune cells may use several recognition mechanisms to identify the process of transformation. Beginning in the 1970s, autologous typing[81] has been used to extensively study individuals' humoral and cellular immune responses to their tumors. The autologous patient's humoral or cellular antitumor immune responses were studied using tumor cell lines derived from a variety of individuals with melanoma or other tumor forms amenable to in vitro growth.[82] Fibroblasts and other autologous normal cell types were used to determine the specificity of the antitumor response. Patients with autologous tumor-recognizing T cells or cell-surface antigen-specific antibodies were distinguished using this strategy.[83] Gene cloning and expression approaches established by Boon and colleagues to identify tumor antigens recognized by CD8+ T lymphocytes and by Pfreundschuh and colleagues to characterize antibody-defined tumor antigens were used to further describe molecular targets revealed through autologous typing.[84] Tumor antigens recognized by CD4+ T lymphocytes that are limited by major histocompatibility complex class II have become much simpler to detect recently.[85]

Since the cloning of the first human tumor antigen in 1991, several cancer antigens have been discovered and roughly grouped into the following five categories: MelanA/MART1, tyrosinase, gp100, and gp100 are examples of differentiation antigens; Her2/neu and mutated p53 are examples of mutational antigens; MAGE and NYESO1 are examples of cancer-testis (CT) antigens; MAGE and NYESO1 are examples of cancer-testis (CT) antigens; and gp100 and gp100 are examples of gp100 (e.g., EBV and HPV). In particular, CT antigens are of interest because of their unique characteristics.[86] In adult normal tissues, their expression is limited to germ cells in the testis; nevertheless, a variable fraction of a wide range of tumor types express CT antigens.[87]

More than 20 CT antigens or antigen families have been characterized as of now,[88] thanks in large part to the serological expression cloning approach (SEREX) established by Pfreundschuh and colleagues to find the humoral response to human cancer. Researchers using the status quo continue their hunt for immunogenic human tumor antigens.

7.5 CANCER-PREVENTING IMMUNE EFFECTOR MOLECULES

The cancer immunosurveillance hypothesis has been bolstered by several landmark studies showing that deficiencies in key immunologic effector molecules enhanced host sensitivity to malignancies, whether they were induced chemically or evolved spontaneously. This research classifies effector mechanisms, such as cytokines and cytotoxic mediators, so that the author may further investigate their functions in driving tumor immunosurveillance.[89]

7.5.1 CYTOKINES

The activation of immune effector systems that suppress tumor development is most likely mediated by cytokines. On the other hand, they could also aid in cancer's initiation, progression, proliferation, invasion, and dissemination. Host immune and stromal cells release cytokines in response to substances produced by cancer cells or as part of the inflammation that often accompanies tumor development. Both healthy and cancer cells respond to stress by releasing cytokines. The local cytokine network in a tumor relies on the types and levels of cytokines generated, as well as the expression patterns of cytokine receptors.[90] Throughout the development of a tumor, alterations in the cytokine environment are to be anticipated. The researcher will be focusing on IFNs because their function in innate tumor immunity has been studied more thoroughly than that of many other cytokines, such as IL-12, IL-18, GM-CSF, Flt3L, and the common g chain cytokines, which may also play an important role in tumor immunosurveillance due to their therapeutic capabilities.

7.5.2 TYPE I INTERFERONS (IFN-A/B)

More than a dozen proteins are produced by the physically similar genes that produce type I interferons; these proteins all engage with the same type I interferon receptor to elicit the same responses in their target cells. Type I IFN is primarily represented by 14 and 12 IFNa proteins in mice and humans, respectively, and a single IFNb protein. Both rodent and human tumors have been proven to be susceptible to IFNa induction, which has been demonstrated to be successful for some time now.[91] It has been hypothesized that IFNa/b induces its antitumor effects primarily through host-dependent mechanisms such as by stimulating NK cell proliferation and cytotoxicity, by increasing the production or secretion of other cytokines by NK cells through an autocrine IFNg loop, or by influencing CD8+ T-cell and B-cell adaptive immune responses by upregulating IL-12. This latter idea is supported, however, by data from studies where exogenous type I IFN was provided.[92] Only lately has it been understood that type I interferons play an important role in triggering a cancer-fighting immune response (immunosurveillance).[93] The rejection of allogeneic or xenogeneic tumors was abolished, while the development of syngeneic tumors transplanted into the animals was sped up, according to preliminary research.[94]

Recent in vitro investigations have connected the expression of the tumor suppressor gene p53 to the suppression of cellular transformation by type I IFNs such as IFNa and IFNb.[95] Further research is required to discover if this IFN activity has a physiologically meaningful impact in reducing cancer growth in vivo[96] since this finding only relates to in vitro transformation experiments.

In recent years, it has become increasingly apparent that type I IFNs play an essential role in suppressing MCA sarcoma formation in mice and in facilitating the rejection of immunogenic transplantable tumors.[97] Mutant versus wild-type (WT) mice differed in their susceptibility to tumorigenesis after MCA treatment.[98] Mice deficient in the IFNAR1 subunit of the type I IFN receptor were more vulnerable. WT mice treated with a blocking IFNAR1 mAb showed the same inability to reject highly immunogenic RAG regressor MCA sarcomas as IFNAR1/animals.[99,100]

The effects of IFNa/b are distinct from those of IFNg, and a search for the functionally relevant targets of type I IFN in these models revealed that tumor cells were not among them. However, type I sensitivity in the host's hematopoietic cells was necessary to establish an effective anti-tumor immune response.[101] Mutant IFNAR1/MCA sarcoma mice had a highly immunogenic phenotype, suggesting that these tumors were not edited throughout development. To fully understand the role of IFNa/b in cancer immunosurveillance and the immunologic pathways that IFNa/b supports leading to cancer immunosurveillance and immunoediting, further research is required.[102]

7.5.3 IMMUNOEDITING SUSTAINS CANCEROUS CELLS

There is still plenty to learn about the connection between immunity and cancer, even if cancer immunoediting fails at the elimination/immunosurveillance stage. Tumors formed in immunodeficient mice are less immunogenic than tumors developed in immunocompetent animals,[103] illustrating the intricacy of the immune system's interaction with malignancies. MCA-generated sarcomas are more immunogenic in immunodeficient hosts (RAG/, SCID, naked, or NKT cell-deficient animals) than tumors arising in immunocompetent hosts. Not only that, but sarcomas extracted from WT mice treated with anti-TRAIL were much more TRAIL-sensitive ex vivo than sarcomas derived from WT animals treated with control Ig. Lymphomas removed from pfp/mice did well when transplanted into animals of the same strain, but were almost always rejected when given to wild-type mice. These findings demonstrate the importance of the local immunological environment in the development and progression of tumors.[104] Tumors that are poorly recognized by the immune system or have acquired mechanisms that limit immune effector responses may arise as a result of this imprinting process, which eliminates tumor cells with high intrinsic immunogenicity and selects for variations with reduced immunogenicity.[105] The author hypothesizes that tumors that survive the immunoediting eradication phase will stabilize and ultimately exit the host.

7.5.4 ESCAPE

Cancer cells that have already passed through the equilibrium phase gain the capacity to multiply in an environment where they are not challenged by the immune system during the escape phase.[106] The tumor either inhibits or tolerates the host's immune system, enabling the tumor to grow and become clinically identifiable, or it obtains resistance to immune detection and eradication via genetic and epigenetic modifications.[107] If the cancer immunosurveillance system is functioning properly, tumors should not be able to grow uncontrolled unless they can elude the adaptive or innate arms of immunity. Tumor cells may use a combination of immunoevasive methods[108] to evade the powerful innate and adaptive anticancer immune responses mounted against their immunogenic progenitors. Therefore, it is believed that a chain of autonomous immunologically driven tumor sculpting processes is required for a malignant cell to reach its ultimate immunogenic phenotype.[109] To what degree a tumor's immunogenicity is influenced by its contact with the host immune system may depend on the identities of the immune editors at work during the equilibrium phase. The mechanisms that promote tumor escape can be different depending on the kind of tissue the tumor started in Ref. [110]. In this case, immunologic sculpting of a metastatic lesion may stand out the most. Recent studies have concentrated on characterizing the molecular basis of cancer escape, even though some pathways implicated in tumor escape remain more speculative than confirmed. Immune responses to tumors may be classified as either intrinsic (involving only tumor cells and tumor-associated antigens) or extrinsic (involving non-tumor cells and non-tumor antigens) (involving the host immune system). For the latter, consider how tumor-associated antigens are surprisingly underexpressed during the first stages of tumor growth. Several examples of the latter include (a) the downregulation or loss of expression of MHC class II molecules and costimulatory molecules, (b) the downregulation or loss of expression of MHC class I molecule proteins, and (c) the downregulation or loss of expression of genes associated with antigen presentation (including transporter associated with antigen processing (TAP), low-molecular-weight protein (LMP), and b2).[111]

Tumor-derived substances that suppress T cells (d, e, f) (including TGFb, IL10, VEGF, FasL, galectin, indoleamine 2,3-dioxygenase (IDO), lymphocyte activation blockade (e.g., NKG2DL) release) include defects in antigen presentation by expert APC and poor APC maturation.[112,113] The formation of tumors expressing the SV40 large T antigen (Tag) in wild-type mice engineered to intermittently release a latent viral oncogene[114] demonstrates that tumor-dependent effects on host immunity may actually contribute to tumor escape.

7.5.5 EQUILIBRIUM

While it is true that a high percentage of mutated cells can be eliminated through this stage of cancer immunoediting, there may be a lag time between the end of the elimination phase and the beginning of the escape phase, at which point clinically recognizable malignant illness would become apparent.[115]

Before a tumor is detected by a doctor, it goes through a period called "equilibrium," during which it interacts with the immune system but does not grow. It's possible that achieving homeostasis will take a very long period, thus it's saved until last.[116] When it comes to solid human malignancies, studies have shown that there may be a lag time of up to 20 years between the first carcinogen exposure and the clinical identification of a tumor. It is speculated that tumor "dormancy"-like behaviors might occur at the equilibrium stage of cancer immunoediting.[117] Other tumor forms keep expanding even after many of the initial tumor cells are eliminated; these tumors have acquired additional mutations that make them more resistant to immune response.[118] Last but not least, in the equilibrium phase, new populations of tumor cells are formed that are better suited to live in the immunocompetent host as a result of the dynamic interaction between immunity and cancer. Cancer cells may be able to elude the host immune system at different periods in the life cycle due to shared genetic and environmental variables that allow them to survive the elimination phase. In a tumor, the immune system is constantly battling the hundreds or thousands of cancer cells that may have accumulated due to genetic mutations. Many kinds of genomic instability have been seen in cancer, and this instability has been connected to the "mutator phenotype" of tumor cells.[119] Examples of such chromosomal abnormalities are microsatellite instability (MIN) and chromosomal instability (NIN)

(CIN). Characteristic of the CIN phenotype is the loss of an estimated 25–50% of alleles in cancer cells due to the addition or loss of entire chromosomes (i.e., aneuploidy). Research on the concept of homeostasis as a liminal state in cancer immunoediting is warranted. To understand the molecular mechanisms behind the emergence of novel tumor phenotypes and the immunological factors that control the maintenance of a stable tumor microenvironment, the researcher needs to construct new "editors" of tumor models. Tumors may be completely destroyed by the immune system, can persist in the equilibrium phase forever due to the immune system's cellular and molecular restraints, or can escape immunological pressure and reach the ultimate escape phase.[120]

7.5.6 HUMAN IMMUNOTHERAPY AND CANCER SURVEILLANCE

Therefore, the rapid development of tumors in patients with mutations in genes that produce immune effector molecules is an outlier. This is because, unlike immunocompromised mice, individuals with impaired immune systems cannot tolerate even brief exposure to non-fatal illnesses and diseases, even in pathogen-free mouse animal facilities.[121] Patients with compromised immune systems are more likely to develop genitourinary and cervical cancers, as well as non-Hodgkin lymphoma, Hodgkin disease, and Kaposi's sarcoma. The various viruses and bacteria that may have infected these people might drastically alter the findings,[122] which could be years before non-viral cancers arise. Despite this, there are three pieces of evidence suggesting that humans have a role in cancer immunosurveillance: Lymphocyte infiltration of tumors may be a predictive indicator of patient survival, and adaptive and innate immune responses to tumors might arise spontaneously in people with cancer.[123,124] Nonetheless (a), nonviral malignancies were more common in immunosuppressed transplant patients than in age-matched, immunocompetent controls.[125,126]

Recent years have seen a rise in interest in transplant patients with compromised immune systems. Finally, research involving thousands of transplant recipients in the Western Hemisphere has found an increased relative risk ratio for various tumors of unknown viral origin.[127] These include cancers of the colon, lung, bladder, pancreas, kidney, ureter, melanomas, squamous cell carcinomas, and endocrine tumors.

So, those with weak immune systems are more prone to acquire cancers of unknown viral origin than people with strong immune systems.[128] This research demonstrates that immunosuppressive therapy increases the risk of tumor development or recurrence in transplant patients.[129] The transmission of cancer between organ donors and recipients is one clinical circumstance where the equilibrium phase is likely to occur in humans.[130] Two kidney transplant recipients who shared a donor were recently diagnosed with metastatic melanoma.[131,132] In the 16 years before the donor's death, primary melanoma was detected; nevertheless, at the time of donation, physicians falsely assumed that no tumors were present.[133]

7.6 CONCLUSION

In this chapter, the author takes a high-level look at the data supporting the existence of an efficient cancer immunosurveillance system capable of halting the progression of cancer in both mice and humans. Cancer suppression by the innate immune system is affected by several factors. The author has also argued that a more accurate way to conceptualize cancer immunosurveillance is as a process known as cancer immunoediting, in which the immunogenic phenotype of the malignant cell is altered by its interaction with the host immune system to evaluate its fitness for sustained survival and development in an immunocompetent environment. Here, the researcher discusses some of the most urgent questions that have yet to be answered about how the immune system can play a role in cancer diagnosis and therapy. Immunotherapy will certainly be utilized to treat and/or eradicate neoplastic sickness once the immunobiology of cancer immunoediting is better known and a molecular description of how tumors are transformed by this process is established.

KEYWORDS

- **immunogenicity**
- **cancer blocking agents**
- **extrinsic factors**
- **cancer immunosurveillance theory**

REFERENCES

1. Burnet, F. M. Immunological Factors in the Process of Carcinogenesis. *Br. Med. Bull.* **1964,** *20*, 154–158.
2. Burnet, F. M. The Concept of Immunological Surveillance. *Prog. Exp. Tumor Res.* **1970,** *13*, 1–27.
3. Burnet, F. M. Immunological Surveillance in Neoplasia. *Transplant Rev.* **1971,** *7*, 3–25.
4. Burnet, M. Cancer: A Biological Approach. III. Viruses Associated with Neoplastic Conditions. IV. Practical applications. *Br. Med. J.* **1957,** *1*, 841–847.
5. Burstein, N. A.; Law, L. W. Neonatal Thymectomy and Non-Viral Mammary Tumours in Mice. *Nature* **1971,** *231*, 450–452.
6. Carayannopoulos, L. N.; Naidenko, O. V.; Fremont, D. H.; Yokoyama, W. M.Cutting Edge: Murine UL16-Binding Protein-Like Transcript 1: A Newly Described Transcript Encoding a High-Affinity Ligand for Murine NKG2D. *J. Immunol.* **2002,** *169*, 4079–4083.
7. Carey, T. E.; Takahashi, T.; Resnick, L. A.; Oettgen, H. F.; Old, L. J. Cell Surface Antigens of Human Malignant Melanoma: Mixed Hem Adsorption Assays for Humoral Immunity to Cultured Autologous Melanoma Cells. *Proc. Natl. Acad. Sci. USA* **1976,** *73*, 3278–3282.
8. Carmeliet, P.; Jain, R. K. Angiogenesis in Cancer and Other Diseases. *Nature* **2000,** *407*, 249–257.
9. Cerwenka, A.; Baron, J. L.; Lanier, L. L. Ectopic Expression of Retinoic Acid Early Inducible-1 Gene (RAE-1) Permits Natural Killer Cell-Mediated Rejection of a MHC Class I-Bearing Tumor In Vivo. *Proc. Natl. Acad. Sci. USA* **2001,** *98*, 11521–11526.
10. Cerwenka, A.; Lanier, L. L. Ligands for Natural Killer Cell Receptors: Redundancy or Specificity. *Immunol. Rev.* **2001,** *181*, 158–169.
11. Chen, Y. T.; Scanlan, M. J.; Sahin, U.; Tureci, O.; Gure, A. O.; Tsang, S.; Williamson, B.; Stockert, E.; Pfreundschuh, M.; Old, L. J. A Testicular Antigen Aberrantly Expressed in Human Cancers Detected by Autologous Antibody Screening. *Proc. Natl. Acad. Sci. USA* **1997,** *94*, 1914–1918.
12. Chin, Y. E.; Kitagawa, M.; Kuida, K.; Flavell, R. A.; Fu, X. Y. Activation of the STAT Signaling Pathway Can Cause Expression of Caspase 1 and Apoptosis. *Mol. Cell. Biol.* **1997,** *17*, 5328–5337.
13. Chin, Y. E.; Kitagawa, M.; Su, W. C.; You, Z. H.; Iwamoto, Y.; Fu, X. Y. Cell Growth Arrest and Induction of Cyclin-Dependent Kinase Inhibitor p21 WAF1/CIP1 Mediated by STAT1. *Science* **1996,** *272*, 719–722.
14. Clemente, C. G.; Mihm, M. C., Jr.; Bufalino, R.; Zurrida, S.; Collini, P.; Cascinelli, N. Prognostic Value of Tumor Infiltrating Lymphocytes in the Vertical Growth Phase of Primary Cutaneous Melanoma. *Cancer* **1996,** *77*, 1303–1310.
15. Coca, S.; Perez-Piqueras, J.; Martinez, D.; Colmenarejo, A.; Saez, M. A.; Vallejo, C.; Martos, J. A.; Moreno, M. The Prognostic Significance of Intratumoral Natural Killer Cells in Patients with Colorectal Carcinoma. *Cancer* **1997,** *79*, 2320–2328.
16. Coley, W. B. The Treatment of Malignant Tumors by Repeated Inoculations of Erysipelas: With a Report of Ten Original Cases 1893. *Clin. Orthop.* **1991,** *262*, 3–11.
17. Conejo-Garcia, J. R.; Benencia, F.; Courreges, M. C.; Kang, E.; Mohamed-Hadley, A.; Buckanovich, R. J.; Holtz, D. O.; Jenkins, A.; Na, H.; Zhang, L.; Wagner, D. S.; Katsaros, D. et al. Tumor- Infiltrating Dendritic Cell Precursors Recruited by a

Beta-Defensin Contribute to Vasculogenesis Under the Influence of Vegf-A. *Nat. Med.* **2004,** *10*, 950–958.

18. Conejo-Garcia, J. R.; Benencia, F.; Courreges, M. C.; Khang, E.; Zhang, L.; Mohamed-Hadley, A.; Vinocur, J. M.; Buckanovich, R. J.; Thompson, C. B.; Levine, B.; Coukos, G. Letal, A Tumor-Associated NKG2D Immunoreceptor Ligand, Induces Activation and Expansion of Effector Immune Cells. *Cancer Biol. Ther.* **2003,** *2*, 446–451.
19. Cory, S.; Huang, D. C.; Adams, J. M. The Bcl-2 Family: Roles in Cell Survival and Oncogenesis. *Oncogene* **2003,** *22*, 8590–8607.
20. Coughlin, C. M.; Salhany, K. E.; Gee, M. S.; LaTemple, D. C.; Kotenko, S.; Ma, X.; Gri, G.; Wysocka, M.; Kim, J. E.; Liu, L.; Liao, F.; Farber, J. M. et al. Tumor Cell Responses to IFN Gamma Affect Tumorigenicity and Response to IL-12 Therapy Andantiangiogenesis. *Immunity* **1998,** *9*, 25–34.
21. Cretney, E.; Takeda, K.; Yagita, H.; Glaccum, M.; Peschon, J. J.; Smyth, M. J.Increased Susceptibility to Tumor Initiation and Metastasis in TNF-Related Apoptosis-Inducing Ligand- Deficient Mice. *J. Immunol.* **2002,** *168*, 1356–1361.
22. Crowe, N. Y.; Smyth, M. J.; Godfrey, D. I. A Critical Role for Natural Killer T Cells in Immunosurveillance of Methyl Cholanthrene-Induced Sarcomas. *J. Exp. Med.* **2002,** 196, 119–127.
23. Cui, J.; Shin, T.; Kawano, T.; Sato, H.; Kondo, E.; Toura, I.; Kaneko, Y.; Koseki, H.; Kanno, M.; Taniguchi, M. Requirement for Valpha14 NKT Cells inIL-12-Mediated Rejection of Tumors. *Science* **1997,** *278*, 1623–1626.
24. Groh, V.; Bahram, S.; Bauer, S.; Herman, A.; Beauchamp, M.; Spies, T. Cell Stress-Regulated Human Major Histocompatibility Complex Class I Gene Expressed in Gastrointestinal Epithelium. *Proc. Natl. Acad. Sci. USA* **1996,** *93*, 12445–12450. Spontaneous Regression of Advanced Cancer: Identification of a Unique Genetically Determined, Age-Dependent Trait in Mice. *Proc. Natl. Acad. Sci. USA 100*, 6682–6687.
25. Curcio, C.; Di Carlo, E.; Clynes, R.; Smyth, M. J.; Boggio, K.; Quaglino, E.; Spadaro, M.; Colombo, M. P.; Amici, A.; Lollini, P. L.; Musiani, P.; Forni, G. Nonredundant roles of Antibody, Cytokines, and Perforin in the Eradication of Established Her-2/neu Carcinomas. *J. Clin. Invest.* **2003,** *111*, 1161–1170.
26. Curiel, T. J.; Coukos, G.; Zou, L.; Alvarez, X.; Cheng, P.; Mottram, P.; Evdemon-Hogan, M.; Conejo-Garcia, J. R.; Zhang, L.; Burow, M.; Zhu, Y.; Wei, S. et al. Specific Recruitment of Regulatory T Cells in Ovarian Carcinoma Fosters Immune Privilege and Predicts Reduced Survival. *Nat. Med.* **2004,** *10*, 942–949.
27. (a) Danial, N. N.; Korsmeyer, S. J. Cell Death: Critical Control Points. *Cell* **2004,** *116*, 205–219. (b) Darnell, R. B. Onconeural Antigens and the Paraneoplastic Neurologic Disorders: At the Intersection of Cancer, Immunity, and the Brain. *Proc. Natl. Acad. Sci. USA* **1996,** *93*, 4529–4536.
28. Darnell, R. B.; DeAngelis, L. M. Regression of Small-Cell Lung Carcinoma Inpatients with Paraneoplastic Neuronal Antibodies. *Lancet* **1993,** *341*, 21–22.
29. (a) Darnell, R. B.; Posner, J. B. Observing the Invisible: Successful Tumor Immunity in Humans. *Nat. Immunol.* **2003,** *4*, 201. (b) Das, H.; Groh, V.; Kuijl, C.; Sugita, M.; Morita, C. T.; Spies, T.; Bukowski, J. F. MICA Engagement by Human Vgamma2Vdelta2 T Cells Enhances Their Antigen-Dependent Effector Function. *Immunity* **2001,** *15*, 83–93.
30. Dave, S. S.; Wright, G.; Tan, B.; Rosenwald, A.; Gascoyne, R. D.; Chan, W. C.; Fisher, R. I.; Braziel, R. M.; Rimsza, L. M.; Grogan, T. M.; Miller, T. P.; LeBlanc, M. et al.

Prediction of Survival in Follicular Lymphoma Based on Molecular Features of Tumor-Infiltrating Immune Cells. *N. Engl. J. Med.* **2004,** *351*, 2159–2169.
31. Davidson, W. F.; Giese, T.; Fredrickson, T. N. Spontaneous Development of Plasmacytoid Tumors in Mice with Defective Fas-Fas Ligand Interactions. *J. Exp. Med.* **1998,** 187, 1825–1838.
32. Dhodapkar, M. V.; Geller, M. D.; Chang, D. H.; Shimizu, K.; Fujii, S.; Dhodapkar, K. M.; Krasovsky, J. A Reversible Defect in Natural Killer T Cell Function Characterizes the Progression of Premalignant to Malignant Multiple Myeloma. *J. Exp. Med.* **2003,** *197*, 1667–1676.
33. Di Carlo, E.; Forni, G.; Lollini, P.; Colombo, M. P.; Modesti, A.; Musiani, P. The Intriguing Role of Polymorphonuclear Neutrophils in Antitumor Reactions. *Blood* **2001,** *97*, 339–345.
34. Affabris, E.; Romeo, G.; Federico, M.; Coccia, E.; Locardi, C.; Belardelli, F.; Rossi, G. B. Molecular Mechanisms of Action of Interferons in the Friend Virus-Induced Leukemia Cell System. *Haematologica* **1987,** *72*, 76–78.
35. Akazawa, T.; Masuda, H.; Saeki, Y.; Matsumoto, M.; Takeda, K.; Tsujimura, K.; Kuzushima, K.; Takahashi, T.; Azuma, I.; Akira, S.; Toyoshima, K.; Seya, T. Adjuvant-Mediated Tumor Regression and Tumor-Specific Cytotoxic Response are Impaired in MyD88-Deficient Mice. *Cancer Res.* **2004,** *64*, 757–764.
36. Albert, M. L.; Austin, L. M.; Darnell, R. B. Detection and Treatment of Activated T Cells in the Cerebrospinal Fluid of Patients with Paraneoplastic Cerebellar Degeneration. *Ann. Neurol.* **2000,** *47*, 9–17.
37. Albert, M. L.; Darnell, J. C.; Bender, A.; Francisco, L. M.; Bhardwaj, N.; Darnell, R. B. Tumor-Specific Killer Cells in Paraneoplastic Cerebellar Degeneration. *Nat. Med. 4*, 1321–1324.
38. Albert, M. L.; Darnell, R. B. Paraneoplastic Neurological Degenerations: Keys to Tumour Immunity. *Nat. Rev. Cancer 4*, 36–44. Algarra, I.; Cabrera, T.; Garrido, F.The HLA Crossroad in Tumor Immunology. *Hum. Immunol.* **2000,** *61*, 65–73.
39. Asahara, T.; Murohara, T.; Sullivan, A.; Silver, M.; van der Zee, R.; Li, T.; Witzenbichler, B.; Schatteman, G.; Isner, J. M. Isolation of Putative Progenitor Endothelial Cells for Angiogenesis. *Science* **1997,** *275*, 964–967.
40. Bach, E. A.; Aguet, M.; Schreiber, R. D. The IFN Gamma Receptor: A Paradigm for Cytokine Receptor Signaling. *Annu. Rev. Immunol.* **1997,** *15*, 563–591.
41. Bancroft, G. J.; Schreiber, R. D.; Unanue, E. R. Natural Immunity: AT-Cell- Independent Pathway of Macrophage Activation, Defined in the Scid Mouse. *Immunol. Rev.* **1991,** *124*, 5–24.
42. Bauer, S.; Groh, V.; Wu, J.; Steinle, A.; Phillips, J. H.; Lanier, L. L.; Spies, T. Activation of NK Cells and T Cells by NKG2D, a Receptor for Stress-Inducible MICA. *Science* **1999,** *285*, 727–729.
43. Baxevanis, C. N.; Voutsas, I. F.; Tsitsilonis, O. E.; Tsiatas, M. L.; Gritzapis, A. D.; andPapamichail, M. Compromised Anti-Tumor Responses in Tumor Necrosis Factor-Alpha Knockout Mice. *Eur. J. Immunol.* **2000,** *30*, 1957–1966.
44. Belardelli, F.; Ferrantini, M.; Proietti, E.; Kirkwood, J. M. Interferon-Alphain Tumor Immunity and Immunotherapy. *Cytokine Growth Factor Rev.* **2002,** *13*, 119–134.
45. Blasius, A.; Vermi, W.; Krug, A.; Facchetti, F.; Cella, M.; Colonna, M. A Cell-Surface Molecule Selectively Expressed on Murine Natural Interferon-Producing Cells That Blocks Secretion of Interferon-Alpha. *Blood* **2004,** *103*, 4201–4206.

46. Boon, T.; van der Bruggen, P. Human Tumor Antigens Recognized by T Lymphocytes. *J. Exp. Med.* **1996,** *183*, 725–729.
47. Borg, C.; Terme, M.; Taieb, J.; Menard, C.; Flament, C.; Robert, C.; Maruyama, K.; Wakasugi, H.; Angevin, E.; Thielemans, K.; Le Cesne, A.; Chung-Scott, V. et al. Novel Mode of Action of C-Kit Tyrosine Kinase Inhibitors Leading to NK Cell-Dependent Antitumor Effects. *J. Clin. Invest.* **2004,** *114*, 379–388.
48. (a) Boshoff, C.; Weiss, R. AIDS-*related malignancies. Nat. Rev. Cancer* 2,373–382. (b) Bromberg, J. F.; Horvath, C. M.; Wen, Z.; Schreiber, R. D.; Darnell, J. E., Jr. Transcriptionally Active Stat1 Is Required for the Anti Proliferative Effects of Both Interferon Alpha and Interferon Gamma. *Proc. Natl. Acad. Sci. USA* **1996,** *93*, 7673–7678.
49. Bronte, V.; Apolloni, E.; Cabrelle, A.; Ronca, R.; Serafini, P.; Zamboni, P.; Restifo, N. P.; Zanovello, P. Identification of a CD11b(þ)/Gr-1(þ)/CD31(þ) Myeloid Progenitor Capable of Activating or Suppressing CD8(þ) T Cells. *Blood* **2000,** *96*, 3838–3846.
50. Ehrlich, P. Uber den jetzigen Stand der Karzinomforschung. *Ned. Tijdschr. Genees.* **1909,** *5*, 273–290.
51. Elder, G. J.; Hersey, P.; Branley, P. Remission of Transplanted Melanoma—Clinical Course and Tumour Cell Characterisation. *Clin. Transplant.* **1997,** *11*, 565–568.
52. Engel, A. M.; Svane, I. M.; Mouritsen, S.; Rygaard, J.; Clausen, J.; Werdelin, O.Methylcholanthrene-Induced Sarcomas in Nude Mice Have Short Induction Times and Relatively Low Levels of Surface MHC Class I Expression. *Apmis* **1996,** *104*, 629–639.
53. Enzler, T.; Gillessen, S.; Manis, J. P.; Ferguson, D.; Fleming, J.; Alt, F. W.; Mihm, M.; Dranoff, G. Deficiencies of GM-CSF and Interferon Gamma Link Inflammation and Cancer. *J. Exp. Med.* **2003,** *197*, 1213–1219.
54. Erianne, G. S.; Wajchman, J.; Yauch, R.; Tsiagbe, V. K.; Kim, B. S.; Ponzio, N. M.B Cell Lymphomas of C57L/J Mice; The Role of Natural Killer Cells and T Helper Cells in Lymphoma Development and Growth. *Leuk. Res.* **2000,** *24*, 705–718.
55. (a) Fridman, J. S.; Lowe, S. W. Control of apoptosis by p53. Oncogene **2003,** *22*, 9030–9040. (b) Gabrilovich, D. Mechanisms and Functional Significance of Tumour-Induced Dendritic-Cell Defects. *Nat. Rev. Immunol.* **2004,** *4*, 941–952.
56. Gabrilovich, D. I.; Chen, H. L.; Girgis, K. R.; Cunningham, H. T.; Meny, G. M.; Nadaf, S.; Kavanaugh, D.; Carbone, D. P. Production of Vascular Endothelial Growth Factor by Human Tumors Inhibits the Functional Maturation of Dendritic Cells. *Nat. Med.* **1996,** *2*, 1096–1103.
57. Gabrilovich, D. I.; Ishida, T.; Nadaf, S.; Ohm, J. E.; Carbone, D. P. Antibodies to Vascular Endothelial Growth Factor Enhance the Efficacy of Cancer Immunotherapy by Improving Endogenous Dendritic Cell Function. Clin. *Cancer Res.* **1999,** *5*, 2963–2970.
58. Gao, Y.; Yang, W.; Pan, M.; Scully, E.; Girardi, M.; Augenlicht, L. H.; Craft, J.; Yin, Z. Gamma Delta T Cells Provide an Early Source of Interferon Gamma in Tumor Immunity. *J. Exp. Med.* **2003,** *198*, 433–442.
59. Gasser, S.; Orsulic, S.; Brown, E. J.; Raulet, D. H. The DNA Damage Pathway Regulates Innate Immune System Ligands of the NKG2D Receptor. *Nature* **2005,** *436*, 1186–1190.
60. Gerosa, F.; Baldani-Guerra, B.; Nisii, C.; Marchesini, V.; Carra, G.; Trinchieri, G. Reciprocal Activating Interaction Between Natural Killer Cells and Dendritic Cells. *J.Exp. Med.* **2002,** *195*, 327–333.

61. Gershwin, M. E.; Ohsugi, Y.; Castles, J. J.; Ikeda, R. M.; Ruebner, B. Anti-mu Induces Lymphoma in Germfree Congenitally Athymic (Nude) But Not in Heterozygous (nu/þ) Mice. *J. Immunol.* **1983,** *131*, 2069–2073.
62. Gilfillan, S.; Ho, E. L.; Cella, M.; Yokoyama, W. M.; Colonna, M. NKG2D Recruits Two Distinct Adapters to Trigger NK Cell Activation and Costimulation. *Nat. Immunol.* **2002,** *3*, 1150–1155.
63. Girardi, M.; Glusac, E.; Filler, R. B.; Roberts, S. J.; Propperova, I.; Lewis, J.; Tigelaar, R. E.; Hayday, A. C. The Distinct Contributions of Murine T Cell Receptor (TCR) Gamma Delta þ and TCRalphabetaþ T Cells to Different Stages of Chemically Induced Skin Cancer. *J. Exp. Med.* **2003,** *198*, 747–755.
64. Girardi, M.; Oppenheim, D.; Glusac, E. J.; Filler, R.; Balmain, A.; Tigelaar, R. E.; andHayday, A. C. Characterizing the Protective Component of the Alphabeta T Cell Response to Transplantable Squamous Cell Carcinoma. *J. Invest. Dermatol.* **2004,** *122*, 699–706.
65. Girardi, M.; Oppenheim, D. E.; Steele, C. R.; Lewis, J. M.; Glusac, E.; Filler, R.; Hobby, P.; Sutton, B.; Tigelaar, R. E.; Hayday, A. C. Regulation of Cutaneous Malignancy by Gammadelta T Cells. *Science* **2001,** *294*, 605–609.
66. Godfrey, D. I.; Hammond, K. J.; Poulton, L. D.; Smyth, M. J.; Baxter, A. G. NKT Cells: Facts, Functions and Fallacies. *Immunol. Today* **2000,** *21*, 573–583.
67. Godfrey, D. I.; Pellicci, D. G.; Smyth, M. J. Immunology: The Elusive NKT Cell Antigen—Is the Search Over? *Science* **2004,** *306*, 1687–1689. Gorelik, L.; Flavell, R. A.Immune-Mediated Eradication of Tumors Through the Blockade of Transforming Growth Factor-Beta Signaling in T Cells. *Nat. Med.* **2001,** *7*, 1118–1122.
68. Gorelik, L.; Flavell, R. A. Transforming Growth Factor-Beta in T-Cell Biology. *Nat. Rev. Immunol.* *2*, 46–53. Grant, G. A.; Miller, J. F. Effect of Neonatal Thymectomy on the Induction of Sarcomata in C57 BL Mice. *Nature* **1965,** *205*, 1124–1125.
69. Graus, F.; Dalmou, J.; Rene, R.; Tora, M.; Malats, N.; Verschuuren, J. J.; Cardenal, F.; Vinolas, N.; Garcia del Muro, J.; Vadell, C.; Mason, W. P.; Rosell, R. et al. Anti-Hu Antibodies in Patients with Small-Cell Lung Cancer: Association with Complete Response to Therapy and Improved Survival. *J. Clin. Oncol.* **1997,** *15*, 2866–2872.
70. (a) Green, D. R.; Kroemer, G. The Pathophysiology of Mitochondrial Cell Death. *Science* **2004,** *305*, 626–629. (b) Gresser, I.; Belardelli, F. Endogenous Type I Interferons as a Defense Against Tumors. *Cytokine Growth Factor Rev.* **2002,** *13*, 111–118.
71. Gresser, I.; Belardelli, F.; Maury, C.; Maunoury, M. T.; Tovey, M. G. Injection of Mice with Antibody to Interferon Enhances the Growth of Transplantable Murine Tumors. *J. Exp. Med.* **1983,** *158*, 2095–2107.
72. Gresser, I.; Maury, C.; Vignaux, F.; Haller, O.; Belardelli, F.; Tovey, M. G. Antibody to Mouse Interferon Alpha/Beta Abrogates Resistance to the Multiplication of Friend Erythroleukemia Cells in the Livers of Allogeneic Mice. *J. Exp. Med.* **1988,** *168*, 1271–1291.
73. Cui, Z.; Willingham, M. C.; Hicks, A. M.; Alexander-Miller, M. A.; Howard, T. D.; Hawkins, G. A.; Miller, M. S.; Weir, H. M.; Du, W.; DeLong, C. J. Spontaneous Regression of Advanced Cancer: Identification of a Unique Genetically Determined, Age-Dependent Trait In Mice. *Proc. Natl. Acad. Sci. USA* **2003,** *100*, 6682–6687.
74. Groh, V.; Rhinehart, R.; Randolph-Habecker, J.; Topp, M. S.; Riddell, S. R.; Spies, T. Costimulation of CD8alphabeta T Cells by NKG2D via Engagement by MIC Induced on Virus- Infected Cells. *Nat. Immunol.* **2001,** *2*, 255–260.

75. Hayashi, T.; Faustman, D. L. Development of Spontaneous Uterine Tumors in Low Molecular Mass Polypeptide-2 Knockout Mice. *Cancer Res.* **2002,** *62*, 24–27.
76. Heidelberger, C. Chemical carcinogenesis. Annu. Rev. Biochem. **1975,** *44*, 79–121.
77. Hodge-Dufour, J.; Noble, P. W.; Horton, M. R.; Bao, C.; Wysoka, M.; Burdick, M. D.; Strieter, R. M.; Trinchieri, G.; Pure, E. Induction of IL-12 and Chemokines by Hyaluronan Requires Adhesion-Dependent Priming of Resident But Not Elicited Macrophages. *J. Immunol.* **1997,** *159*, 2492–2500.
78. Huang, A. Y.; Golumbek, P.; Ahmadzadeh, M.; Jaffee, E.; Pardoll, D.; Levitsky, H. Role of Bone Marrow-Derived Cells in Presenting MHC Class I-Restricted Tumor Antigens. *Science* **1994,** *264*, 961–965.
79. Humbert, P.; Russell, S.; Richardson, H. Dlg, Scribble and Lgl in Cell Polarity, Cell Proliferation and Cancer. *Bioessays* **2003,** *25*, 542–553.
80. Hunig, T.; Bevan, M. J. Specificity of Cytotoxic T Cells from Athymic Mice. *J. Exp. Med.* **1980,** *152*, 688–702.
81. Iinuma, T.; Homma, S.; Noda, T.; Kufe, D.; Ohno, T.; Toda, G. Prevention of Gastrointestinal Tumors Based on Adenomatous Polyposis Coli Gene Mutation by Dendritic Cell Vaccine. *J. Clin. Invest.* **2004,** *113*, 1307–1317.
82. Ikehara, S.; Pahwa, R. N.; Fernandes, G.; Hansen, C. T.; Good, R. A. Functional T Cells in Athymic Nude Mice. *Proc. Natl. Acad. Sci. USA* **1984,** *81*, 886–888.
83. Ishigami, S.; Natsugoe, S.; Tokuda, K.; Nakajo, A.; Che, X.; Iwashige, H.; Aridome, K.; Hokita, S.; Aikou, T. Prognostic Value of Intratumoral Natural Killer Cells in Gastric Carcinoma. *Cancer* **2000,** *88*, 577–583.
84. Liyanage, U. K.; Moore, T. T.; Joo, H. G.; Tanaka, Y.; Herrmann, V.; Doherty, G.; Drebin, J. A.; Strasberg, S. M.; Eberlein, T. J.; Goedegebuure, P. S.; Linehan, D. C.Prevalence of Regulatory T Cells Is Increased in Peripheral Blood and Tumor Microenvironment of Patients with Pancreas or breast adenocarcinoma. *J. Immunol.* **2002,** *169*, 2756–2761.
85. Locksley, R. M.; Killeen, N.; Lenardo, M. J. The TNF and TNF Receptor Super Families: Integrating Mammalian Biology. *Cell* **2001,** *104*, 487–501. Loeb, L. A. Mutator Phenotype May Be Required for Multistage Carcinogenesis. *Cancer Res.* **1991,** *51*, 3075–3079.
86. Loeb, L. A.; Loeb, K. R.; Anderson, J. P. Multiple Mutations and Cancer. *Proc. Natl. Acad. Sci. USA* **2003,** *100*, 776–781.
87. Loeffler, C. M.; Smyth, M. J.; Longo, D. L.; Kopp, W. C.; Harvey, L. K.; Tribble, H. R.; Tase, J. E.; Urba, W. J.; Leonard, A. S.; Young, H. A. et al. Immunoregulation in Cancer-Bearing Hosts. Down-Regulation of Gene Expression and Cytotoxic Function inCD8þ T Cells. *J. Immunol.* **1992,** *149*, 949–956.
88. (a) Lowe, S. W.; Cepero, E.; Evan, G. Intrinsic Tumour Suppression. *Nature* **2004,** *432*, 307–315. (b) Ludwig, A. T.; Moore, J. M.; Luo, Y.; Chen, X.; Saltsgaver, N. A.; O'Donnell, M. A.; Griffith, T. S. Tumor Necrosis Factor-Related Apoptosis-Inducing Ligand: A Novel Mechanism for Bacillus Calmette-Guerin-Induced Antitumor Activity. *Cancer Res.* **2004,** *64*, 3386–3390.
89. Luster, A. D.; Leder, P. IP-10, a -C-X-C- Chemokine, Elicits a Potent Thymus-Dependent Antitumor Response In Vivo. *J. Exp. Med.* **1993,** *178*, 1057–1065.
90. Lyden, D.; Hattori, K.; Dias, S.; Costa, C.; Blaikie, P.; Butros, L.; Chadburn, A.; Heissig, B.; Marks, W.; Witte, L.; Wu, Y.; Hicklin, D. et al. Impaired Recruitment of Bone-Marrow-Derived Endothelial and Hematopoietic Precursor Cells Blocks Tumor Angiogenesis and Growth. *Nat. Med.* **2001,** *7*, 1194–1201.

91. MacKie, R. M.; Reid, R.; Junor, B. Fatal Melanoma Transferred in a Donated Kidney 16 Years After Melanoma Surgery. *N. Engl. J. Med.* **2003,** *348*, 567–568.
92. Macleod, K. Tumor Suppressor Genes. *Curr. Opin. Genet. Dev.* **2000,** *10*, 81–93.
93. MacMicking, J.; Xie, Q. W.; Nathan, C. Nitric Oxide and Macrophage Function. *Annu. Rev. Immunol.* **1997,** *15*, 323–350.
94. Maleckar, J. R.; Sherman, L. A. The Composition of the T Cell Receptor Repertoire in Nude Mice. *J. Immunol.* **1987,** *138*, 3873–3876.
95. Marincola, F. M.; Jaffee, E. M.; Hicklin, D. J.; Ferrone, S. Escape of Human Solid Tumors from T-Cell Recognition: Molecular Mechanisms and Functional Significance. *Adv. Immunol.* **2000,** *74*, 181–273.
96. Mattes, J.; Hulett, M.; Xie, W.; Hogan, S.; Rothenberg, M. E.; Foster, P.; Parish, C.Immunotherapy of Cytotoxic T Cell-Resistant Tumors by T Helper 2 Cells: Aneotaxin and STAT6- Dependent Process. *J. Exp. Med.* **2003,** *197*, 387–393.
97. Matzinger, P. Tolerance, Danger, and the Extended Family. *Annu. Rev. Immunol.* **1994,** *12*, 991–1045. Medawar, P. The Behaviour and Fate of Skin Autografts and Skin Homografts in Rabbits. *J. Anat.* **1944,** *78*, 176–199.
98. Melani, C.; Chiodoni, C.; Forni, G.; Colombo, M. P. Myeloid Cell Expansion Elicited by the Progression of Spontaneous Mammary Carcinomas in c-erbB-2 Transgenic BALB/c Mice Suppresses Immune Reactivity. *Blood* **2003,** *102*, 2138–2145.
99. Mihara, M.; Erster, S.; Zaika, A.; Petrenko, O.; Chittenden, T.; Pancoska, P.; Moll, U. M. p53 Has a Direct Apoptogenic Role at the Mitochondria. *Mol. Cell* **2003,** *11*, 577–590.
100. Mihm, M. C., Jr.; Clemente, C. G.; Cascinelli, N. Tumor Infiltrating Lymphocytes in Lymph Node Melanoma Metastases: A Histopathologic Prognostic Indicator and an Expression of Local Immune Response. *Lab. Invest.* **1996,** *74*, 43–47.
101. Pardoll, D. Does the Immune System See Tumors as Foreign or Self? *Annu. Rev. Immunol.* **2003,** *21*, 807–839.
102. Pende, D.; Rivera, P.; Marcenaro, S.; Chang, C. C.; Biassoni, R.; Conte, R.; Kubin, M.; Cosman, D.; Ferrone, S.; Moretta, L.; Moretta, A. Major Histocompatibility Complex Class I-Related Chain A and UL16-Binding Protein Expression on Tumor Cell Lines of Different Histotypes: Analysis of Tumor Susceptibility to NKG2D-Dependent Natural Killer Cell Cytotoxicity. *Cancer Res.* **2002,** *62*, 6178–6186.
103. Peng, S. L.; Robert, M. E.; Hayday, A. C.; Craft, J. Atumor-Suppressor Function for Fas (CD95) Revealed in T Cell-Deficient Mice. *J. Exp. Med.* **1996,** *184*, 1149–1154.
104. Penn, I. Malignant Melanoma in Organ Allograft Recipients. *Transplantation* **1996,** *61*, 274–278.
105. Penn, I.; Starzl, T. E. Malignant Lymphomas in Transplantation Patients: A Review of the World Experience. *Int. Z. Klin. Pharmakol. Ther. Toxikol.* **1970,** *3*, 49–54.
106. Pestka, S.; Krause, C. D.; Walter, M. R. Interferons, Interferon-Like Cytokines, and Their Receptors. *Immunol. Rev*. **2004,** *202*, 8–32.
107. Peter, M. E.; Krammer, P. H. The CD95(APO-1/Fas) DISC and Beyond. *Cell Death Differ.* **2003,** *10*, 26–35.
108. Qin, Z.; Blankenstein, T. CD4þ T Cell-Mediated Tumor Rejection Involves Inhibition of Angiogenesis That Is Dependent on IFN Gamma Receptor Expression by Non Hematopoietic Cells. *Immunity* **2000,** *12*, 677–686.
109. Radosavljevic, M.; Cuillerier, B.; Wilson, M. J.; Clement, O.; Wicker, S.; Gilfillan, S.; Beck, S.; Trowsdale, J.; Bahram, S. A Cluster of Ten Novel MHC Class I Related Genes on Human Chromosome 6q24.2–q25. *Genomics* **2002,** *79*, 114–123.

110. Rafii, S.; Lyden, D.; Benezra, R.; Hattori, K.; Heissig, B. Vascular and Haematopoietic Stem Cells: Novel Targets for Anti-Angiogenesis Therapy? *Nat. Rev. Cancer* **2002,** *2*, 826–835.
111. (a) Raulet, D. H. Roles of the NKG2D Immuno Receptor and Its Ligands. *Nat. Rev. Immunol.* **2003,** *3*, 781–790. Rosenberg, S. A. A New Era for Cancer Immunotherapy Based on the Genes That Encode Cancer Antigens. *Immunity* **2003,** *10*, 281–287.
112. Rubinstein, N.; Alvarez, M.; Zwirner, N. W.; Toscano, M. A.; Ilarregui, J. M.; Bravo, A.; Mordoh, J.; Fainboim, L.; Podhajcer, O. L.; Rabinovich, G. A. Targeted Inhibition of Galectin-1 Gene Expression in Tumor Cells Results in Heightened T Cell-Mediated Rejection; A Potential Mechanism of Tumor-Immune Privilege. *Cancer Cell* **2004,** *5*, 241–251.
113. Ruggeri, L.; Capanni, M.; Urbani, E.; Perruccio, K.; Shlomchik, W. D.; Tosti, A.; Posati, S.; Rogaia, D.; Frassoni, F.; Aversa, F.; Martelli, M. F.; Velardi, A. Effectiveness of Donor Natural Killer Cell Alloreactivity in Mismatched Hematopoietic Transplants. *Science* **2002,** *295*, 2097–2100.
114. (a) Rygaard, J.; Povlsen, C. O. Is Immunological Surveillance Not a Cell-Mediated Immune Function? *Transplantation* **1974a,** *17*, 135–136. (b) Rygaard, J.; Povlsen, C. O. The Mouse Mutant Nude Does Not Develop Spontaneous Tumours: An Argument Against Immunological Surveillance. *Acta Pathol. Microbiol. Scand. Microbiol. Immunol.* **1974b,** *82*, 99–106.
115. Sahin, U.; Tureci, O.; Schmitt, H.; Cochlovius, B.; Johannes, T.; Schmits, R.; Stenner, F.; Luo, G.; Schobert, I.; Pfreundschuh, M. Human Neoplasms Elicit Multiple Specific Immune Responses in the Autologous Host. *Proc. Natl. Acad. Sci. USA* **1995,** *92*, 11810–11813.
116. Sakaguchi, S. Regulatory T Cells: Key Controllers of Immunologic Self-Tolerance. *Cell* **2000,** *101*, 455–458.
117. Sakaguchi, S. Immunologic Tolerance Maintained by Regulatory T Cells: Implications for Autoimmunity, Tumor Immunity and Transplantation Tolerance. *Vox Sang.* **2002,** *83* (Suppl. 1), 151–153.
118. Sakaguchi, S. Naturally Arising CD4þ Regulatory T Cells for Immunologic Self-Tolerance and Negative Control of Immune Responses. *Annu. Rev. Immunol.* **2004,** *22*, 531–562.
119. Takeda, K.; Hayakawa, Y.; Smyth, M. J.; Kayagaki, N.; Yamaguchi, N.; Kakuta, S.; Iwakura, Y.; Yagita, H.; Okumura, K. Involvement of Tumor Necrosis Factor-Related Apoptosis- Inducing Ligand in Surveillance of Tumor Metastasis by Liver Natural Killer Cells. *Nat. Med.* **2001,** *7*, 94–100.
120. Takeda, K.; Smyth, M. J.; Cretney, E.; Hayakawa, Y.; Kayagaki, N.; Yagita, H.; Okumura, K. Critical Role for Tumor Necrosis Factor-Related Apoptosis-Inducing Ligand in Immune Surveillance Against Tumor Development. *J. Exp.Med.* **2002,** *195*, 161–169.
121. Tawara, I.; Take, Y.; Uenaka, A.; Noguchi, Y.; Nakayama, E. Sequential Involvement of Two Distinct CD4þ Regulatory T Cells During the Course of Transplantable Tumor Growth and Protection from 3-Methylcholanthrene-Induced Tumorigenesis by CD25-Depletion. *Jpn. J. Cancer Res.* **2002,** *93*, 911–916.
122. Terabe, M.; Matsui, S.; Noben-Trauth, N.; Chen, H.; Watson, C.; Donaldson, D. D.; Carbone, D. P.; Paul, W. E.; Berzofsky, J. A. NKT cell-Mediated Repression of Tumor

Immunosurveillance by IL-13 and the IL-4R-STAT6 Pathway. *Nat. Immunol.* **2000,** *1,* 515–520.

123. Terabe, M.; Matsui, S.; Park, J. M.; Mamura, M.; Noben-Trauth, N.; Donaldson, D. D.; Chen, W.; Wahl, S. M.; Ledbetter, S.; Pratt, B.; Letterio, J. J.; Paul, W. E. et al. Transforming Growth Factor-Beta Production and Myeloid Cells Are an Effect or Mechanism Through Which CD1d-Restricted T Cells Block Cytotoxic T Lymphocyte-Mediated Tumor Immunosurveillance: Abrogation Prevents Tumor Recurrence. *J. Exp. Med.* *198*, 1741–1752.
124. Uhr, J. W.; Scheuermann, R. H.; Street, N. E.; Vitetta, E. S. Cancer Dormancy: Opportunities for New Therapeutic Approaches. *Nat. Med.* **1997,** *3*, 505–509.
125. Uyttenhove, C.; Pilotte, L.; Theate, I.; Stroobant, V.; Colau, D.; Parmentier, N.; Boon, T.; Van den Eynde, B. J. Evidence for a Tumoral Immune Resistance Mechanism Based on Tryptophan Degradation by Indoleamine 2,3-Dioxygenase. *Nat. Med.* **2003,** *9*, 1269–1274.
126. van den Broek, M. E.; Kagi, D.; Ossendorp, F.; Toes, R.; Vamvakas, S.; Lutz, W. K.; Melief, C. J.; Zinkernagel, R. M.; Hengartner, H. Decreased Tumor Surveillance in Perforin- Deficient Mice. *J. Exp. Med.* **1996,** *184*, 1781–1790.
127. van den Broek, M. F.; Kagi, D.; Zinkernagel, R. M.; Hengartner, H. Perforin Dependence of Natural Killer Cell-Mediated Tumor Control In Vivo. *Eur. J. Immunol.* **1995,** *25*, 3514–3516.
128. van der Bruggen, P.; Traversari, C.; Chomez, P.; Lurquin, C.; De Plaen, E.; Vanden Eynde, B.; Knuth, A.; Boon, T. A Gene Encoding an Antigen Recognized by Cytolytic T Lymphocytes on a Human Melanoma. *Science* **1991,** *254*, 1643–1647.
129. Vetter, C. S.; Groh, V.; Thor Straten, P.; Spies, T.; Brocker, E. B.; Becker, J. C. Expression of Stress-Induced MHC Class I Related Chain Molecules on Human Melanoma. *J. Invest. Dermatol.* **2002,** *118*, 600–605.
130. Vicari, A. P.; Caux, C. Chemokines in Cancer. *Cytokine Growth Factor Rev.* **2002,** *13*, 143–154.
131. Villegas, F. R.; Coca, S.; Villarrubia, V. G.; Jimenez, R.; Chillon, M. J.; Jareno, J.; Zuil, M.; Callol, L. Prognostic Significance of Tumor Infiltrating Natural Killer cells Subset CD57 in Patients with Squamous Cell Lung Cancer. *Lung Cancer* **2002,** *35*, 23–28.
132. Wang, R. F.; Rosenberg, S. A. Human Tumor Antigens for Cancer Vaccine Development. *Immunol. Rev.* *170*, 85–100.
133. Wang, T.; Niu, G.; Kortylewski, M.; Burdelya, L.; Shain, K.; Zhang, S.; Bhattacharya, R.; Gabrilovich, D.; Heller, R.; Coppola, D.; Dalton, W.; Jove, R. et al. Regulation of the Innate and Adaptive Immune Responses by Stat-3 Signaling in Tumor Cells. *Nat. Med.* **2004,** *10*, 48–54.

CHAPTER 8

The Immune Stimulating Effects of Conventional Chemotherapy

ABSTRACT

This chapter examines how gemcitabine alters the immune systems of experimental animals. Researchers have examined how gemcitabine affects the immune system in patients with pancreatic cancer. Chemotherapy for cancer not only kills off the tumor cells but also increases the immune system's contact with tumor antigens. It is possible that the process of apoptosis itself acts as a signal that revs up the immune system. One way in which chemotherapy affects the immune system is by eradicating or rendering nonfunctional immune regulatory T-cells (Tregs) and triggering the body's homeostatic T-cell growth. It is possible that both may stimulate immune responses that will help eliminate tumors. The results imply that immunological therapy may be implemented at different stages. Traditional approaches to measuring the immunological effects of chemotherapy, such as monitoring changes in lymphocyte subset numbers and DTH reactions, have been shown to be insufficient. Developments in tetramer technology, fluorescence-activated cell sorting, and the identification of tumor-specific antigens have facilitated better human investigations (FACS). The great clinical success of checkpoint blockers shows that reactivating inactive immune surveillance has intriguing uses in cancer therapy. Despite mounting evidence that it is essential to the success of both conventional and targeted anticancer therapies, the (re)activation of tumor-targeting immune responses is often disregarded in clinical practice. Chemotherapy, by increasing the immunogenicity of malignant cells or by blocking the immunosuppressive circuitries created by developing tumors, may induce such responses. Understanding these immunological "side effects" of chemotherapy is essential for creating novel combinatorial regimens with increased clinical efficacy.

Advances in Immunological Treatment of Cancer: Cellular Mechanism and Regulation.
Rishabha Malviya, Arun Kumar Singh, and Deepika Yadav (Authors)

8.1 INTRODUCTION

The goal of standard cancer chemotherapy is to permanently eradicate the tumor or stop its cells from proliferating. Cytotoxic drugs work by halting the replication of tumor cells by interfering with DNA synthesis or by causing chemical damage to the DNA of tumor cells, killing them.[1–3] The host may also have a role in the curative benefits of chemotherapy, but this viewpoint ignores this possibility. The success rates of some chemotherapeutic treatments against tumors implanted in immunocompetent hosts are higher than those of immunodeficient hosts, and this trend is expected to continue.[4,5]

Responses from the immune system to conventional chemotherapy might fall into two categories. As a result of being exposed to different treatments, tumor cells undergo morphological changes that make them easier for the immune system to recognize as antigens and destroy.[6–8] Medications may temporarily deplete lymph nodes (LNs), while others can activate or increase immune effector cells or processes. Targeted chemotherapeutic medications, which are chemicals aimed at attacking biochemical pathways crucial to tumor cell survival and/or growth, are also credited with immunomodulatory effects.[9,10] These results raise the intriguing possibility that immunomodulatory chemotherapeutic drugs might be successfully integrated with immune-based treatment methods.[11] Because they inhibit metabolic pathways or mutant proteins vital to cancer cell growth and survival,[12] targeted medicines have proven helpful in combating malignant tumors. Drug-resistant mutations limit the therapeutic usefulness of these therapies over the long term,[13] despite the fact that many targeted medicines considerably decrease tumor growth. Several targeted medicines and standard chemotherapeutic drugs have immunomodulatory effects. TK inhibitors, such as sunitinib and imatinib, reduce the immunosuppressive effects of regulatory T-cells (Treg) and myeloid-derived suppressor cells (MDSCs) and increase the synthesis of indoleamine 2,3-dioxygenase (IDO) in myeloid cells. The angiogenic activity is decreased by mTOR kinase inhibitors (e.g., everolimus and temsirolimus) because they increase autophagy, which in turn inhibits vascular cell proliferation and, by extension, growth factor synthesis. Janus kinase inhibitors are a type of small-molecule cytokine signaling inhibitors that may play a significant role in facilitating dendritic cell (DC) maturation and priming of antigen-specific T lymphocytes.[14] Bortezomib, an approved treatment for multiple myeloma, acts by making tumor cells more vulnerable to apoptosis caused by tumor necrosis factor (TNF)-related apoptosis-inducing ligand (TRAIL) and natural killer (NK) cells. Recently licensed for therapy against certain melanomas, the particular BRAF inhibitor vemurafenib causes an increase in the expression of several tumor antigens, facilitating their immunological detection by T-cells.[15,16] Trastuzumab and cetuximab are two monoclonal antibodies that target the tumor-associated receptors TK HER-2 and epidermal growth factor receptor (EGFR), respectively, and which enhance T-cell-mediated immune responses by enhancing antigen presentation through the creation of immune complexes. These findings open the door for more clinical research into the complementary effects of targeted therapies and immunotherapies.[17]

The purpose of this chapter is to describe the immune-based mechanisms employed by various cytotoxic medications in the quest for generating more effective, rationally driven therapies against cancer,[18] ultimately aiming to uncover situations for successful conjunction with immunotherapy techniques.

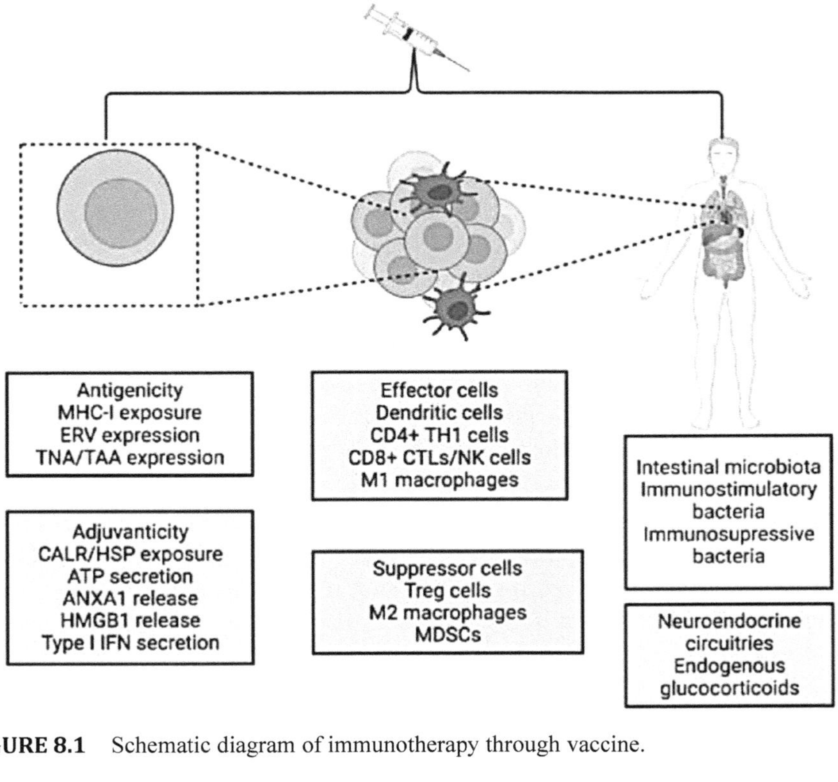

FIGURE 8.1 Schematic diagram of immunotherapy through vaccine.

8.2 EFFECTS ON THE INNATE IMMUNE SYSTEM

Cytotoxic medications have been found to have direct effects on macrophages, DCs, and NK cells.[19] However, the current literature on the effects of cytotoxic treatment on NK-cell function in cancer patients has shown contradictory results. Studies reveal that individuals with locally progressed or metastatic breast cancer had reduced NK-cell responses after undergoing cytotoxic therapy regimens. It has been shown in several rodent models that metronomic (i.e., low dosage for a long period) cyclophosphamide (CTX) therapy rapidly attracts DCs, macrophages, and NK cells to the tumor site, and greatly increases NK-dependent antitumor immunity in terminal cancer patients. Surprisingly, in a model of pancreatic cancer, adding interferon (IFN)- to 5-fluorouracil (5-FU) boosted both the number of invading NK cells and their lethality.[20]

Macrophage death as a result of chemotherapy has also been examined. There are two types of macrophages that may develop from blood monocytes: M1 macrophages have effector functions, whereas M2 macrophages have suppressive functions. Macrophages called tumor-associated macrophages (TAMs) are sometimes seen within solid tumors.[21] Similarly to M2 macrophages, TAMs inhibit NK and T-cell function, which is a protumorigenic action. There are two possible approaches to doing this (through cytokine production). There is a correlation between TAMs and an increased risk of tumor progression in cancer patients (TAMs). Numerous studies have been undertaken to determine whether or not cytotoxic treatment may reduce macrophages' tumor-promoting abilities.[22] Several lines of evidence show that low-dose CTX may convert M2 macrophages to M1 in vivo, leading to increased production of oxygen radicals, interleukin 6 (IL-6), and IL-12 and a fortification of innate responses. Treatment of mice with B16.F10 melanoma with vincristine, CTX, and doxorubicin simultaneously dramatically increased a TAM fraction that may be M1-polarized.[23] Coculturing human primary ovarian tumor cells with macrophages increased their sensitivity to cisplatin and carboplatin but decreased their sensitivity to 5- FU. Similar to docetaxel, paclitaxel has the ability to induce tumor regression by activating DCs, NK cells, and tumor-specific CTL via upregulation of IL-12, TNF-, and inducible nitric oxide synthase (iNOS) production. In both mouse mammary carcinoma and human breast cancer chemotherapy patients, paclitaxel suppressed the TAM response.[24]

Chemotherapy's potential to affect myeloid cell mobilization means it may interfere with hematopoiesis in the bone marrow (BM). These results suggest that a single, low-dose injection of CTX may sustain DC precursors in the BM, leading to greater numbers of DCs and improved DC differentiation in the periphery.[25] The most notable impact of CTX was a reduction in lymphoid organ CD8$^+$ DCs, followed by an excessive recovery of these cells after drug withdrawal.[26,27] Our results demonstrated that CD8$^+$ DCs moved to the tumor site after CTX injection and cross-presented tumor-associated antigen.[28] Platinum derivative cisplatin increased immune-based effector responses in melanoma-bearing mice by increasing DCs and decreasing MDSCs. Patients with advanced pancreatic cancer who were given gemcitabine also showed a higher frequency of CD14+ monocytes, CD11c$^+$ myeloid DCs, and CD123$^+$ plasmacytoid DCs than those who had not been given the drug.

It's also possible that cytotoxic drugs might directly stimulate DC immune function. Using a functional objective screen,[29] the impact of 54 different chemotherapeutic medications on DCs and their proliferation and flourishing was assessed. Methylating medicines (mechlorethamine, diaziquone) and topoisomerase inhibitors (etoposide, mitoxantrone, and doxorubicin) have been shown to promote DC maturation at doses that generate minimal DC mortality.[30] Studies in vitro have shown that DCs generated from murine BM that express more costimulatory molecules, IL-12p70, and antigen-processing machinery gene components are more effective in delivering antigens to T-cells.[31] Human DCs made from monocytes performed similarly. Vinblastine promotes DCs' phenotypic and functional maturation in vitro and in vivo by stimulating humoral and cellular immunological responses and inducing in situ maturation of skin-resident DCs at modest dosages (0.1–1 M).[32] The proportion of 6-sulfo LacNAc$^+$ (slan) DCs in human blood is rather high. The capacity of slan DCs to generate proinflammatory cytokines and activate T lymphocytes and NK cells has been shown to be preserved by methotrexate and paclitaxel, but suppressed by doxorubicin, according to recent research.[33]

Indirect mechanisms may regulate the impact of cytotoxic therapy on DC function. Innovating research[34,35] has revealed that heat shock proteins (HSPs) promote the engulfment of cell debris by human DCs, which in turn increases the cross-presentation of tumor antigens to T-cells. 5-FU and doxorubicin (2-AD) trigger the production of HSPs in response to treatment (dox). Injecting syngeneic mice with doxorubicin-killed tumor cells triggered tumor-specific immune responses by increasing DC phagocytosis and

CD8 T-cell responses, according to recent research. This chapter provides a concise overview of research performed by L. Zitvogel's team, which establishes the molecular route linking the immunogenic apoptosis induced by chemotherapy with DC activation.[36]

8.3 EFFECTS ON THE ADAPTIVE IMMUNE SYSTEM

Radiation and chemotherapy both reduce lymphocytic populations, including B lymphocytes. In patients with breast cancer, treatment with 5-FU, doxorubicin (5-FU), or both, has been shown to increase the number of cytotoxic T and NK cells in the blood while drastically decreasing the proportion of B-cells.[37] Gemcitabine improved T-cell responses, reduced B-cell counts, and significantly suppressed antigen-specific IgG antibody responses in a mouse model of malignant mesothelioma. It is not yet known, however, whether this process holds true in vivo or in the case of other chemotherapeutic medicines. Perhaps this disparate effect on the two lymphocyte subsets can be explained by the fact that B-cells are more susceptible to gemcitabine's antiproliferative activities in vitro than T-cells. A single dosage of gemcitabine (120 mg/kg) retained T- and B-cells in the spleens of mice with big tumors, and 5- FU-based adjuvant treatment elicited significant tumor-specific antibody responses in patients with colon cancer.[38]

There are many different takes on what constitutes a good CTX joke. Although CTX may reduce humoral responses at low levels while enhancing cellular responses, studies have demonstrated that B-cells are more vulnerable to the toxin's cytotoxic effects.[39] Finally, it has been shown that low-dose CTX increases the relative percentages of B- and T-cells in mice with SW1C melanoma,[40] and that it enhances cellular and antibody responses in patients with advanced disease. CTX has been shown in many trials to have short-lived lymphodepleting effects, with the lymphocytic pool quickly recovering to pretreatment levels after drug cessation.[41,42] These results show that humoral and cellular responses may be inhibited following pharmaceutical administration, although in distinct ways and at distinct times. During a drug-induced "cytokine storm," several hematological and homeostatic variables, danger signals, pattern recognition receptors, inflammatory mediators, and growth factors are upregulated, leading to increased peripheral B- and T-cell expansion.[43] Myelo- and lymphotoxicity caused by drugs may be detected by these mediators working together to trigger a DNA damage response and further enhance immunological activation. Surprisingly, genotoxic stress induces immunological responses remarkably comparable to those generated by viral infections by activating the production of IFN- and IFN- genes.[44] In addition, combining CTX and fludarabine with adoptive transplantation of tumor-specific B lymphocytes in a mouse model of melanoma metastasis greatly improved outcomes.[45] The humoral response to an in vivo cancer vaccine was extended when patients were treated with high doses of CTX and doxorubicin at the same time.[46]

Chemotherapy may enhance T-cell-mediated immunological responses, as has been established in a number of studies. Ovarian cancer cell-immunized mice treated with doxorubicin or cisplatin had improved antitumor immunity and a longer lifetime.[47] Used in combination with low doses of cisplatin and paclitaxel, it is very effective against platinum-resistant ovarian tumors in mice and people.[48] The production of IL-2 and IFN- by the combo is what makes it so effective in eliciting strong tumor-specific CD8 T-cell responses. T-cell-dependent anticancer responses are improved by 5-FU because of its ability to eliminate MDSCs in vivo and stimulate IFN-production by tumor-specific CD8 T-cells in the tumor microenvironment.[49] Treatment with 5-FU and cisplatin as neoadjuvant treatment in mice models of patients with esophageal squamous cell carcinoma increased the trafficking of CD4 and CD8 T-cells into tumors.[50,51] In experimental carcinogen-induced adenocarcinomas and fibrosarcomas, doxorubicin treatment increased tumor-specific proliferation of CD8 T-cells in tumor-draining LNs and facilitated tumor infiltration of activated, IFN—producing CD8 T-cells.[52] Doxorubicin's therapeutic activity in this scenario relied on the presence of T-cells and IL-1 and IL-17.[53]

Tumor-bearing mice with an enhanced anticancer response to a single dose of CTX after adoptive transfer of tumor-reactive T-cells.[54] Drug-induced bystander proliferation of memory CD4 and CD8 T-cells,[55] drug-induced stimulation of CD8 T-cells proliferation and IFN-production,[41] and antigen-specific CD4 T-cell differentiation into activated polyfunctional T helper cells with potent antitumor activity[56] are all proposed mechanisms to explain this effect. Therefore, the therapeutic effectiveness of combined CTX and adoptive transfer of tumor-specific spleen cells was shown to be highly dependent on donor CD4 T-cells.[57] CD8-mediated anticancer effects were enhanced when anti-4-1BB was coupled with CTX in a mouse model of B16 melanoma.[58]

Cancer vaccines were used in a small pilot clinical study, and the results confirmed that chemotherapy boosts antitumor immune responses. Gene expression analysis of PBMCs taken the day after chemotherapy showed that melanoma patients who received both dacarbazine and a peptide-based immunization had higher levels of tumor antigen-specific CD8 T-cell responses than those who received the vaccine alone.[59] Because of these adjustments, the antigenic repertoire grew, and T-cells became more sensitive to tumor antigens.[60]

Disruption of the balance between the Th1 and Th2 arms of the immune system is a common side effect of cancer treatment, although various anticancer medications have been proven to be successful at restoring this balance. Paclitaxel boosts Th1 cellular immunity by raising the number of circulating IFN—secreting CD8 T-cells and IL-2–secreting CD4 T-cells; for example, in patients with advanced nonsmall cell lung cancer.[61] Additionally, CTX lowers Th2 cytokines (IL-4 and IL-10) and increases Th1 polarizing cytokines (IL-2 and IFN-) in tumor-bearing mice41 and rats.[62] In tumor-bearing mice treated with CTX,[63] the frequency of splenic T helper 1 (Th1) and T helper 17 (Th17) cells increased, and these cell types recovered from drug-induced lymphodepletion more quickly than Treg cells. There is substantial evidence that the Th17 T-cell subset plays a major role in the pathophysiology of autoimmune and inflammatory diseases. Despite this, their function in cancer development and therapy is still a matter of some controversy.[64,65] Recent studies have shown that chemotherapy medicines such as gemcitabine and 5- FU might reduce the effectiveness of anticancer therapies by increasing IL-1 synthesis, which in turn promotes IL-17 release by CD4 T-cells.[66] However, previous studies have shown that IL-17 generating T-cells are crucial for efficient anticancer responses after doxorubicin therapy, hence our findings are at odds with other studies. According to the same research, CD4 T-cells accounted for the vast majority of the IL-17 generated in response to T-cell receptor activation in metronomic CTX patients. CTX dramatically suppressed CD4 T-cell proliferation and/or differentiation when given to tumor-free and tumor-bearing rats. If CTX-induced Th17 helps CTX's anticancer effect is still an unsettled subject.[67]

8.4 CONSTITUTIVE CROSS-PRESENTATION OF TUMOR ANTIGENS

Tumor cells are notoriously bad at presenting antigens to T lymphocytes. However, cross-presentation, a mechanism similar to how the immune system recognizes antigens generated by virally infected cells, may allow the immune system to recognize cancer antigens. Cross-presentation requires an APC, such as a DC, to collect antigen from a target cell and convert it into tiny peptides so that the peptides may be offered to T-cells through the T-cells' own class I presentation pathway.[68] Therefore, the major histocompatibility complex class I route is used to introduce foreign antigens to the immune system. $CD11c^+$ $CD8^+$ DCs[69] are a subset of DCs that are mostly responsible for this. However, how and from which other cells DCs receive antigens remains unknown: (I) taking in antigens from living organisms, (II) taking in foreign antigens (phagocytosis), and (III) heat-shock protein-mediated antigen or peptide transfer, and intercellular transfer. All of these theories on antigen transfer through exosomes are now gaining traction. The results demonstrate that each of them plays an important part.[70] Proteins, rather than peptides, seem to be the source of transmitted antigens since signal sequence CD8+ T-cell epitopes are cleaved as peptides shortly after synthesis. The role of HSPs in phenotypic plasticity has been the subject of much debate.[71] Phagocytosis of apoptotic debris seems to play a key part in the cross-presentation of tumor antigens that occurs after chemotherapy, as shown by a considerable body of research. Tolerance and immune response activation are both affected by antigen cross-presentation. CD8+ DC may be either immunogenic or tolerogenic, depending on the circumstances. A subset of microbial compounds known as pathogen-associated molecular patterns (PAMPs) is known to promote an immunogenic or proinflammatory milieu. PAMPs include viral components like double-stranded RNA, bacterial components such as lipopolysaccharide (LPS) and flagellin, and unmethylated CpG DNA patterns. These structures are responsible for initiating robust inflammatory responses[72] by exchanging information with Toll-like receptors (TLRs).

Type I IFNs have a role in the inflammatory response caused by TLR-stimulation (IFN- and IFN-). Type I IFNs seem to play a critical role in eliciting CD8 T-cell responses in response to cross-presented antigens.[73] Since tumor antigens are often cross-presented, type I IFNs may play a critical role in reversing anticancer tolerance and triggering potent antitumor immune responses. The ability of DCs to recognize tumor antigens for cross-presentation and subsequently transmit these antigens to T-cells in an immunogenic environment is crucial for a successful immune response to tumors. Chemotherapy is likely to impact both processes due to the variability in toxicity of individual medications toward immune system cells and the variety of routes by which they induce cell death.[74]

8.5 CHEMOTHERAPY AUGMENTS ANTIGEN CROSS-PRESENTATION

It was shown that a larger fraction of tumor antigen is cross-presented in relation to the initial tumor size when apoptosis in tumor cells is induced by chemotherapy. Nowak et al.'s[75] study lend credence to this claim. In mice with subcutaneously developing tumors, the antimetabolite gemcitabine was used to lower tumor burden by inducing tumor cell death. To measure antigen cross-presentation, researchers used adoptive transfer of tumor-specific CD8 T-cells.[76] For the purpose of measuring T-cell proliferation in a fluorescence-activated cell sorter (FACS),[77] these cells were fluorescently labeled with 5,6-carboxyfluorescein succinimidyl (CFSE). To infer antigen presentation, the researchers examined cell proliferation. These studies showed that antigen cross-presentation in the LNs draining the tumor was significantly boosted when tumor cells were killed. In this way, chemotherapy has the potential to change the antigenic composition of cross-presented antigens. The findings also suggest that a substantial number of antigens useful for cross-presentation are present in apoptotic cells. The LNs draining the damaged region may produce an immune response if the overall quantity of cross-presented antigen increases since this will create a rise in the concentration of other antigens. A medication that induces apo ptosis may also enhance antigen cross-presentation, although this is not yet known.[78]

8.6 PATHWAYS TO IMMUNOGENICITY

8.6.1 IMMUNOGENIC APOPTOSIS

The possibility that chemotherapy raises levels of cross-presented antigens raises two obvious concerns.[79] In the first place, how immunogenic are these tumor antigens in the view of the immune system? Is it fair to say that all cytotoxic medications have the ability to stimulate an immunological response, or do some work better than others at this? The absence of typical TLR-ligands (such as PAMPs) in the tumor microenvironment makes chemotherapy-induced tumor cell death unlikely to stimulate an immune response. After a cell has died through apoptosis, a molecule called phosphatidylserine (PS) is transferred from the inner to the outside leaflet of the plasma membrane. This PS acts as a molecular flag instructing the immune system to disregard the dead cell. In contrast to DCs, macrophages are more adept at absorbing PS, and the molecule also boosts the synthesis of anti-inflammatory factors including TGF- and IL-10 while suppressing IL-12 production.[80] Regulating the daily death of billions of cells essential in maintaining tissue homeostasis, tolerogenic apoptosis avoids excessive autoimmunity. Necrotic cell death, on the other hand, has been related to infections and tissue degeneration, which may cause an immune response.

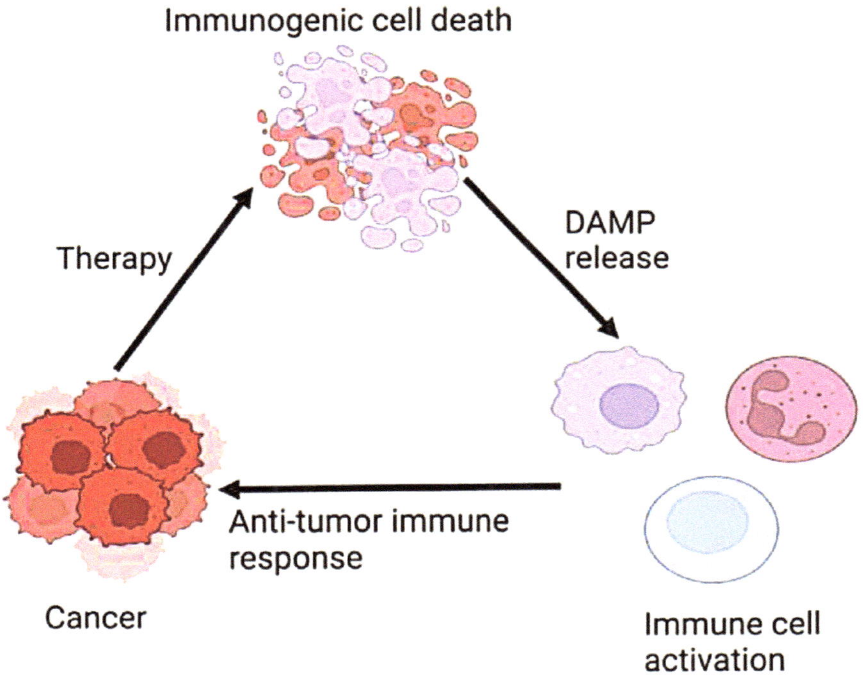

FIUGRE 8.2 Schematic diagram of immune cell activation.

This view is supported by some evidence, but it will be demonstrated that there are just as many reasons to reject it. The disparity between the apoptosis and necrosis argument may be explained, according to the writers of this chapter, by the fact that not all apoptosis is equal.[81] Apoptosis is used to describe a wide range of processes that may be induced by different inputs. In contrast to inflammatory apoptosis, which causes an immune response, noninflammatory apoptosis results in immunological apathy or tolerance.[82] Using recombinant alphaviruses, scientists have shown beyond a reasonable doubt that apoptotic cell death is required for immunogenicity.[83,84] Alphaviruses are a class of RNA viruses that are very brief and cause cell death.[85,86] By employing a recombinant alphavirus expressing the tumor antigen tyrosine-related protein-1 (TRP-1) and a plasmid harboring the antiapoptotic gene BclXL,[87–90] Restifo and colleagues demonstrated that apoptotic cell death was necessary to make the vaccination entirely immunogenic. Production of Bcl-XL, an inhibitor of apoptosis, dampened T-cell responses to TRP-1. However, coimmunization with a plasmid expressing Bcl-XL improved the immunogenicity of a normal DNA plasmid vaccine by reducing the amount of apoptosis that occurred. Leitner and Restifo (2003) proposed that apoptosis increases the immunogenicity of somatic cells by enhancing antigen transport to the cross-presentation pathway, but attenuates the response when apoptosis also occurs in transfected DCs. Therefore, adding antiapoptotic genes would have different effects on different cell types.[91] An increase in Bcl-XL under these conditions might lead to a reduction in cross-presented antigens from both dead somatic cells and activated DCs, as proposed by this idea.[92] Plasmid DNA vaccines, on the other hand, may rely more on the long-term viability of transfected DCs for their immunogenicity. These findings[93–96] highlight the ability of recombinant alphavirus vectors to mimic immunogenic cell death when combined with cross-presented antigens.

The need to trigger immunogenic apoptosis was a potential obstacle to this study's progress. Given that TLR activation often coincides with viral infection during apoptotic cell death, this makes it reasonable. Treatment-induced immunogenic cell death is not a given.[97–99] Many clinical tests have looked at the immunological implications of doxorubicin- and mitomycin-induced cancer cell death. These findings provide support to the idea that various forms of apoptosis may be induced by chemotherapy. Mice vaccinated with cells killed with doxorubicin were better protected from a tumor challenge than mice vaccinated with mitomycin-killed cells.[100] Local tumor reduction after intralesional doxorubicin administration. Immune response to doxorubicin and mitomycin is distinct because the two drugs destroy cells in distinct ways.[101] DCs rapidly ingested antigens produced by doxorubicin-killed cells, but mitomycin-killed cell antigens were ignored. Genealogical material is altered in different ways by each drug. Oxarubicin and mitomycin inhibit DNA replication by crosslinking and intercalating, respectively. However, the precise mechanism by which doxorubicin poisons topoisomerase II to induce cell death is not fully understood.[102] That DC can take in antigen and deliver it in a form that T-cells can recognize and respond to makes this unique immunologically. Doxorubicin, but not mitomycin, is thought to cause the formation or release of a molecular flag that instructs DCs to phagocytose dying tumor cells.[103] Despite the fact that cross-presentation does not guarantee immunogenicity,[104] it does play a role in maintaining tolerance. Gemcitabine seems to have the opposite effect of doxorubicin by promoting DC absorption and cross-presentation. These studies show that not all apoptosis is the same from an immunological standpoint, but they also raise more questions than they answer. One such puzzle is the nature of the inflammatory signals that have been postulated to play a role in chemo-induced cell death. That is, how does the immune system understand that a cell has died? There are a variety of endogenous danger signals, including uric acid, heat-shock proteins, DNA damage, and cytokines.[105]

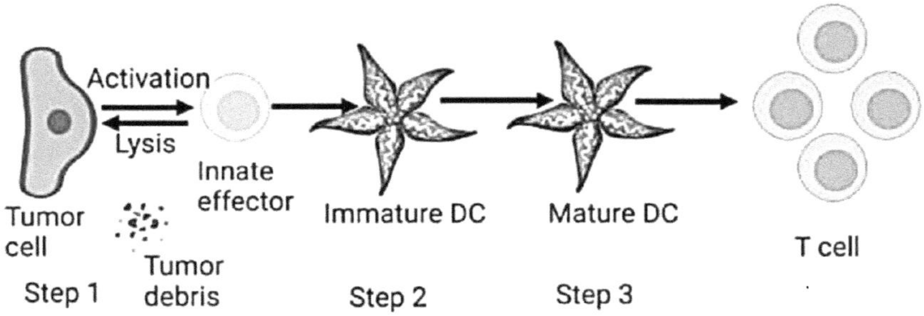

FIGURE 8.3 Schematic diagram of T-cell.

8.6.2 EFFECTS ON REGULATORY SUBSETS AND PATHWAYS

It is possible that cytotoxic treatment may increase immunopotency by directly stimulating effector cells or by reducing tumor-induced immune suppression. Since their initial identification, many classes of immunoregulatory cells in cancer patients have been defined. The buildup of suppressive-function myeloid cells, such as MDSCs and TAMs, and Tregs, such as $CD4^+$ $CD25^+$ Tregs.[106]

Several methods have been identified for therapeutically eliminating suppressive cell subsets.

Gemcitabine has been shown to preferentially destroy MDSCs in both in vitro and in vivo tests. With their selective eradication and the removal of MDSCs, CD8 T-cells and NK cells showed increased antitumor activity.[107] Inhibiting STAT6's regulation of the synthesis of programmed death ligand-2, platinum-based medicines lessen immunosuppression by both DCs and tumor cells. To temporarily and selectively reduce Tregs and the production of immune-suppressing cytokines such as IL-4, IL-10, and IL-13,[108] low-dose CTX is used. Metronomic CTX regimens have been shown to block Tregs for a longer duration and more efficiently than standard CTX regimens[109] in patients with advanced solid tumors or metastatic breast cancer. This dacarbazine derivative decreased the number of circulating Tregs and their suppressive efficacy,[110] despite the fact that metronomic temozolomide did not prevent tumor growth in glioma-bearing rats.

Communication about CTX and MDSCs is ongoing. Preliminary findings seem to support the hypothesis that CTX promotes the development of natural suppressor cells. The administration of CTX and doxorubicin enhanced the amount of circulating MDSC in B16 melanoma-bearing mice and human breast cancer patients.[111] Inflammatory mediators such as granulocyte-macrophage colony-stimulating factor (GM-CSF), IL-1β, IL-10, IFN-, and TNF- accumulated in skin tumors and metastatic LNs of mice with spontaneously developing melanomas treated with low doses of CTX. Several lines of evidence suggest that CTX generates early myeloid effector cells that may restrict cancer cell proliferation through nitric oxide (NO) generation and that this effect is enhanced when CTX is combined with gemcitabine. In contrast to FOXP3^{+}CD4 effector cells, paclitaxel (a kind of taxone) specifically reduces the survival and cytokine production of FOXP3+Treg cells.[112,113] Docetaxel, like paclitaxel, has been shown to polarize MDSCs toward an M1-like phenotype and to selectively kill MDSCs while sparing the M-like cells.[114] Recent research has shown that by selectively eliminating MDSCs in the lab and in the wild, 5-FU enhances IFN- by tumor-infiltrating CD8 T-cells and T-cell-mediated anti-cancer responses.

Suppressor cells utilize a variety of strategies to dampen the efforts of effector lymphocytes to eliminate tumors. Tumor growth and immune suppression have been linked to a number of enzymes and peptides.[115] These include arginine kinase I (AKI), IDO, and iNOS, as well as latency-associated peptide (LAP) and CD124. The suppression of these mediators by chemotherapy seems crucial for the tumor-induced breakdown of tolerance. Patients with advanced nonsmall cell lung cancer who were treated with platinum-based drugs had reduced levels of iNOS, IDO, and CD124 expression in their blood samples compared to those who were not. It has recently been discovered that the cancer medicines gemcitabine and 5-FU are not as successful as formerly believed since they reduce immune suppression and enhance the production of protumorigenic cytokines through inflammasome pathways.[116]

8.6.3 EFFECTS ON TUMOR CELLS AND ON TUMOR MICROENVIRONMENT

Tumors in the brain and spinal cord respond very well to treatment (TCD). Only lately have cytotoxic medicines, which kill cancer cells and alter the body's immunological response, shown their full potential. There is now growing evidence that the immune infiltrate's composition, which often outnumbers neoplastic cells, is critical for cancer prognosis.[117] So, cancer treatments that alter the immune infiltrate's makeup might be very useful. When the conditions are right, chemotherapy that induces TCD may pave the way for a strong antitumor immune response. Some anticancer medications, for instance, upregulate genes that play key roles in cell death, such as fas, TNF receptors, and TRAIL receptors. Cytochrome c is released from mitochondria and acts as a trigger for apoptosis in certain medicines.[118] In a number of tumor settings, the severity of tumor-associated stroma (TCD) is linked to clinical prognosis.[119]

Anthracyclines, oxaliplatin, and CTX have the unique ability to induce an immunogenic kind of TCD by converting apoptotic tumor cells into adjuvanted-endogenous vaccines.[120] According to vaccination theory, tumor antigens should be taken up by activated DCs to trigger CD4+ and CD8+ T-cell-mediated adaptive immunological responses. The death of tumor cells releases antigens into the surrounding environment, stimulating DCs and the immune system. I the early surface exposure of calreticulin (ecto-CRT) on stressed cells, (ii) the following secretion of ATP, and (iii) the release of high-mobility group box-1 (HMGB1) and HSPs by dying cancer cells; these molecular processes distinguish immunogenic from nonimmunogenic cell death.[121] However, HMGB1 and ATP block DC-mediated tumor antigen cross-presentation and T-cell polarization, while ecto-CRT encourages DC absorption of apoptotic materials. Recent studies in mice have demonstrated that ATP may influence the recruitment and differentiation of myeloid DC to the cancer site. Human colon cancer cells were killed extensively by gemcitabine, oxaliplatin, leucovorin, and 5- FU, which triggered strong in vitro antigen-specific CTL responses. Antigen cross-presentation by DC and CD8 T-cell cross-priming are both facilitated by the immunogenic TCD produced by some alkylating medications, such as anthracyclines.[122]

Tumors in the brain and spinal cord respond very well to treatment (TCD). Only lately have cytotoxic medicines, which kill cancer cells and alter the body's immunological response, shown their full potential. There is now growing evidence that the immune infiltrate's composition, which often outnumbers neoplastic cells, is critical for cancer prognosis.[123] So, cancer treatments that alter the immune infiltrate's makeup might be very useful. When the conditions are right, chemotherapy that induces TCD may pave the way for a strong antitumor immune response.[124] Some anticancer medications, for instance, upregulate genes that play key roles in cell death, such as fas, TNF receptors, and TRAIL receptors. Cytochrome c is released from mitochondria and acts as a trigger for apoptosis in certain medicines. In a number of tumor settings, the severity of tumor-associated stroma (TCD) is linked to clinical prognosis.

Anthracyclines, oxaliplatin, and CTX have the unique ability to induce an immunogenic kind of TCD by converting apoptotic tumor cells into adjuvanted-endogenous vaccines.[125] The vaccination hypothesis suggests that activated DCs will take up tumor antigens and then trigger $CD4^+$ and $CD8^+$ T-cell-mediated adaptive immune responses. When tumor cells die, they leak antigens into the surrounding environment, which in turn stimulates DCs and the immune system.[126] (i) The early surface exposure of calreticulin (ecto-CRT) on stressed cells, (ii) the following secretion of ATP, and (iii) the release of high-mobility group box-1 (HMGB1) and HSPs by dying cancer cells; these three molecular processes distinguish immunogenic from nonimmunogenic cell death. However, HMGB1 and ATP block DC-mediated tumor antigen cross-presentation and T-cell polarization, while ecto-CRT encourages DC absorption of apoptotic materials.[127] Recent studies in mice have demonstrated that ATP influences the recruitment and differentiation of myeloid DC to the cancer site. Human colon cancer cells were killed extensively by gemcitabine, oxaliplatin, leucovorin, and 5- FU, which triggered strong in vitro antigen-specific CTL responses. There is evidence to indicate that certain alkylating medications, including anthracyclines, generate an immunogenic TCD that encourages antigen cross-presentation by DC and CD8 T-cell cross-priming.[128,129]

8.7 IMPLICATIONS FOR COMBINATION WITH OTHER TREATMENTS

These results have important implications for how future clinical studies of combination chemotherapy and immunotherapy should be planned. To eliminate immune suppressor cells, it may be necessary to combine adoptive immunotherapy with cytotoxic drugs that induce immunogenic TCD.[130] This might turn the tumor into an endogenous vaccine by causing sufficient DC activation to be created by the transmission of danger signals. Ipilimumab is a human anti-CTLA-4 monoclonal antibody that blocks the CTLA-4 inhibitory signal on T-cells, and it has been shown to significantly increase survival rates in phase-III clinical studies on patients with metastatic melanoma and phase-II clinical trials on patients with lung cancer.[131,132] As chemotherapy raises the immune visibility of surviving or damaged tumor cells by upregulating HLA molecules and NK-related or TRAIL ligands, the co-administration of immunostimulatory molecules (such as TLR ligands, cytokines, etc.) may further facilitate immune recognition by DCs and NK, thereby enhancing antitumor T-cell responses. One promising use of type I IFN in conjunction with chemotherapy is its synergy with intertumoral injections of CTX21, which enhances the cross-presentation of tumor-derived antigens by DCs.[133] Third, several anticancer medications or regimens (such as CTX or gemcitabine) target inhibitory subsets, creating an ideal setting for combining with active vaccination techniques to augment preexisting tumor-reactive immune responses (such as metronomic). This field has recently benefited from the growing use of targeted medications such as imatinib (an IDO inhibitor) and sunitinib (an MDSC and Treg antagonist).[134,135] A fourth need for synergy with active or adoptive immunotherapy is for some medications to have bystander effects on the host.[136,137] Considering how long the components should be mixed is essential for good outcomes. Evidence from both animal and human research suggests that the synergy between immunotherapy and chemotherapy may be maximized if the two medicines are provided close to one another (1–2 days).

8.8 CONCLUSION

More and more studies point to the possibility that typical chemo drugs have profound effects on tumors and the patient's immune system. To the best of our knowledge, there has been no exhaustive examination of how the immune-based effects of typical chemotherapeutic medications differ by cancer histology or stage. There have been fresh insights into the mechanisms of therapeutic synergy and antagonism thanks to recent developments in the field of cancer immunology. Taken together, these findings point the way toward mechanism-based rather than empirical explanations for why combining specialized chemotherapeutic drugs with individualized immunotherapeutic treatment may lead to more successful care for cancer patients. There is growing evidence that combining conventional cancer treatment with immune system modulation improves patient outcomes. The immunological contribution might be created in two different ways by conventional treatment. Treatments for cancer that induce biological reactions other than apoptosis may render the death of tumor cells immunogenic. Bortezomib and HSP90 upregulate major histocompatibility complex components, tumor-specific antigens, and death receptors; anthracyclines and oxaliplatin promote the release of HMGB1 after apoptosis; and these changes all have immunogenic consequences. Different methods may be used by certain immunogenic cancer chemotherapies. On the other hand, certain drugs may have immune-boosting side effects (beyond their effect on the tumor). Results may include short-term lymphodepletion, disruption of immunosuppressive mechanisms, and direct or indirect activation of immunological effectors.

Animal and human studies show that immunization against cancer-specific antigens makes tumors more responsive to later chemotherapeutic therapy. The author hypothesizes that vaccination-induced increases in the frequency of primed T-cells, which are unrelated to tumor formation, may give a significant advantage if the tumor is assaulted by cytotoxic drugs. However, the underlying processes that produce this additive effect are still unknown.

KEYWORDS

- **immunogenicity**
- **cancer blocking agents**
- **extrinsic factors**
- **cancer immunosurveillance theory**

REFERENCES

1. Apetoh, L.; Ghiringhelli, F.; Tesniere, A.; Obeid, M.; Ortiz, C.; Criollo, A. et al. Toll-like receptor 4-dependent contribution of the immune system to anticancer chemotherapy and radiotherapy. *Nat Med.* **2007,** *13*, 1050–1059.
2. Mackay, I. R.; Goodyear, M. D.; Riglar, C.; Penschow, J. Effect on Natural Killer and Antibody-Dependent Cellular Cytotoxicity of Adjuvant Cytotoxic Chemotherapy Including Melphalan in Breast Cancer. *Cancer Immunol. Immunother.* **1983,** *16*, 98–100.

3. Braun, D. P.; Harris, J. E. Effect of Chemotherapy on NK Function in the Peripheral Blood of Cancer Patients. *Cancer Immunol. Immunother.* **1986,** *21*, 240–245.
4. Brenner, B. G.; Margolese, R. G. The Relationship of Chemotherapeutic and Endocrine Intervention on Natural Killer Cell Activity in Human Breast Cancer. *Cancer* **1991,** *68*, 482–488.
5. Sewell, H. F.; Halbert, C. F.; Robins, R. A.; Galvin, A.; Chan, S.; Blamey, R. W. Chemotherapy-Induced Differential Changes in Lymphocyte Subsets and Natural-Killer-Cell Function in Patients with Advanced Breast Cancer. *Int. J. Cancer* **1993,** *55*, 735–738.
6. Ghiringhelli, F.; Menard, C.; Puig, P. E.; Ladoire, S.; Roux, S.; Martin, F. et al. Metronomic Cyclophosphamide Regimen Selectively Depletes CD4 + CD25 + Regulatory T-Cells and Restores T and NK Effector Functions in End Stage Cancer Patients. *Cancer Immunol. Immunother.* **2007,** *56*, 641–648.
7. Liu, P.; Jaffar, J.; Hellstrom, I.; Hellstrom, K. E. Administration of Cyclophosphamide Changes the Immune Profile of Tumor-Bearing Mice. *J. Immunother.* **2010,** *33*, 53–59.
8. Doloff, J. C.; Waxman, D. J. VEGF Receptor Inhibitors Block the Ability of Metronomically Dosed Cyclophosphamide to Activate Innate Immunity-Induced Tumor Regression. *Cancer Res.* **2012,** *72*, 1103–1115.
9. Khallouf, H.; Marten, A.; Serba, S.; Teichgraber, V.; Buchler, M. W.; Jager, D. et al. 5-Fluorouracil and Interferon-Alpha Immunochemotherapy Enhances Immunogenicity of Murine Pancreatic Cancer Through Upregulation of NKG2D Ligands and MHC Class, I. *J. Immunother.* **2012,** *35*, 245–253.
10. Martinez, F. O.; Sica, A.; Mantovani, A.; Locati, M. Macrophage Activation and Polarization. *Front Biosci.* **2008,** *13*, 453–461.
11. Mills, C. D.; Kincaid, K.; Alt, J. M.; Heilman, M. J.; Hill, A. M. M-1/M-2 Macrophages and theTh1/Th2 Paradigm. *J. Immunol.* **2000,** *164*, 6166–6173.
12. Heusinkveld, M.; van der Burg, S. H. Identification and Manipulation of Tumor Associated Macrophages in Human Cancers. *J. Transl. Med.* **2011,** *9*, 216.
13. Bryniarski, K.; Szczepanik, M.; Ptak, M.; Zemelka, M.; Ptak, W. Influence of Cyclophosphamide and Its Metabolic Products on the Activity of Peritoneal Macrophages in Mice. *Pharmacol Rep.* **2009,** *61*, 550–557.
14. Buhtoiarov, I. N.; Sondel, P. M.; Wigginton, J. M.; Buhtoiarova, T. N.; Yanke, E. M.; Mahvi, D. A. et al. Anti-Tumour Synergy of Cytotoxic Chemotherapy and Anti-CD40 plus CpG-ODN Immunotherapy Through Repolarization of Tumour-Associated Macrophages. *Immunology* **2011,** *132*, 226–239.
15. Potapov, S. L.; Korman, D. B.; Shamaev, V. I.; Ershova, R. B.; Makarov, O. V. Sensitivity of Clonogenic Cells of Human Ovarian Ascitic Cancer to Antitumor Drugs. *Arch. Geschwulstforsch* **1988,** *58*, 99–104.
16. Park, S.; Kang, S.; Chen, X.; Kim, E. J.; Kim, J.; Kim, N. et al. Tumor Suppression via Paclitaxel-Loaded Drug Carriers That Target Inflammation Marker Upregulated in Tumor Vasculature and Macrophages. *Biomaterials* **2013,** *34*, 598–605.
17. Javeed, A.; Ashraf, M.; Riaz, A.; Ghafoor, A.; Afzal, S.; Mukhtar, M. M. Paclitaxel and Immune System. *Eur. J. Pharm Sci.* **2009,** *38*, 283–290.
18. DeNardo, D. G.; Brennan, D. J.; Rexhepaj, E.; Ruffell, B.; Shiao, S. L.; Madden, S. F. et al. Leukocyte Complexity Predicts Breast Cancer Survival and Functionally Regulates Response to Chemotherapy. *Cancer Discov.* **2011,** *1*, 54–67.

19. Shree, T.; Olson, O. C.; Elie, B. T.; Kester, J. C.; Garfall, A. L.; Simpson, K. et al. Macrophages and Cathepsin Proteases Blunt Chemotherapeutic Response in Breast Cancer. *Genes Dev.* **2011,** *25*, 2465–2479.
20. Nakahara, T.; Uchi, H.; Lesokhin, A. M.; Avogadri, F.; Rizzuto, G. A.; Hirschhorn-Cymerman, D. et al. Cyclophosphamide Enhances Immunity by Modulating the Balance of Dendritic Cell Subsets in Lymphoid Organs. *Blood* **2010,** *115*, 4384–4392.
21. Schiavoni, G.; Sistigu, A.; Valentini, M.; Mattei, F.; Sestili, P.; Spadaro, F. et al. Cyclophosphamide Synergizes with Type I Interferons Through Systemic Dendritic Cell Reactivation and Induction of Immunogenic Tumor Apoptosis. *Cancer Res.* **2011,** *71*, 768–778.
22. Chen, J.; Huang, X.; Huang, G.; Chen, Y.; Chen, L.; Song, H. Preconditioning Chemotherapy with Cisplatin Enhances the Antitumor Activity of Cytokine-Induced Killer Cells in a Murine Melanoma Model. *Cancer Biother. Radiopharm.* **2012,** *27*, 210–220.
23. Soeda, A.; Morita-Hoshi, Y.; Makiyama, H.; Morizane, C.; Ueno, H.; Ikeda, M. et al. Regular Dose of Gemcitabine Induces an Increase in CD14+ Monocytes and CD11c+ Dendritic Cells in Patients with Advanced Pancreatic Cancer. *Jpn. J. Clin. Oncol.* **2009,** *39*, 797–806.
24. Tanaka, H.; Matsushima, H.; Mizumoto, N.; Takashima, A. Classification of Chemotherapeutic Agents Based on Their Differential In Vitro Effects on Dendritic Cells. *Cancer Res.* **2009,** *69*, 6978–6986.
25. Shurin, G. V.; Tourkova, I. L.; Kaneno, R.; Shurin, M. R. Chemotherapeutic Agents in Noncytotoxic Concentrations Increase Antigen Presentation by Dendritic Cells via an IL-12-Dependent Mechanism. *J Immunol.* **2009,** *183*, 137–144.
26. Kaneno, R.; Shurin, G. V.; Tourkova, I. L.; Shurin, M. R. Chemomodulation of Human Dendritic Cell Function by Antineoplastic Agents in Low Noncytotoxic Concentrations. *J. Transl. Med.* **2009,** *7*, 58.
27. Tanaka, H.; Matsushima, H.; Nishibu, A.; Clausen, B. E.; Takashima, A. Dual Therapeutic Efficacy of Vinblastine as a Unique Chemotherapeutic Agent Capable of Inducing Dendritic Cell Maturation. *Cancer Res.* **2009,** *69*, 6987–6994.
28. Wehner, R.; Bitterlich, A.; Meyer, N.; Kloss, A.; Schakel, K.; Bachmann, M. et al. Impact of Chemotherapeutic Agents on the Immunostimulatory Properties of Human 6-sulfoLacNAc+ (slan) Dendritic Cells. *Int. J. Cancer* **2013,** *132*, 1351–1359.
29. Buttiglieri, S.; Galetto, A.; Forno, S.; De Andrea, M.; Matera, L. Influence of Drug-Induced Apoptotic Death on Processing and Presentation of Tumor Antigens by Dendritic Cells. *Int. J. Cancer* **2003,** *106*, 516–520.
30. Galetto, A.; Buttiglieri, S.; Forno, S.; Moro, F.; Mussa, A.; Matera, L. Drug- and Cell-Mediated Antitumor Cytotoxicities Modulate Cross-Presentation of Tumor Antigens by Myeloid Dendritic Cells. *Anticancer Drugs* **2003,** *14*, 833–843.
31. Casares, N.; Pequignot, M. O.; Tesniere, A.; Ghiringhelli, F.; Roux, S.; Chaput, N. et al. Caspase-Dependent Immunogenicity of Doxorubicin-Induced Tumor Cell Death. *J Exp. Med.* **2005,** *202*, 1691–1701.
32. Mackall, C. L.; Fleisher, T. A.; Brown, M. R.; Magrath, I. T.; Shad, A. T.; Horowitz, M. E. et al. Lymphocyte Depletion During Treatment with Intensive Chemotherapy for Cancer. *Blood* **1994,** *84*, 2221–2228.

33. Proietti, E.; Moschella, F.; Capone, I.; Belardelli, F. Exploitation of the Propulsive Force of Chemotherapy for Improving the Response to Cancer Immunotherapy. Mol. Oncol. **2012,** *6,* 1–14.
34. Wijayahadi, N.; Haron, M. R.; Stanslas, J.; Yusuf, Z. Changes in Cellular Immunity During Chemotherapy for Primary Breast Cancer with Anthracycline Regimens. J. Chemother. **2007,** *19,* 716–723.
35. Ladoire, S.; Mignot, G.; Dabakuyo, S.; Arnould, L.; Apetoh, L.; Rebe, C. et al. In Situ Immuneresponse After Neoadjuvant Chemotherapy for Breast Cancer Predicts Survival. *J. Pathol.* 224:389–400.
36. Suzuki, E.; Kapoor, V.; Jassar, A. S.; Kaiser, L. R.; Albelda, S. M. Gemcitabine Selectively Eliminates Splenic Gr-1+/CD11b+ Myeloid Suppressor Cells in Tumor-Bearing Animals and Enhances Antitumor Immune Activity. *Clin. Cancer Res.* **2005,** *11,* 6713–6721.
37. Shebzukhov, Y. V.; Koroleva, E. P.; Khlgatian, S. V.; Lagarkova, M. A.; Meshcheryakov, A. A.; Lichinitser, M. R. et al. Humoral Immune Response to Thymidylate Synthase in Colon Cancer Patients After 5-FU Chemotherapy. *Immunol. Lett.* **2005,** *100,* 88–93.
38. Otterness, I. G.; Chang, Y. H. Comparative Study of Cyclophosphamide, 6-Mercaptopurine, Azathiopurine and Methotrexate: Relative Effects on the Humoral and the Cellular Immuneresponse in the Mouse. *Clin Exp Immunol.* **1976,** *26,* 346–354.
39. Smith, S. R.; Terminelli, C.; Kipilman, C. T.; Smith, Y. Comparative Effects of Azathioprine, Cyclophosphamide and Frentizole on Cellular Immunity in Mice. *J. Immunopharmacol.* **1981,** *3,* 133–170.
40. Berd, D.; Maguire, H. C., Jr, Mastrangelo, M. J. Potentiation of Human Cell-Mediated and Humoral Immunity by Low-Dose Cyclophosphamide. *Cancer Res.* **1984,** *44,* 5439–5443.
41. Bracci, L.; Moschella, F.; Sestili, P.; La Sorsa, V.; Valentini, M.; Canini, I. et al. Cyclophosphamide Enhances the Antitumor Efficacy of Adoptively Transferred Immune Cells Through the Induction of Cytokine Expression, B-Cell and T-Cell Homeostatic Proliferation, and Specific Tumor Infiltration. *Clin. Cancer Res.* **2007,** *13,* 644–653.
42. Proietti, E.; Greco, G.; Garrone, B.; Baccarini, S.; Mauri, C.; Venditti, M. et al. Importance of Cyclophosphamide-Induced by Stander Effect on T-Cells for a Successful Tumor Eradication in Response to Adoptive Immunotherapy in Mice. J Clin. Invest. **1998,** *101,* 429–441.
43. Moschella, F.; Valentini, M.; Arico, E.; Macchia, I.; Sestili, P.; MT, et al. Unraveling Cancer Chemoimmunotherapy Mechanisms by Gene and Protein Expression Profiling of Responses to Cyclophosphamide. *Cancer Res.* **2011,** *71,* 3528–3539.
44. Brzostek-Racine, S.; Gordon, C.; Van Scoy, S.; Reich, N. C. The DNA Damage Response Induces IFN. *J. Immunol.* **2011,** *187,* 5336–5345.
45. Li, Q.; Teitz-Tennenbaum, S.; Donald, E. J.; Li, M.; Chang, A. E. In Vivo Sensitized and In Vitro Activated B Cells Mediate Tumor Regression in Cancer Adoptive Immunotherapy. *J. Immunol.* **2009,** *183,* 3195–3203.
46. Montero, E.; Valdes, M.; Avellanet, J.; Lopez, A.; Perez, R.; Lage, A. Chemotherapy Induced Transient B-Cell Depletion Boosts Antibody-Forming Cells Expansion Driven by an Epidermal Growth Factor-Based Cancer Vaccine. *Vaccine* **2009,** *27,* 2230–2239.
47. Kim, J. E.; Jang, M. J.; Lee, J. I.; Chung, Y. H.; Jeong, J. H.; Hung, C. F. et al. Cancer Cells Containing Nanoscale Chemotherapeutic Drugs Generate Antiovarian Cancer-Specific CD4+ T-Cells in Peritoneal Space. *J. Immunother.* **2012,** *35,* 1–13.

48. Spisek, R.; Charalambous, A.; Mazumder, A.; Vesole, D. H.; Jagannath, S.; Dhodapkar, M. V. Bortezomib Enhances Dendritic Cell (DC)-Mediated Induction of Immunity to Human Myelomavia Exposure of Cell Surface Heat Shock Protein 90 on Dying Tumor Cells: Therapeutic Implications. *Blood* **2007,** *109*, 4839–4845.
49. Guerriero, J. L.; Ditsworth, D.; Catanzaro, J. M.; Sabino, G.; Furie, M. B.; Kew, R. R. et al. DNA Alkylating Therapy Induces Tumor Regression Through an HMGB1-Mediated Activation of Innate Immunity. *J. Immunol.* **2011,** *186*, 3517–3526.
50. Wolchok, J. D.; Chiarion-Sileni, V.; Gonzalez, R.; Rutkowski, P.; Grob, J. J.; Cowey, C. L. et al. Overall survival with combined nivolumab and ipilimumab in advanced melanoma. *N. Engl. J. Med.* **2017,** *377*, 1345–1356.
51. Bulkley, G. B.; Cohen, M. H.; Banks, P. M.; Char, D. H.; Ketcham, A. S. Long-Term Spontaneous Regression of Malignant Melanoma with Visceral Metastases: Report of a Case with Immunologic Profile. *Cancer* **1975,** *36*, 485–494.
52. Nelson, D. J.; Mukherjee, S.; Bundell, C.; Fisher, S.; van Hagen, D.; Robinson, B. Tumor Progression Despite Efficient Tumor Antigen Cross-Presentation and Effective "Arming" of Tumor Antigen-Specific CTL. *J. Immunol.* **2001,** *166*, 5557–5566.
53. Garbe, A. I.; Vermeer, B.; Gamrekelashvili, J.; von Wasielewski, R.; Greten, F. R.; Westendorf, A. M. et al. Genetically Induced Pancreatic Adenocarcinoma Is Highly Immunogenic and Causes Spontaneous Tumor-Specific Immune Responses. *Cancer Res.* **2006,** *66*, 508–516.
54. Kraman, M.; Bambrough, P. J.; Arnold, J. N.; Roberts, E. W.; Magiera, L.; Jones, J. O. et al. Suppression of Antitumor Immunity by Stromal Cells Expressing Fibroblast Activation Protein-Alpha. *Science* **2010,** *330*, 827–830.
55. Snyder, A.; Makarov, V.; Merghoub, T.; Yuan, J.; Zaretsky, J. M.; Desrichard, A. et al. Genetic Basis for Clinical Response to CTLA-4 Blockade in Melanoma. *N. Engl. J. Med.* **2014,** *371*, 2189–2199.
56. Rizvi, N. A.; Hellmann, M. D.; Snyder, A.; Kvistborg, P.; Makarov, V.; Havel, J. J. et al. Cancer Immunology. Mutational Landscape Determines Sensitivity to PD-1 Blockade in Non-Small Cell Lung Cancer. *Science 348*, 124–128.
57. Gandara, D. R.; Paul, S. M.; Kowanetz, M.; Schleifman, E.; Zou, W.; Li, Y. et al. Blood-Based Tumor Mutational Burden as a Predictor of Clinical Benefit in Non-Small-Cell Lung Cancer Patients Treated with a Tezolizumab. *Nat. Med.* **2018,** *24*, 1441–1448.
58. van der Bruggen, P.; Traversari, C.; Chomez, P.; Lurquin, C.; De Plaen, E.; Van den Eynde, B. et al. A Gene Encoding an Antigen Recognized by Cytolytic T Lymphocytes on a Human Melanoma. *Science* **1991,** *254*, 1643–1647.
59. Jass, J. R. Lymphocytic Infiltration and Survival in Rectal Cancer. *J. Clin. Pathol.* **1986,** *39*, 585–589.
60. Zhang, L.; Conejo-Garcia, J. R.; Katsaros, D.; Gimotty, P. A.; Massobrio, M.; Regnani, G. et al. Intratumoral T Cells, Recurrence, and Survival in Epithelial Ovarian Cancer. *N. Engl. J. Med.* **2003,** *348*, 203–213.
61. Galon, J.; Costes, A.; Sanchez-Cabo, F.; Kirilovsky, A.; Mlecnik, B.; Lagorce-Pages, C. et al. Type, Density, and Location of Immune Cells Within Human Colorectal Tumors Predict Clinical Outcome. *Science* **2006,** *313*, 1960–1964.
62. Shimizu, K.; Nakata, M.; Hirami, Y.; Yukawa, T.; Maeda, A.; Tanemoto, K. Tumor-Infiltrating Foxp3 + Regulatory T Cells Are Correlated with Cyclooxygenase-2 Expression and Are Associated With Recurrence in Resected Non-Small Cell Lung Cancer. *J. Thorac. Oncol.* **2010,** *5*, 585–590.

63. Hwang, W. T.; Adams, S. F.; Tahirovic, E.; Hagemann, I. S.; Coukos, G. Prognostic Significance of Tumor-Infiltrating T Cells in Ovarian Cancer: A Meta-Analysis. *Gynecol. Oncol.* **2012,** *124*, 192–198.
64. Ledys, F.; Klopfenstein, Q.; Truntzer, C.; Arnould, L.; Vincent, J.; Bengrine, L. et al. RAS Status and Neoadjuvant Chemotherapy Impact CD8+ Cells and Tumor HLA Class I Expression in Liver Metastatic Colorectal Cancer. *J. Immunother. Cancer* **2018,** *6*, 123.
65. Pages, F.; Mlecnik, B.; Marliot, F.; Bindea, G.; Ou, F. S.; Bifulco, C. et al. International Validation of the Consensus Immunoscore for the classification of Colon Cancer: A Prognostic and Accuracy Study. *Lancet* **2018,** *391*, 2128–2139.
66. Tumeh, P. C.; Harview, C. L.; Yearley, J. H.; Shintaku, I. P.; Taylor, E. J.; Robert, L. et al. PD-1 Blockade Induces Responses by Inhibiting Adaptive Immune Resistance. *Nature* **2014,** *515*, 568–571.
67. Hamid, O.; Schmidt, H.; Nissan, A.; Ridolfi, L.; Aamdal, S.; Hansson, J. et al. A Prospective Phase II Trial Exploring the Association Between Tumor Microenvironment Biomarkers and Clinical Activity of Ipilimumab in Advanced Melanoma. *J. Transl. Med.* **2011,** *9*, 204.
68. Kim, J. M.; Chen, D. S. Immune Escape to PD-L1/PD-1 Blockade: Seven Steps to Success (or Failure). *Ann Oncol.* **2016,** *27*, 1492–1504.
69. Galon, J.; Bruni, D. Approaches to Treat Immune Hot, Altered and Cold Tumours with Combination Immunotherapies. *Nat. Rev. Drug Discov. 18*, 197–218.
70. Demaria, S.; Volm, M. D.; Shapiro, R. L.; Yee, H. T.; Oratz, R.; Formenti, S. C. et al. Development of Tumor-Infiltrating Lymphocytes in Breast Cancer After Neoadjuvant Paclitaxel Chemotherapy. *Clin. Cancer Res.* **2001,** *7*, 3025–3030.
71. Bracci, L.; Moschella, F.; Sestili, P.; La Sorsa, V.; Valentini, M.; Canini, I. et al. Cyclophosphamide Enhances the Antitumor Efficacy of Adoptively Transferred Immune Cells Through the Induction of Cytokine Expression, B-Cell and T-Cell Homeostatic Proliferation, and Specific Tumor Infiltration. *Clin. Cancer Res.* **2007,** *13* (2 Pt 1), 644–653.
72. Hong, M.; Puaux, A. L.; Huang, C.; Loumagne, L.; Tow, C.; Mackay, C. et al. Chemotherapy Induces Intratumoral Expression of Chemokines in Cutaneous Melanoma, Favoring T-Cell Infiltration and Tumor Control. *Cancer Res.* **2011,** *71*, 6997–7009.
73. Mattarollo, S. R.; Loi, S.; Duret, H.; Ma, Y.; Zitvogel, L.; Smyth, M. J. Pivotal Role of Innate and Adaptive Immunity in Anthracycline Chemotherapy of Established Tumors. *Cancer Res.* **2011,** *71*, 4809–4820. DOI: 10.1158/0008-5472.CAN-11-0753
74. Peng, J.; Hamanishi, J.; Matsumura, N.; Abiko, K.; Murat, K.; Baba, T. et al. Chemotherapy Induces Programmed Cell Death-Ligand 1 Overexpression via the Nuclear Factor-κB to Foster an Immunosuppressive Tumor Microenvironment in Ovarian Cancer. *Cancer Res.* **2015,** *75*, 5034–5045.
75. Nowak, A. K.; Robinson, B. W.; Lake, R. A. Gemcitabine Exerts a Selective Effect on the Humoral Immune Response: Implications for Combination Chemo-Immunotherapy. *Cancer Res.* **2002,** *62*, 2353–2358.
76. Mikucki, M. E.; Fisher, D. T.; Matsuzaki, J.; Skitzki, J. J.; Gaulin, N. B.; Muhitch, J. B. et al. Non-Redundant Requirement for CXCR3 Signalling During Tumoricidal T-Cell Trafficking Across Tumour Vascular Checkpoints. *Nat. Commun.* **2015,** *6*, 7458. DOI: 10.1038/ncomms8458

77. Sistigu, A.; Yamazaki, T.; Vacchelli, E.; Chaba, K.; Enot, D. P.; Adam, J. et al. Cancer Cell-Autonomous Contribution of Type I Interferon Signaling to the Efficacy of Chemotherapy. *Nat. Med.* **2014,** *20*, 1301–1309. DOI: 10.1038/nm.3708
78. Liu, W. M.; Fowler, D. W.; Smith, P.; Dalgleish, A. G. Pre-Treatment with Chemotherapy Can Enhance the Antigenicity and Immunogenicity of Tumours by Promoting Adaptive Immune Responses. *Br. J. Cancer* **2010,** *102*, 115–123. DOI: 10.1038/sj.bjc.6605465
79. Marigo, I.; Bosio, E.; Solito, S.; Mesa, C.; Fernandez, A.; Dolcetti, L. et al. Tumor-Induced Tolerance and Immune Suppression Depend on the C/EBP Beta Transcription Factor. *Immunity* **2010,** *32*, 790–802. DOI: 10.1016/j.immuni.2010.010
80. Fleming, V.; Hu, X.; Weber, R.; Nagibin, V.; Groth, C.; Altevogt, P. et al. Targeting Myeloid-Derived Suppressor Cells to Bypass Tumor-Induced Immunosuppression. *Front. Immunol.* **2018,** *9*, 398. DOI: 10.3389/fimmu.2018.00398
81. Sato, K.; Sato, N.; Xu, B.; Nakamura, Y.; Nagaya, T.; Choyke, P. L. et al. Spatially Selective Depletion Of Tumor-Associated Regulatory T Cells with Near-Infrared Photo Immunotherapy. *Sci. Transl. Med.* **2016,** *8*, 352ra110. DOI: 10.1126/scitranslmed.aaf6843
82. Zhang, L.; Dermawan, K.; Jin, M.; Liu, R.; Zheng, H.; Xu, L. et al. Differential Impairment of Regulatory T Cells Rather Than Effector T Cells by Paclitaxel-Based Chemotherapy. *Clin Immunol.* **2008,** *129*, 219–229. DOI: 10.1016/j.clim.2008.013
83. Ghiringhelli, F.; Larmonier, N.; Schmitt, E.; Parcellier, A.; Cathelin, D.; Garrido, C. et al. CD4 + CD25 + Regulatory T Cells Suppress Tumor Immunity But Are Sensitive to Cyclophosphamide Which Allows Immunotherapy of Established Tumors to be Curative. *Eur. J. Immunol.* **2004,** *34*, 336–344. DOI: 10.1002/eji.200324181
84. Apetoh, L. et al. Toll-Like Receptor 4-Dependent Contribution of the Immune System to Anticancer Chemotherapy and Radiotherapy. *Nat. Med.* **2007,** *13*, 1050–1059. The First Demonstration of a Role for TLR4 and HMGB1 in the Antitumour Effects That Are Mediated by Chemotherapy and Radiotherapy.
85. Apetoh L et al. The Interaction Between HMGB1 and TLR4 Dictates the Outcome of Anticancer Chemotherapy and Radiotherapy. *Immunol. Rev.* **2007,** *220*, 47–59.
86. Shiratsuchi, A., Watanabe, I., Takeuchi, O., Akira, S.; Nakanishi, Y. Inhibitory Effect of Toll-Like Receptor 4 on Fusion Between Phagosomes and Endosomes/Lysosomes in Macrophages. *J. Immunol.* **2004,** *172*, 2039–2047.
87. Antonia, S. J. et al. Combination of p53 Cancer Vaccine with Chemotherapy in Patients with Extensive Stage Small Cell Lung Cancer. *Clin. Cancer Res.* **2006,** *12*, 878–887.
88. Gribben, J. G. et al. Unexpected Association Between Induction of Immunity to the Universal Tumor Antigen CYP1B1 and Response to Next Therapy. *Clin. Cancer Res.* **2005,** *11*, 4430–4436.
89. Arlen, P. M. et al. A Randomized Phase II Study of Concurrent Docetaxel Plus Vaccine Versus Vaccine Alone in Metastatic Androgen-Independent Prostate Cancer. *Clin. Cancer Res.* **2006,** *12*, 1260–1269.
90. Noguchi, M. et al. Immunological Evaluation of Individualized Peptide Vaccination with a Low Dose of Estramustine for HLA-A24+ HRPC Patients. *Prostate* **2005,** *63*, 1–12.
91. Sun, T. et al. FASL -844C Polymorphism Is Associated with Increased Activation-Induced T Cell Death and Risk of Cervical Cancer. *J. Exp. Med.* **2005,** *202*, 967–974.
92. Lech-Maranda, E. et al. Interleukin-10 Gene Promoter Polymorphisms Influence the Clinical Outcome of Diffuse Large B-Cell Lymphoma. *Blood* **2004,** *103*, 3529–3534.

93. Bushley, A. W. et al. Polymorphisms of Interleukin (IL)-1α, IL-1β, IL-6, IL-10, and IL-18 and the Risk of Ovarian Cancer. *Gynecol. Oncol.* **2004,** *95*, 672–679.
94. Nimmerjahn, F.; Ravetch, J. V. Antibodies, Fc receptors and cancer. *Curr. Opin. Immunol.* **2007,** *19*, 239–245.
95. Kroemer, G. et al. Classification of Cell Death: Recommendations of the Nomenclature Committee on Cell Death. *Cell Death Differ.* **2005,** *12* (Suppl. 2), 1463–1467.
96. Zitvogel, L. et al. Immune Response Against Dying Tumor Cells. *Adv. Immunol.* **2004,** *84*, 131–179.
97. Claude, L. et al. Lymphopenia: A New Independent Prognostic Factor for Survival in Patients Treated with Whole Brain Radiotherapy for Brain Metastases from Breast Carcinoma. *Radiother. Oncol.* **2005,** *76*, 334–339.
98. Schuerwegh, A. J., van Offel, J. F., Bridts, C. H., Stevens, W. J.; De Clerck, L. S. Influence of Long Term Therapy with Methotrexate and Low Dose Corticosteroids on Type 1 and Type 2 Cytokine Production in CD4+ and CD8+ T Lymphocytes of Patients with Rheumatoid Arthritis. *J. Rheumatol.* **2001,** *28*, 1793–1799.
99. Panici, P. B. et al. Systematic Aortic and Pelvic Lymphadenectomy Versus Resection of Bulky Nodes Only in Optimally Debulked Advanced Ovarian Cancer: A Randomized Clinical Trial. *J. Natl. Cancer Inst.* **2005,** *97*, 560–566.
100. Chan, O. T.; Yang, L. X. The Immunological Effects of Taxanes. *Cancer Immunol. Immunother.* **2000,** *49*, 181–185.
101. Chakraborty, M. et al. External Beam Radiation of Tumors Alters Phenotype of Tumor Cells to Render Them Susceptible to Vaccine-Mediated T-Cell Killing. *Cancer Res.* **2004,** *64*, 4328–4337.
102. Cho, B. K., Rao, V. P., Ge, Q., Eisen, H. N.; Chen, J. Homeostasis-Stimulated Proliferation Drives Naive T Cells to Differentiate Directly Into Memory T Cells. *J. Exp. Med.* **2000,** *192*, 549–556.
103. Ge, Q., Hu, H., Eisen, H. N.; Chen, J. Different Contributions of Thymopoiesis and Homeostasis-Driven Proliferation to the Reconstitution of Naive and Memory T Cell Compartments. *Proc. Natl. Acad. Sci. USA* **2002,** *99*, 2989–2994.
104. Ghiringhelli, F. et al. CD4 + CD25 + Regulatory T Cells Suppress Tumor Immunity But Are Sensitive to Cyclophosphamide Which Allows Immunotherapy of Established Tumors to be Curative. *Eur. J. Immunol.* **2004,** *34*, 336–344.
105. Mullins, D. W., Burger, C. J.; Elgert, K. D. Paclitaxel Enhances Macrophage IL-12 Production in Tumor Bearing Hosts Through Nitric Oxide. *J. Immunol.* **1999,** *162*, 6811–6818.
106. Yuan, L. et al. Restoration of Macrophage Tumoricidal Activity by Bleomycin Correlates with the Decreased Production of Transforming Growth Factor β in Rats Bearing KDH-8 Hepatoma Cells. *Cancer Immunol. Immunother.* **1997,** *45*, 71–76.
107. Hornung, R. L., Young, H. A., Urba, W. J.; Wiltrout, R. H. Immunomodulation of Natural Killer Cell Activity by Flavone Acetic Acid: Occurrence via Induction of Interferon α/β. *J. Natl. Cancer Inst.* **1988,** *80*, 1226–1231.
108. Keilholz, U. et al. Dacarbazine, Cisplatin, and Interferon-α2b with or Without Interleukin-2in Metastatic Melanoma: A Randomized Phase III Trial (18951) of the European Organisation for Research and Treatment of Cancer Melanoma Group. *J. Clin. Oncol.* **2005,** *23*, 6747–6755.

109. Bajetta, E. et al. Multicenter Phase III Randomized Trial of Poly Chemotherapy (CVD Regimen) Versus the Same Chemotherapy (CT) Plus Subcutaneous Interleukin-2 and Interferon-α2b in Metastatic Melanoma. *Ann. Oncol.* **2006,** *17*, 571–577.
110. Massacesi, C., Burattini, L., Marcucci, F.; Bonsignori, M. Short Communication: The Efficacy of Fixed Dose Rate Infusion of Gemcitabine Combined with IFN-α2a in Patients with Advanced Refractory Renal Cell Carcinoma. *J. Interferon Cytokine Res.* **2005,** *25*, 165–168.
111. Atzpodien, J. et al. Interleukin-2- and Interferon α2a-Based Immunochemotherapy in Advanced Renal Cell Carcinoma: A Prospectively Randomized Trial of the German Cooperative Renal Carcinoma Chemoimmunotherapy Group (DGCIN). *J. Clin. Oncol.* **2004,** *22*, 1188–1194.
112. Parra, H. S. et al. Combined Regimen of Cisplatin, Doxorubicin, and α-2b Interferon in the Treatment of Advanced Malignant Pleural Mesothelioma: A Phase II Multicenter Trial of the Italian Group on Rare Tumors (GITR) and the Italian Lung Cancer Task Force (FONICAP). *Cancer* **2001,** *92*, 650–656.
113. Ishida, A. et al. Intrapleural Cisplatin and OK432 Therapy for Malignant Pleural Effusion Caused by Non-Small Cell Lung Cancer. *Respirology* **2006,** *11*, 90–97.
114. Kasamon, Y. L. et al. Phase I Study Of Low-Dose Interleukin-2, Fludarabine, and Cyclophosphamide for Previously Untreated Indolent Lymphoma and Chronic Lymphocytic Leukemia. *Clin. Cancer Res.* **2005,** *11*, 8413–8417.
115. Gomez, G. G., Hutchison, R. B.; Kruse, C. A. Chemo-Immunotherapy and Chemo-Adoptive Immunotherapy of Cancer. *Cancer Treat. Rev.* **2001,** *27*, 375–402.
116. Ehrlich, P. Über den jetztigen Stand der Karzinomforschung. *Ned. Tijdschr. Geneeskd.* **1909,** *5*, 273–290 (in German).
117. Burnet, M. Cancer; A Biological Approach: I—The Processes of Control. *BMJ* **1957,** *1*, 779–786.
118. van den Broek, M. E. et al. Decreased Tumor Surveillance in Perforin-Deficient Mice. *J. Exp. Med.* **1996,** *184*, 1781–1790.
119. Shankaran, V. et al. IFNγ and Lymphocytes Prevent Primary Tumour Development and Shape Tumour Immunogenicity. *Nature* **2001,** *410*, 1107–1111.
120. Galon, J. et al. Type, Density, and Location Of Immune Cells Within Human Colorectal Tumors Predict Clinical Outcome. *Science* **2006,** *313*, 1960–1964.
121. Salem, M. L.; El-Naggar, S. A.; Cole, D. J. Cyclophosphamide Induces Bone Marrow to Yield Higher Numbers of Precursor Dendritic Cells In Vitro Capable of Functional Antigen Presentation to T-Cells In Vivo. *Cell Immunol.* **2010,** *261*, 134–143.
122. Sefc, L.; Psenak, O.; Sykora, V.; Sulc, K.; Necas, E. Response of Hematopoiesis to Cyclophosphamide Follows Highly Specific Patterns in Bone Marrow and Spleen. *J. Hematother. Stem Cell Res.* **2003,** *12*, 47–61.
123. Salem, M. L.; Kadima, A. N.; El-Naggar, S. A.; Rubinstein, M. P.; Chen, Y.; Gillanders, W. E. et al. Defining the Ability of Cyclophosphamide Preconditioning to Enhance the Antigen-Specific CD8+T-Cell Response to Peptide Vaccination: Creation of a Beneficial Host Microenvironment Involving Type I IFNs and Myeloid Cells. *J. Immunother.* **2007,** *30*, 40–53.
124. Eberhard, A.; Kahlert, S.; Goede, V.; Hemmerlein, B.; Plate, K. H.; Augustin, H. G. Heterogeneity of Angiogenesis and Blood Vessel Maturation in Human Tumors: Implications for Antiangiogenic Tumor Therapies. *Cancer Res.* **2000,** *60*, 1388–1393.

125. Miller, K. D.; Sweeney, C. J.; Sledge, G. W., Jr. Redefining the Target: Chemotherapeutics as Antiangiogenics. *J. Clin. Oncol.* **2001,** *19,* 1195–1206.
126. Bertolini, F.; Paul, S.; Mancuso, P.; Monestiroli, S.; Gobbi, A.; Shaked, Y. et al. Maximum Tolerable Dose and Low-Dose Metronomic Chemotherapy Have Opposite Effects on the Mobilization and Viability of Circulating Endothelial Progenitor Cells. *Cancer Res.* **2003,** *63,* 4342–4346.
127. Loeffler, M.; Kruger, J. A.; Reisfeld, R. A. Immunostimulatory Effects of Low-Dose Cyclophosphamide Are Controlled by Inducible Nitric Oxide Synthase. *Cancer Res.* **2005,** *65,* 5027–5030.
128. Hodi, F. S.; O'Day, S. J.; McDermott, D. F.; Weber, R. W.; Sosman, J. A.; Haanen, J. B. et al. Improved Survival with Ipilimumab in Patients with Metastatic Melanoma. *N. Engl. J. Med.* **2010,** *363,* 711–723.
129. Robert, C.; Thomas, L.; Bondarenko, I.; O'Day, S.; DJ, M.; Garbe, C. et al. Ipilimumab Plus Dacarbazine for Previously Untreated Metastatic Melanoma. *N. Engl. J. Med.* **2011,** *364,* 2517–2526.
130. Zielinski, C.; Knapp, S.; Mascaux, C.; Hirsch, F. Rationale for Targeting the Immune System Through Checkpoint Molecule Blockade in the Treatment of Non-Small-Cell Lung Cancer. *Ann. Oncol.* **2013,** *24,* 1170–1179.
131. Lorenzi, S.; Mattei, F.; Sistigu, A.; Bracci, L.; Spadaro, F.; Sanchez, M. et al. Type I IFNs Control Antigen Retention and Survival of CD8{alpha}+ Dendritic Cells After Uptake of Tumor Apoptotic Cells Leading to Cross-Priming. *J. Immunol.* **2011,** *186,* 5142–5150132.
132. Vanneman, M.; Dranoff, G. Combining Immunotherapy and Targeted Therapies in Cancer Treatment. *Nat. Rev. Cancer* **2012,** *12,* 237–251.
133. Farsaci, B.; Higgins, J. P.; Hodge, J. W. Consequence of Dose Scheduling of Sunitinib on Host Immune Response Elements and Vaccine Combination Therapy. *Int. J. Cancer* **2012,** *130,* 1948–1959.
134. Ko, J. S.; Zea, A. H.; Rini, B. I.; Ireland, J. L.; Elson, P.; Cohen, P. et al. Sunitinib Mediates Reversal of Myeloid-Derived Suppressor Cell Accumulation in Renal Cell Carcinoma Patients. *Clin Cancer Res.* **2009,** *15,* 2148–2157.
135. Balachandran, V. P.; Cavnar, M. J.; Zeng, S.; Bamboat, Z. M.; Ocuin, L. M.; Obaid, H. et al. Imatinib Potentiates Antitumor T-Cell Responses in Gastrointestinal Stromal Tumor Through the Inhibition of Ido. *Nat Med.* **2011,** *17,* 1094–1100.
136. Nefedova, Y.; Nagaraj, S.; Rosenbauer, A.; Muro-Cacho, C.; Sebti, S. M.; Gabrilovich, D. I. Regulation of Dendritic Cell Differentiation and Antitumor Immune Response in Cancer by Pharmacologic-Selective Inhibition of the Janus-Activated Kinase 2/ Signal Transducers and Activators of Transcription 3 Pathway. *Cancer Res.* **2005,** *65,* 9525–9535.
137. Seeger, J. M.; Schmidt, P.; Brinkmann, K.; Hombach, A. A.; Coutelle, O.; Zigrino, P. et al. The Proteasome Inhibitor Bortezomib Sensitizes Melanoma Cells Toward Adoptive CTL Attack. *Cancer Res.* **2010,** *70,* 1825–1834.
138. Boni, A.; Cogdill, A. P.; Dang, P.; Udayakumar, D.; Njauw, C. N.; Sloss, C. M. et al. Selective BRAFV600E Inhibition Enhances T-Cell Recognition of Melanoma Without Affecting Lymphocyte Function. *Cancer Res.* **2010,** *70,* 5213–5219.
139. Correale, P.; Botta, C.; Cusi, M. G.; Del Vecchio, M. T.; De Santi, M. M.; Gori Savellini, G. et al. Cetuximab +/− Chemotherapy Enhances Dendritic Cell-Mediated Phagocytosis

of Colon Cancer Cells and Ignites a Highly Efficient Colon Cancer Antigen-Specific Cytotoxic T-Cell Response In Vitro. *Int. J. Cancer* **2012,** *130*, 1577–1589.

140. Kim, P. S.; Armstrong, T. D.; Song, H.; Wolpoe, M. E.; Weiss, V.; Manning, E. A. et al. Antibody Association with HER-2/neu-Targeted Vaccine Enhances CD8 T-Cell Responses in Mice Through Fc-Mediated Activation of DCs. *J. Clin. Invest.* **2008,** *118*, 1700–1711.

CHAPTER 9

The Role of Innate and Adaptive Antitumor Immunity in Immune Surveillance

ABSTRACT

Determined by a complex interaction between innate and adaptive antitumor immunity, the effectiveness and power of the body's natural defenses against cancer are ultimately established. Cancer cells' production of chemicals in response to stress activates the innate immune system, while the adaptive immune system responds to tumor-associated antigens. It is possible that the rate at which a tumor grows is affected by a combination of variables that are affected by both innate and adaptive anticancer responses. Tumor development may be slowed by endogenous responses, but this might potentially lead to escape variations being selected. Tumor cells may be combated by certain host responses, such as the production of inflammatory mediators and cytokines, but these responses can also foster tumor development, invasion, and metastasis. Recent research has uncovered a communication network connecting tumor cells, immunological elements, and stromal components of the tumor microenvironment (TME) that together regulate the anticancer response's trajectory, quality, and dynamics. When a virus enters the body, the immune system responds by mounting a defense. The germline-encoded pattern recognition receptors in this system swiftly identify infected or injured cells and initiate robust effector responses that successfully block the progression of infection. These pattern-sensing devices have recently been shown to be useful for determining tumor-induced stress variables. It has been shown that toll-like receptors (TLRs) and cytosolic sensors for DNA and RNA recognition activate innate signaling pathways and produce protective immune

Advances in Immunological Treatment of Cancer: Cellular Mechanism and Regulation.
Rishabha Malviya, Arun Kumar Singh, and Deepika Yadav (Authors)

responses against nascent tumors in both animal models and people. This article provides a concise overview of the current state of knowledge about the therapeutic use of innate immune signaling in the never-ending war against tumors.

9.1 INTRODUCTION

As immunotherapy, the use of the immune system to treat preexisting cancers was first proposed by Paul Ehrlich in the early 20th century.[1] He came up with the concept of the antibody, the protein molecule presently used to deliver poisons directly to cancer cells. It was in the 1950s when Burnet and Thomas improved upon Ehrlich's "magic bullet" strategy.[2] According to the immune surveillance hypothesis,[3] the immune system may also protect the body against cancer by removing cancer-causing cells before they can multiply and become tumors.[4] Research done since the 1970s has shown that the immune system may be manipulated to prevent cancer from developing and to reject preexisting tumors. With its rediscovery, immune surveillance was rethought to include immunoediting. Many malignant and premalignant cells are eliminated by the immune system during immunoediting, but other cells are able to avoid the onslaught and give rise to malignant tumors that continue to spread.[5,6] After a tumor has taken root, the host's immune response may still influence its phenotype via a process called immunoediting. Both innate and adaptive immunity have been shown to have potential "dark sides" that can promote tumor growth or aid in its elimination.[7] Consistent with the well-established link between chronic inflammation and an increased risk of getting cancer, this polarization of immunity toward effectors that promote tumor formation.[8] Achieving a satisfactory equilibrium between these competing immune system functions is critical for the efficacy of immune surveillance and immunotherapy in cancer, since the immune system may either speed up or halt tumor growth and progression (Fig. 9.1). Following a short review of the evidence for immune surveillance, this article dives into the cells that drive pro- and anti-tumor immunity,[9] highlighting how inflammation may polarize both innate and adaptive immunity toward a pro- or antitumor phenotype.[10,11]

By demonstrating that sarcomas develop more rapidly in immunodeficient mice that were either exposed to carcinogens or received transplantation of syngeneic tumor cells,[12] this work lends more credence to the hypothesis that the immune system offers protection against nascent malignant cells.

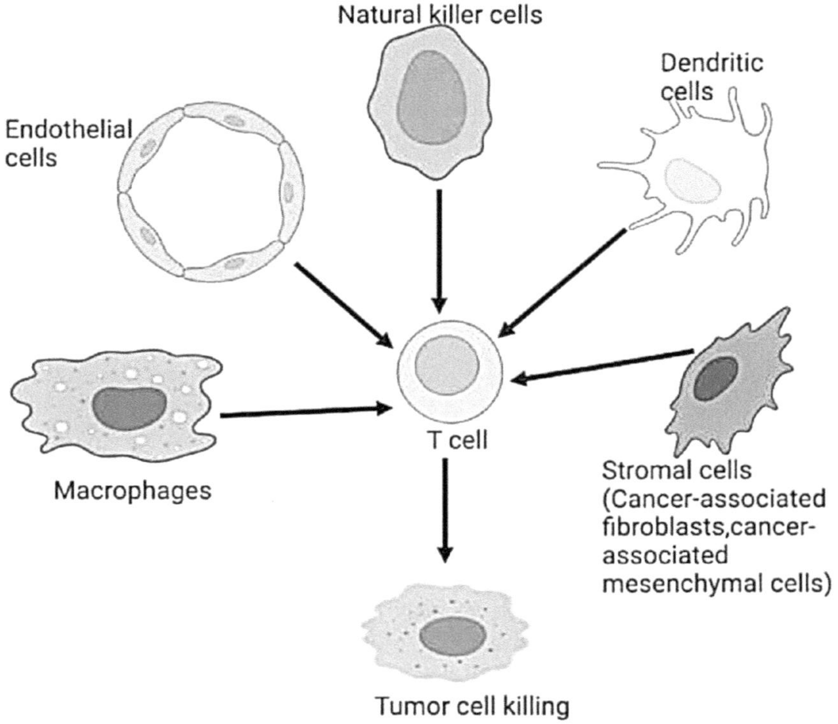

FIGURE 9.1 Schematic representation of the destruction of cancer cells through natural killer cells.

Many cases of low B and/or T-cell counts may be traced back to a lack of the recombination-activating gene-2 (RAG2), which is necessary for immunoglobulin and T-cell receptor gene rearrangements. Mice lacking innate immune effector cells such as natural killer (NK), or natural killer T (NKT) cells develop tumors more quickly after receiving a xenograft than do wild-type or immunocompetent mice.[13,14] Perforin is a protein essential for cell-mediated cytotoxicity that is used by the majority of effector cells of the innate and adaptive immune systems.

Circumstantial evidence suggests that immune surveillance and immunoediting may also occur in cancer patients. Viral and other infectious cancers are more prevalent in people whose immune systems are compromised in some other way, whether by genetics or environmental toxins. People who have had organ transplants and are being maintained following the procedure while receiving immune suppressive drugs are 3–8 times more likely to have cancer than healthy controls, despite the fact that cancers are not disseminated uniformly throughout all anatomical sites. Patients with ovarian, colorectal,[15] and melanoma who have tumor-infiltrating lymphocytes tend to do better than those without them. Evidence from both animal and human studies, as well as clinical trials, suggests that the immune system may have a role in preventing cancer from developing and causing existing tumors to go into remission.[16]

9.2 INNATE ANTITUMOR RESPONSES

Errors in DNA replication, oxidative damage, microbial infection, inflammation, and other stresses may all negatively affect normal cells, but these cells have a robust defense mechanism to help them survive. Failure of the DNA damage response to repair single- or double-stranded DNA breaks[17] increases the risk of malignant transformation. The immune system's job is to provide defense against genotoxic stress that comes from the outside.[18] The NKG2D ligands MICA, MICB, and six UL16 binding proteins in humans and RAE and H60 in rodents are activated after DNA damage through a cascade involving ATM, ATR, Chk-1, and Chk-2. When these ligands are expressed on the surfaces of stressed cells, NKG2D-dependent activation of NK, NKT, and gamma-delta (γδ; gd) T cells [and the cluster of differentiation 8 (CD8) T-cells] suppresses tumor growth through cytotoxicity and interferon-gamma (IFN)-γ production. Adaptive immunity is triggered by TLR binding by the cytoplasmic stress response molecules heat-shock protein 70 (HSP-70), high-mobility group box 1 (HMGB1), and uric acid.[19] Inappropriate stimulation of these innate reactions, as in autoimmune illness or chronic inflammation, may have deleterious effects but may aid in tumor control.[20] By increasing endocytosis, endogenous NKG2D ligand expression reduces NKG2D surface expression and dampens immunological responses.[21] NKG2D ligands are removed from tumor cells and stromal cells in the TME by the isomerase ERp5 and the metalloproteinases ADP-ribosyl adenine dinucleotide phosphate metalloproteinases.[22] This may contribute to immune suppression. Stromal cytokines, such as transforming growth factor beta (TGF-β), may play a role in facilitating suppression via their direct inhibitory impact. In addition, a low-grade inflammation might emerge, which could fuel tumor development.[23] Some research suggests that stress ligand synthesis either triggers cytotoxic antitumor responses or helps tumors elude the immune system.

9.3 INNATE IMMUNE CELLS

9.3.1 NK CELLS

The production of IFN-gamma (IFN-γ) and other effector mechanisms such as perforin and death receptor ligands by NK cells are essential to the innate antitumor response.[24] NK cells are distinguished by their capacity to lyse major histocompatibility complex (MHC) class I-deficient tumor cells even in the absence of any external stimuli.[25] NK-cell activity may be influenced by numerous factors present in the TME. These include cytokines such as interleukin 2 (IL-2) and IL-15, inhibitory and activating receptors, and costimulatory molecules including CD80, CD86, CD40, CD70, and ICOS.[26] NK cell depletion makes animals more susceptible to methylcholanthrene-induced tumors,[27] lending credence to the hypothesis that NK cells are essential to tumor defense.[28] Injections of antibodies targeting the membrane proteins NK1.1 or asiago-GM1 produce this deficiency.

NK cells provide negative regulatory signals through several types of inhibitory receptors after attaching to MHC class I molecules on their targets. Proteins such as killer cell immunoglobulin-like receptors (KIRs) and C-type lectin-like molecules (CD94 and NKG2A/E) are found in primates (primates and rodents). Because of this heterogeneity in inhibitory receptor expression, NK cells are especially vulnerable to the loss of a single MHC class I allele.[29]

Other Ly49 proteins, NKG2D, and the natural cytotoxicity receptors (NKp46, NKp44, NKp30, and NKp80) are also part of the NK cell activating receptor family. Recognizing and eliminating cancer cells, as well as stopping their spread, are only two of the numerous roles that natural cytotoxicity receptors play.[30] An intriguing fact about the activating receptor NKp44 is that it may also have inhibitory properties for cancer immunity,[31] and this may depend on the ligand it binds to. However, it is not yet known how much protection these receptors really provide against tumors in general.[32]

The NKG2D pathway, in contrast to natural cytotoxicity receptors,[33] has been demonstrated to be critical in NK cell tumor detection. Transformed cells produce NKG2D ligands in response to DNA damage, indicating that NKG2D-mediated responses provide a crucial link between genotoxic stress and the eradication of a target cell by the immune system. Lack of NKG2D (NK2GD) in mice makes them more vulnerable to many different types of cancer and increases the tumor burden in those animals who get cancer.[34] Tumors induced by chemical carcinogenesis in wild-type mice often lack NKG2D ligand expression, suggesting that escape of NKG2D monitoring is adaptive. Antibody injections into mice that inhibit NKG2D increase their susceptibility to chemically produced tumors,[35] revealing new insights into the potent NKG2D-dependent mechanism of tumor suppression. Perforin has a crucial role in NK cell anticancer effects, despite interferon-g and tumor necrosis factor-related death-inducing ligand (TRAIL)[36] synthesis being more important in other contexts. This theory is corroborated by research showing that sarcomas derived from chemically produced tumors in perforin-deficient animals express the NKG2D ligand Rae-1 and are rejected after transplantation into wild-type mice. Recent animal studies show that NK cells may develop memory-like responses to various antigens (viral, haptens)[37] and IL-12, IL-15, and IL-18 cytokine stimulation.[38] There needs to be further investigation into whether or if individuals have NK-cell responses and the possible role they play in tumor immunity.

9.3.2 B LYMPHOCYTES

Subcutaneous injections of tumor-reactive monoclonal antibodies in cancer patients might produce powerful antitumor effects. Most cancer vaccines and other drugs designed to generate tumor-reactive antibodies fall short of their intended goal of promoting tumor rejection.[39] Carcinogenesis has been related to several soluble products of activated B cells. Researchers showed that B cells promote a chronic inflammatory microenvironment that recruits innate immune cells and factors to the tumor site, thereby establishing a stromal environment that supports de novo carcinogenesis, using a transgenic mouse model in which the human keratin promoter drives expression of early region genes of human papillomavirus. Taken together, these results raise the possibility that humoral immunity promotes cancer growth by increasing innate immune system activation.[40,41]

9.3.3 MACROPHAGES

Macrophages have a role in every area of defense, making them an important part of the immune system. All sorts of different kinds of cells make up this population. Macrophages may either help in cancer killing or encourage tumor growth and metastasis, but to a lower amount than $CD4^+$ T-cells.

Tumor-killing nitric oxide and type-1 cytokines and chemokines are released by "classically activated" macrophages in response to interferon and bacterial lipopolysaccharides. The macrophages serve as both antigen-presenting cells and cytotoxic $CD8^+$ T stimulators.[42] Activated macrophages stimulate tumor development by using an "alternative" route (with IL-4, IL-13, and/or TGF-) that leads to enhanced angiogenesis and the production of type-2 cytokines and chemokines.[43] Mills came up with the labels "M1-M2" to describe the two kinds of activated macrophages based on similarities between the Th1/Th2 paradigm[44] and his findings of the cytokine profiles of conventionally and alternatively activated macrophages. Despite the fact that M1 and M2 macrophages are simply two ends of a continuum, Mantovani and colleagues.[45,46]nonetheless came up with this nomenclature.

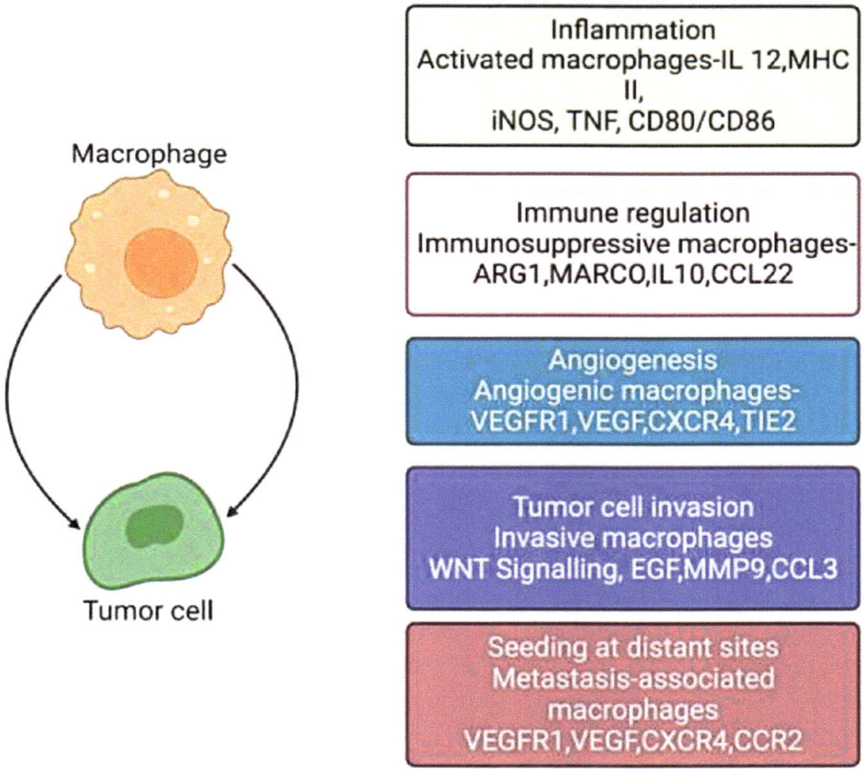

FIGURE 9.2 Schematic diagram of macrophage during inflammation activation.

Most quickly growing tumors are invaded by swarms of macrophages. Tumor-associated macrophage swarming (TAMS) are essential for tumor growth-sustaining mechanisms such as angiogenesis and matrix remodeling and make up a significant portion of the tumor stroma. Researchers using a spontaneous mouse mammary tumor model discovered that neoangiogenesis was decreased by removing TAMS and that angiogenesis was increased by infiltrating TAMS early in the process of tumor development. Intravital multiphoton imaging of in vivo breast tumors provides visible evidence that TAMS facilitates metastasis by allowing tumor cells to invade local blood vessels.[47,48] TAMS enhances tumor development, as seen in human ovarian cancer.[49] TAMS do this by expressing B7-H4, a negative regulator of T-cell activation.

TAMs are often categorized as M2 macrophages due to their role in fostering tumor growth. Gene profiling has shown that TAMS share several components with M2 macrophages, but also exhibit features of M1 macrophages, such as IFN-inducible genes.[50,51] This suggests that TAMS occupy a middle ground between the two extremes of the macrophage phenotypic spectrum.

9.3.4 MYELOID-DERIVED SUPPRESSOR CELLS (MDSC)

Myeloid-derived suppressor cells (MDSC), a morphologically and functionally varied population of cells, are seen in higher numbers in almost all cancer patients and animals with cancer developed in experimental models.[52] They suppress antitumor immunity at both the innate and adaptive levels by reducing the number of $CD8^+$ and $CD4^+$ T-cells, NK cells, and NKT cells, and by preventing DC maturation.[53,54] Arginase and/or reactive oxygen species (ROS) secreted by TAMS and stromal cells (MDSC) suppress T-cell proliferation.[55,56] However, the need for a variety of mediators is consistent across all cancers. Some MDSCs lack CD80 expression but are nevertheless capable of performing their suppressive roles (Refs. [57, 58]), whereas in other cases CD80 expression is essential. MDSC derived from either IL-4R-deficient or wild-type mice exhibited equal levels of inhibition. The IL-4 receptor (IL-4R) similarly has a similar function for IL-13 to activate certain MDSC. By cleaving tyrosines on $CD8^+$ T-cell TCRs,[59] MDSC suppresses $CD8^+$ T-cell activity, rendering CD8+ T-cells unable to react to antigen presentation through peptide-MHC I complexes. Suppression occurs as a result of interactions between MDSC and T-cells.

MDSC inhibits T-cell activation and produces $CD4^+$ T regs through an IL-10 and IFN-dependent, ROS-independent mechanism. They polarize immunity toward a tumor-promoting type-2 phenotype by inhibiting macrophage production of the type-1 cytokine IL-12 and secreting large amounts of IL-10. Improved IL-10 production by MDSCs is a result of macrophages' facilitation of tumor development.[60]

Like other immune system cells, prolonged inflammation promotes the protumorigenic activity of MDSC. There is a correlation between elevated levels of IL-1 and IL-6 and a subsequent decrease in MDSCs.[61,62] Coculture of c-kit$^+$ mice bone marrow stem cells with PGE2 results in the production of Gr1$^+$CD11b$^+$ MDSC that inhibits the immune system,[63] while arginase expression in human MDSC is up-regulated by cyclooxygenase-2 (COX-2) generated by human lung cancer cells.[64]

9.4 DENTRIC CELLS PRIME T-CELLS AND OFFER ANTIGENS

One kind of innate immune cell called dendritic cells (DCs) is commonly thought to have a pivotal role in setting off anticancer immunity. In the TME, DCs act as APCs to initiate the cancer–immunity cycle by cross-presenting tumor-associated antigens to naive T-cells.[65] The control of T-cell priming in mice is dependent on conventional dendritic cells (cDCs). Different from the CD103$^+$ cDC1 lineage, which is specialized for priming CD8$^+$ cytotoxic T lymphocytes, the CD11b$^+$ cDC2 lineage may have a function in priming helper CD4$^+$ T-cells (CTLs). However, despite their critical role in orchestrating T-cell infiltration into tumors, circulating DC-1 (cDC1) cells are surprisingly few. In a mouse model of melanoma, Batf3$^+$ CD103$^+$ cDC1s were shown to play an essential role in CTL recruitment to the TME. By secreting chemoattractants including CXCL9/10, these cDC1s aid in the recruitment of T lymphocytes.[66] T-cell immunosurveillance in cancer may be enhanced by techniques that increase cDC1 production and activation at tumor locations.[67] By increasing the number of circulating DC-1s (cDC1s), such as by the use of FMS-like tyrosine kinase 3 ligands (FLT3L) and a TLR3 agonist, tumors become more susceptible to PDL1 immune checkpoint inhibition.[68,69] Antigens from apoptotic and necrotic tumor cells are picked up by a small but sizable population of CD103$^+$ DCs inside the tumor bed. From the primary tumor site to the secondary lymphoid organs, CD103$^+$ DCs cross-present antigens to naive T-cells.[70,71] Lymph nodes (LN) collect soluble antigens from draining tissues (such as tumors) and present them for CD8$^+$ cDC1s to recognize.

The function of cDC2s in regulating cancer immunosurveillance is just now becoming appreciated. There are many subsets of coordination DC2s (cDC2s) that work together to coordinate CD4$^+$ T-cell priming. Analyses of individual cells have shown that distinct developmental processes and transcriptional regulators give rise to the varied subsets of cDC2.[72] Distinct subsets of cDC2s show different capacities to promote differentiation of conventional $CD4^+$ T-cell subsets, suggesting they may play distinct roles in regulating anti- and proinflammatory immune responses, in contrast to their shared ability to promote regulatory T-cell (Treg) differentiation.[73–75] Tumors have 25-fold more cDC2s after granulocyte macrophage–colony-stimulating factor (GM–CSF) therapy. The presence of cDC2s in a melanoma mouse model was negatively related to the presence of Tregs. Genetic decrease of Tregs has been shown to improve antigen-specific CD4+ T-cell priming, cDC2 migration to tumor-draining LNs, and $CD4^+$ T-cell differentiation in vivo. Finally, it was shown that a higher frequency of cDC2 and a lower frequency of Treg was associated with better progression-free survival (PFS) in newly diagnosed cases of head and neck cancer. 24 Patients with melanoma who benefit from anti-PD1 treatment may have higher numbers of cDC2s in their tumors. This data implies that cDC2s and Tregs may communicate with one another to influence the immune response against cancer and the success of immunotherapy.[76]

Under the correct circumstances, antigen-loaded DCs may cause extreme excitement in activated T-cells. However, DCs need to go through a process termed "licensing" to gain the ability to excite naive T-cells before they can induce the formation of antigen-specific CTLs.[77] Accreditation of DCs requires the CD40–CD40L pathway. It is extensively expressed on APCs such DCs, macrophages, and B cells since it is a member of the TNF receptor superfamily. Following antigen recognition, $CD4^+$ T helper cells expressing CD40L migrate to DCs expressing the matching CD40 receptor. Priming of T cells is facilitated by upregulation of MHC class II, CD80 (B7-1), and CD86 (B7-2) expression on DC surfaces through activation of the CD40 pathway.[78] New studies confirm that when $CD4^+$ T-cells license DCs and promote CCL5 production, CCR5-naive $CD8^+$ T-cells are attracted to the region. DCs work in concert with other cells of the adaptive immune system to orchestrate the development of antigen-specific T lymphocytes.[79]

9.5 MACROPHAGE PHENOTYPIC PLASTICITY REGULATES T-CELL PRIMING

Macrophages are another kind of innate immune cell that, like DCs, may promote or inhibit the expansion of effector T-cells. Still, macrophages are the most common kind of myeloid-derived immune invading cell in the great majority of solid tumors. Because of their tremendous phenotypic plasticity and functional flexibility, they may assume several activation modes. Macrophages are highly plastic cells that may alter their appearance and behavior in response to their immediate environment.[80]

Macrophages have several functions in tissue homeostasis and may respond physiologically to their surroundings. Depending on the stimulus, macrophages may switch from pro- to anticancer activity,[81] as indicated by in vitro investigations of macrophage polarity. IFN- and TLR agonists are two examples of substances that might be employed to promote the expansion of M1-phenotype antitumor macrophages.[82] IL-10 and IL-17 suppress the expansion of M1 phenotype macrophages, whereas IL-4 and IL-13 encourage the development of M2 phenotype macrophages, which are associated with protumor qualities. The M1-M2 distinction is simplistic, but it might provide a starting point for considering macrophage polarization. "Classically" activated M1 macrophages, for example, play a pivotal role in coordinating the immune response to exogenous infections.[83] Thus, M1 macrophages demonstrate differential up-regulation of the expression of MHC II molecules and secrete a potent profile of proinflammatory cytokines (including IL-6, IL-12, and IL-23).[84] Anti-inflammatory cytokines, such as IL-4 and IL-10,[85] are secreted by "alternatively" activated M2 macrophages, which repair tissue damage left over after the initial M1 response. To help in adaptive immunosurveillance, macrophages are generally stimulated by cancer and carcinogenesis to adopt an M1-like phenotype, whereas polarization toward an M2-like state is hypothesized to infuse suppressive T-cell antitumor immunity.[86]

Despite the widespread belief that macrophages dampen the immune system's ability to fight cancer, several studies have shown that macrophage invasion of tumors may actually stimulate the growth of a T-cell-mediated immune response. The bulk of APCs are migratory and lymphoid-resident DCs[87]; however, M1-like macrophages in the TME may phagocytose malignant cells and cross-present tumor antigens to trigger $CD8^+$ T-cells. Recent studies have shown that DCs' ability to stimulate and activate CTLs is enhanced by the presence of macrophages in extra-tumoral tissue. The spleen's MZMs may cross-prime phagocytosed antigens to help generate an adaptive immune response.[88] This suggests that under perfect conditions, macrophages might improve anticancer effectiveness by encouraging the expansion of T-cells that recognize tumor antigens.

Although macrophages have been shown to boost cancer immunity, malignancies often use them to sabotage adaptive immunosurveillance. By inhibiting the TCR from attaching to its CD8 coreceptor, M2-like macrophages limit T-cell activation.[89] Increasing the number of branches in the N-glycans on the surface of effector T-cells is one method for achieving this goal. The capacity of $CD8^+$ T lymphocytes to bind peptide-loaded MHC might be inhibited by reactive nitrogen species (RNS) produced by macrophages if these RNS nitrate tyrosine residues on the TCR. Metabolic changes enabled by macrophages allow T-cells to become completely activated when their TCR attaches to their matching antigen. Immunological checkpoint proteins like PDL1 and B7-H4 are only two examples of the variety of inhibitory signals sent by tumor-infiltrating macrophages that dampen T-cell activation. Due to their ability to regulate T-cell function, macrophages may impede the formation of an adaptive immune response to cancer.[90]

Macrophages' employment of indirect interference strategies might lead to T-cell malfunction. More and more research indicates that macrophages are responsible for swallowing cancer antigens and secreting them away from APCs.[91] One thing to keep in mind is that macrophages (CD169+) from the subcapsular sinus (SCS) can end up in the LNs that drain tumors. Macrophages from the SCS populate the space surrounding the LN's interstitium. The region containing LNs and lymph fluid.[92] Macrophages in the splenic choroid plexus (SCS) are among the first immune cells to encounter tumor-associated antigens and apoptotic cellular debris.[93] MZMs that express CD169 in the spleen phagocytose tumor-associated antigens and cross-present them to $CD8^+$ T-cells, which in turn increases anticancer immunity in mice models.[94] A recent body of research, however, suggests that $CD169^+$ macrophages may also have a secondary purpose in limiting the adaptive immune response. Those effects may be immunosuppressive in the case of splenic MZMs, preventing the host from gaining any advantage from a T-cell-directed response to auto- or self-antigens.[95,96] It is possible that $CD169^+$ macrophages are at the root of this finding since they are the ones that eat antigens before they can set off an immune response. Cancer-related LN SCS macrophages may "hoard" tumor-associated antigens to prevent them from being presented to T-cells.[97]

9.6 INNATE IMMUNE CELLS CREATE HOT AND COLD T-CELL MALIGNANCIES

To reach the tumor bed and eliminate cancer cells, tumor-antigen-specific activated T-cells must enter the TME (TME). Consequently, a significant number of T lymphocytes must infiltrate the TME before the effector function of these cells may be activated.[98] Tumors may differ substantially in both the immunological context of their microenvironment (TME) and the composition of the immune cells that infiltrate them. Infiltrated tumors with T-cells are called "inflamed" tumors, and a high concentration of $CD8^+$ T-cells is a consistent indicator of improved prognosis and survival outcomes across a variety of malignancies.[99] Using transcriptomic analysis, researchers have shown that immunoreactive tumor beds increase cellular expression of IFN signaling genes.[100] Low survival rates are associated with T-cell "cold" malignancies, which are characterized by an overabundance of Tregs and immunosuppressive myeloid cells.[101] These results provide an intriguing idea, namely, that the presence of "hot" or "cool" T-cells in a tumor is determined by the degree to which the innate immune system regulates TME phenotypic plasticity.

Crucial for antitumor immunity is the capacity of activated T-cells to penetrate the tumor microenvironment, which is mostly attributable to the participation of innate immune cells (TME).[102] T cell-inflamed tumor development is critically dependent on cytokines and soluble substances produced by innate immune cells, according to research.[103] The innate immune system facilitates the cytolytic activity of T-cells by hastening their migration from the secondary lymphoid organs (where T-cells are primed and activated) to the tumor bed. $CXCR3^+$ T-cell recruitment into tumors is facilitated, for instance, by IFN-activation of APCs, which results in increased production of CXCL9.[104] The production of CXCL10 from Batf3-dependent CD8+ DCs has been shown to induce the infiltration of $CXCR3^+$-activated T lymphocytes into the tumor nest.[105] For anticancer immunity to develop, myeloid-derived cytokines are essential for encouraging the infiltration of tumor antigen-specific T-cells into the tumor bed.[106]

However, other data suggest the innate immune system may contribute to the development of T-cell-cold tumors.[107] Tumors with less infiltrating T-cells also have a larger myeloid infiltration, as indicated by studies of patient tissue.[108,109] Patients with pancreatic ductal adenocarcinoma had a worse prognosis and a worse response to neoadjuvant immunotherapy if their tumors included a high number of tumor-infiltrating myeloid cells, such as $CD163^+$ macrophages and $CD66b^+$ granulocytes.[110] (PDAC). The presence of myeloid cells is negatively related to $CD8^+$ T-cell infiltrates in breast cancer.[111] Intriguingly, these converging findings suggest that the myeloid compartment may play a critical role in driving T-cell exclusion in cancer.

Initiating T-cell access to the tumor microenvironment has been shown in preliminary investigations to be blocked by innate immune cells. Therapeutic reduction of tumor-associated macrophages (TAMs) by blocking cytokine-stimulating factor-1R dramatically increases the fraction of tumor-infiltrating $CD8^+$ T-cells in a mouse model of breast cancer (CSF-1R). The presence of macrophages in the TME may contribute to the development of the malignancy by preventing CD8+ T lymphocytes from accessing and directly engaging malignant cells. $CD8^+$ T-cell penetration into tumors is also enhanced by decreasing Ly6CloF4/80^+ extratemporal macrophages in a mouse model of spontaneous PDAC.[112] Myeloid cell exclusion from the immune system may be aided by RNS secretion, which may start post-translational alterations in T-cell chemoattractants. Myeloid-derived RNS limits T-cell access to the tumor nest by nitrating CCL2. T lymphocytes may be discouraged from accessing the tumor nidus by cytokines produced by stromal cells in the area. Recent research suggests that stromal leukocytes, in collaboration with cancer-associated fibroblasts, may increase TGF- signaling to block T-cell entrance. Pharmacologic suppression of TGF-signaling coupled with blockage of PDL1 increased T-cell infiltration into tumors in animal tumor models.[113] These results demonstrate that the innate immune response to cancer is a key factor in the efficacy of T-cell immunosurveillance. Thus, macrophages, granulocytes, and DCs, which are not tumor cells, may have a major effect on cancer immunogenicity.[114]

9.7 IMMUNE ESCAPE MECHANISMS IN TUMORS THREATEN PRODUCTIVE CANCER IMMUNOSURVEILLANCE

Tumors are immunogenic enough to trigger immune surveillance, which is how the cancer–immunity cycle may function. New studies suggest that cancer cells may hide from the body's natural anticancer defenses. There are two basic ways cancers evade the immune system and grow. To begin, cancer cells may undergo phenotypic changes that favor less immunogenic tumor forms via Darwinian evolution. Second, modulating the host immune system may aid cancers in evading the body's natural elimination process.[115] When tumor cells are present, they might hinder adaptive immune surveillance or reroute the immune system, both of which can hasten the course of the illness. This means that cancer not only increases the tumorigenic potential of the immune system but also selectively suppresses its anticancer properties. Immunoediting requires cancer cells and a healthy host immune system.

Tumor cells may take on disguises to evade the immune system's destruction by adopting new characteristics. Antigen-presenting machinery (APM) deficiencies and downregulation of MHC class I molecule expression are common phenotypic manifestations of this in malignant cells.[116] Downregulation of genes involved in APM production[117] has been seen in a variety of malignancies as determined by transcriptomic analysis of patient tumors. These findings suggest that neoplastic cells modify the translocation of intracellular components for presentation on the MHC complex to evade immune recognition. Loss of heterozygosity (LOH) at the MHC locus is another genetic alteration linked to cancer protection. The frequency of certain MHC haplotypes in both the germline and cancer genomes of a given patient may be counted to estimate LOH occurrences,[118] thanks to recent improvements in bioinformatics. Malignant cells from individuals with treatment-naive non-small-cell lung cancer show deletion of one parental MHC haplotype in approximately 40% of tumor samples using this method.[119] LOH at the MHC locus allows malignant cells to present only the neoepitopes that bind most strongly to the conserved haplotype, even though haplotype diversity is essential for the synthesis of different MHC molecules with variable antigen binding affinities. For this reason, it is believed that the lack of MHC haplotype favors the selection of less immunogenic tumor variants that are able to evade host immune clearance by downregulating the presentation of tumor-associated antigens.[120]

9.8 THE MATURATION OF T-CELLS IS INHIBITED BY PROCESSES INTERNAL TO CANCER CELLS

Cancers may hide from the body's adaptive immune system in a variety of ways. Several places throughout the intricate multistep process of making tumor antigen-specific T lymphocytes are vulnerable to interference by malignant cells. It may be challenging for $CD103^+$ DCs to acquire antigens and cross-present them when malignant cells are present in a tumor microenvironment (TME). In a Pten-deficient BRAFV600E melanoma model, downregulating CCL4 synthesis by stimulating Wnt/-catenin signaling[121] decreased tumor invasion by $CD103^+$ DCs. As an added layer of defense, tumor-derived cholesterol metabolites such as "liver X receptor-" may suppress CCR7 expression on DCs, preventing DCs loaded with tumor antigens from migrating to LNs that drain tumors.[122] Other soluble chemicals produced by tumors have been demonstrated to impair DC biology as well. To illustrate, tumors promote

DC tolerogenic by the secretion of IL-10 and IL-6, which dampens adaptive immune response to tumor-specific antigens. The production of vascular endothelial growth factor (VEGF) inhibits HSC differentiation into DCs. So, it is probable that the functional dysregulation of key events in the cancer-immunity cycle is due to mechanisms inherent to cancer cells.[123]

Tumors may achieve immunologic acquiescence by dampening the activity of effector T-cells that have made their way into the TME. Canonical immunological checkpoint activation pathways play a crucial role in this system. PDL1 on tumor and myeloid cells directly inhibits T-cell effector activity by binding to its corresponding receptor, PD1, on the surface of activated T-cells.[124] Exosomes isolated from metastatic melanoma tumors have been shown to contain PDL1,[125] suggesting that tumor-derived cellular cargo may further worsen this immunoregulatory response. More so, tumors may alter the biochemistry of the TME to promote the expansion of defective T-cells and an enzyme called indoleamine 2,3-dioxygenase (IDO)[126] is produced by cancer cells and myeloid cells to help them accomplish just that. The enzyme IDO is responsible for the conversion of tryptophan to kynurenine. Since tryptophan is essential for T-cell proliferation, alterations in TME nutritional homeostasis suppress T-cell function.[127] Myeloid cells and tumor cells work together to release a plethora of cytokines that suppress the cytolytic effector function of T-cells and slow down tumor metabolism. Inhibiting gene expression that controls proteins such as perforin and granzyme, which together generate T-cell-mediated cytotoxicity, is one mechanism by which tumor-derived TGF may compromise immune system function. Immune-suppressing macrophages ($CCR2^+$ and $CSF1R^+$) and tumor-feeding granulocytes (CXCR2+) are recruited to the tumor site by chemokines secreted by certain tumor cells and myeloid cells. Therefore, cancer cells and innate immunity both contribute to the TME learning to use many pathways that may decrease T-cell immunity.[128]

Malignant cells take advantage of cancer's tumor-initiating capacity by dominating the immune system. The recruitment of myeloid cell subsets that, once established in the TME, severely impede adaptive immunosurveillance and enhance cancer's proliferation capacity is mostly responsible for this phenomenon. Therefore, tumors take advantage of the TME's adaptability to develop a hospitable niche, or zone of "immune privilege," that shields tumor cells from destruction by the immune system.[129–131]

Cancers create a growth-friendly microenvironment by luring immune-suppressing cells to their periphery. Cancer cells recruit specific immune cells to the tumor core by secreting cytokines (such as CXCL1, GM-CSF,

IL-1, IL-4, and IL-13) through paracrine signaling pathways.[132] An invasion-competent $CD11b^{+}Gr1^{+}$ myeloid cell population is promoted in a familial pancreatic cancer model by a tumor-derived GM–CSF. Neutrophils, monocytes, and other innate immune cells, such as $CD11b^{+}Gr1^{+}$ cells,[133] contribute to cancer development. By secreting the enzyme arginase-1, which catalyzes the catabolic degradation of arginine,[134] CD11b+Gr1+ neutrophils deplete resources in the local microenvironment. Low CD3 expression and poor T-cell proliferation due to arginine deprivation impede T-cell activation through the TCR. $CD11b^{+}Gr1^{+}$ myeloid cells may provide an inflammatory microenvironment that aids in tumor growth. The proinflammatory cytokine IL-1 induces nuclear factor kappa B (NF-κB) signaling in $CD11b^{+}Gr1^{+}$ cells, which in turn recruits myeloid cells and increases the onset of gastric neoplasia, as revealed in preclinical investigations.[134] IL-1 production and the creation of an immunosuppressive environment have been shown to come from the activation of TLR4 on epithelial cells, through the NLRP3 inflammasome.[135] Tumor cell death signals, such as ATP, may stimulate resident immune cells. Put another way, extracellular ATP is a DAMP that acts as a chemoattractant, attracting circulating neutrophils (ATP). ATP is converted into immunosuppressant adenosine by ectonucleotidases such as CD39 and CD73.[136] Adenosine may have immunosuppressive effects by, for example, inhibiting T-cell biology or increasing immunosuppressive myeloid cells.[137–140] Since the TME permits innate immunity to hamper adaptive immunosurveillance,[141] cancer cells inside the TME orchestrate tumor-promoting pleiotropic effects.

Because tumors alter the cytokine profile of the TME, these cells may become immunosuppressive or acquire protumorigenic features.[142] Macrophages stand out among innate immune cells because of their remarkable phenotypic diversity and plasticity. Tumor-secreted cytokines such as IL-4, IL-13, CCL2, and CSF have been shown to polarize macrophages, causing them to develop traits that cancer cells may use to their advantage.[143] The possibility that TAMs reduce the immune system's capacity to combat tumors is widely known. Mice with breast cancer are unable to mount an adaptive T-cell response because macrophage-derived IL-10 inhibits DC-produced IL-12.[144] Tumor-associated macrophages (TAMs) could encourage tumor development by boosting circulation to the malignant outgrowth. TGF and low oxygen levels in the TME cause infiltrating macrophages to increase their production of the angiogenesis-regulating protein VEGF.[145,146] Since TAMs have the potential to suppress T-cell immunity as well as nourish tumor growth and metastasis, they may be an important factor in both processes.

9.9 THE EVOLUTION OF TUMORS IMMUNOEDITING REQUIRES A NEW CANCER IMMUNOTHERAPY APPROACH

Tumor cells can evade detection and elimination by the body's defenses by altering their antigenicity (i.e., the expression and presentation of tumor antigens), immunogenicity (i.e., the potential to alert tumor-specific immunity), the immunosuppressive microenvironment, and the functional capacity of the immune system (i.e., the ability of the immune system to produce antitumor activity). Immune-evading tumor development pathways are cancer-causing and detrimental to T-cell immunosurveillance.[147] Examples of these characteristics include the low antigenicity of pancreatic cancer due to the small number of mutated neo-proteins, the poor T-cell response due to low immunogenicity, the immunosuppressive microenvironment directed by myeloid cells, fibroblasts, and Tregs, and the systemic inflammatory reaction associated with T-cell dysfunction.[148] This biology may shed light on why immune checkpoint suppression, either alone or in combination, has not been successful in treating pancreatic cancer (ICB).[149] In the case of glioblastoma and other very aggressive brain tumors, immunotherapy has not shown to be a successful treatment option. These T-cell "cool" malignancies[150] are an example of how important criteria like the immunologic state of the TME are in determining the therapeutic potential of immunotherapy.

Immunotherapy, however, is ineffective against the vast majority of solid tumors. If we want to maximize the production of tumor-specific T-cells and the efficacy of immunotherapy, we may need to "train" tumors to an immunostimulatory state.[151]. Techniques that aim to "keep" cancer immunogenicity may be necessary for long-lasting responses,[152] given that tumors may adapt to immunological pressure. We elaborate on our proposed two-pronged strategy for cancer immunotherapy in the following. The TME is first transformed from a tolerogenic to an immunostimulatory environment via a process we term training. The goal of this conditioning procedure is to increase the efficacy of immunotherapy in the treatment of tumors.[153] This priming may be accomplished in several ways, including I the elimination or suppression of myeloid cell populations often seen in tumors and (ii) the repolarization of immune cells that have shown anticancer characteristics.[154] The second stage, dubbed "maintenance," involves preventing tumors from switching to an immunosuppressive TME by inhibiting possible channels of acquired resistance.[155,156] When using T-cell-directed immunotherapy, it is crucial to begin the maintenance phase as soon as possible. It is hoped that this training and maintenance method may spark new combinatorial strategies that increase the therapeutic efficacy of immunotherapy.[157] Not all tumors need the same "conditioning" and "maintenance" treatment. For instance, in many people with melanoma, ICB is sufficient to initiate a robust and long-lasting anticancer immune response.[158,159] The results of this investigation demonstrate that melanoma often has enough antigenicity and that T lymphocytes have substantial potential for revitalized antitumor action. Furthermore, the results demonstrate immunotherapy's potential for influencing the immunologic status of the TME around melanoma cancer cells.[160–162] Some patients with solid tumors may have dramatic and long-lasting responses to T-cell-directed immunotherapies such as antiprogramed death 1/programed death ligand 1 (anti-PD-1/PD-L1) and anti-cytotoxic T lymphocyte antigen 4 (anti-CTLA-4).[163] Tumor sensitization efforts will be necessary for the vast majority of cancer types. Here, we talk about ways to increase the efficacy of immunotherapy by conditioning and keeping tumors in an immunostimulatory state.[164]

9.10 CONCLUSION

In the TME, communication between tumor cells and immune cells is feasible. Complex two-way interactions between the immune system and cancer cells inside the TME may control the progression of the illness. To better understand the significance of the immunoreactive state of cancer, we investigated how innate and adaptive immunity work together to promote or suppress tumor development. In this article, we examine how tumors alter the phenotype of the TME to evade adaptive immunosurveillance and take advantage of the immune system's tumorigenic potential. Furthermore, we argue that it may be crucial to exploit the TME's natural flexibility to overcome resistance mechanisms that are driven by cancer cells. As evidence mounts that methods that preserve the early immune response to cancer immunotherapy might boost the effectiveness of immunotherapy in this context, they may come to be emphasized more in future cancer treatment. The complex interplay between tumor cells, stroma, and immune system components has recently been shown through research into endogenous anticancer responses.

When cancer cells release stress-inducible signals, the body's innate immune system responds by trying to suppress the illness. An adaptive immune response, with its improved specificity and memory, might develop as a result of such successful countermeasures. Tumor cell proliferation, survival, invasion, angiogenesis, and metastasis may all be aided by the interplay between innate and adaptive immune components during extended immune activation. The success of immunotherapeutic treatments depends on researchers' ability to identify and characterize the molecular processes that underlie the induction of either tumor-protective or tumor-promoting host responses.

KEYWORDS

- **innate immunity**
- **adaptive immunity**
- **tumor**
- **malignancies**
- **antitumor responses**

REFERENCES

1. Swann, J. B.; Smyth, M. J. Immune surveillance of tumors. *J. Clin. Invest.* **2007,** *117*, 1137–1146.
2. Reiman, J. M.; Kmieciak, M.; Manjili, M. H.; Knutson, K. L. Tumor Immunoediting and Immunosculpting Pathways to Cancer Progression. *Semin. Cancer Biol.* **2007,** *17*, 275–287.
3. Dunn, G. P.; Koebel, C. M.; Schreiber, R. D. Interferons, Immunity and Cancer Immunoediting. *Nat. Rev. Immunol.* **2006,** *6*, 836–848.
4. Smyth, M. J.; Dunn, G. P.; Schreiber, R. D. Cancer Immunosurveillance and Immunoediting: The Roles of Immunity in Suppressing Tumor Development and Shaping Tumor Immunogenicity. *Adv. Immunol.* **2006,** *90*, 1–50.
5. Galon, J.; Costes, A.; Sanchez-Cabo, F.; Kirilovsky, A.; Mlecnik, B.; Lagorce-Pages, C.; Tosolini, M.; Camus, M.; Berger, A.; Wind, P. et al. Type, Density, and Location of Immune Cells within Human Colorectal Tumors Predict Clinical Outcome. *Science* **2006,** *313*, 1960–1964.
6. Zou, W. Regulatory T Cells, Tumour Immunity and Immunotherapy. *Nat. Rev. Immunol.* **2006,** *6*, 295–307.
7. Baratelli, F.; Lin, Y.; Zhu, L.; Yang, S. C.; Heuze-Vourc'h, N.; Zeng, G.; Reckamp, K.; Dohadwala, M.; Sharma, S.; Dubinett, S. M. Prostaglandin E2 Induces FOXP3 Gene Expression and T Regulatory Cell Function in Human CD4+ T Cells. *J. Immunol.* **2005,** *175*, 1483–1490.
8. Sharma, S.; Zhu, L.; Yang, S. C.; Zhang, L.; Lin, J.; Hillinger, S.; Gardner, B.; Reckamp, K.; Strieter, R. M.; Huang, M. et al. Cyclooxygenase 2 Inhibition Promotes

IFN-Gamma-Dependent Enhancement of Antitumor Responses. *J. Immunol.* **2005,** *175*, 813–819.
9. Ralainirina, N.; Poli, A.; Michel, T.; Poos, L.; Andres, E.; Hentges, F.; Zimmer, J. Control of NK Cell Functions by CD4+CD25+ Regulatory T Cells. *J. Leukoc. Biol.* **2007,** *81*, 144–153.
10. Wei, S.; Kryczek, I.; Zou, L.; Daniel, B.; Cheng, P.; Mottram, P.; Curiel, T.; Lange, A.; Zou, W. Plasmacytoid Dendritic Cells Induce CD8+ Regulatory T Cells in Human Ovarian Carcinoma. *Cancer Res.* **2005,** *65*, 5020–5026.
11. Castellino, F.; Germain, R. N. Cooperation Between CD4+ and CD8+ T cells: when, where, and how. *Annu. Rev. Immunol.* **2006,** *24*, 519–540.
12. Steinman, L. A Brief History of T(H)17, the First Major Revision in the T(H)1/T(H)2 Hypothesis of T Cell-Mediated Tissue Damage. *Nat. Med.* **2007,** *13*, 139–145.
13. Langowski, J. L.; Zhang, X.; Wu, L.; Mattson, J. D.; Chen, T.; Smith, K.; Basham, B.; McClanahan, T.; Kastelein, R. A.; Oft, M. IL-23 Promotes Tumour Incidence and Growth. *Nature* **2006,** *442*, 461–465.
14. Bi, Y.; Liu, G.; Yang, R. Th17 Cell Induction and Immune Regulatory Effects. *J. Cell Physiol.* **2007,** *211*, 273–278.
15. Langowski, J. L.; Kastelein, R. A.; Oft, M. Swords into Plowshares: IL-23 Repurposes Tumor Immunesurveillance. *Trends Immunol.* **2007,** *28*, 207–212.
16. Numasaki, M.; Fukushi, J.; Ono, M.; Narula, S. K.; Zavodny, P. J.; Kudo, T.; Robbins, P. D.; Tahara, H.; Lotze, M. T. Interleukin-17 Promotes Angiogenesis and Tumor Growth. *Blood* **2003,** *101*, 2620–2627.
17. Benchetrit, F.; Ciree, A.; Vives, V.; Warnier, G.; Gey, A.; Sautes-Fridman, C.; Fossiez, F.; Haicheur, N.; Fridman, W. H.; Tartour, E. Interleukin-17 Inhibits Tumor Cell Growth by Means of a T-Cell-Dependent Mechanism. *Blood* **2002,** *99*, 2114–2121.
18. Bettelli, E.; Carrier, Y.; Gao, W.; Korn, T.; Strom, T. B.; Oukka, M.; Weiner, H. L.; Kuchroo, V. K. Reciprocal Developmental Pathways for the Generation of Pathogenic Effector TH17 and Regulatory T Cells. *Nature* **2006,** *441*, 235–238.
19. Nanni, P.; Nicoletti, G.; De Giovanni, C.; Landuzzi, L.; Di Carlo, E.; Cavallo, F.; Pupa, S. M.; Rossi, I.; Colombo, M. P.; Ricci, C. et al. Combined Allogeneic Tumor Cell Vaccination And Systemic Interleukin 12 Prevents Mammary Carcinogenesis in HER-2/neu Transgenic Mice. *J. Exp. Med.* **2001,** *194*, 1195–1205.
20. De Visser, K. E.; Korets, L. V.; Coussens, L. M. De Novo Carcinogenesis Promoted by Chronic Inflammation Is B Lymphocyte Dependent. *Cancer Cell* **2005,** *7*, 411–423.
21. Tan, T. T.; Coussens, L. M. Humoral Immunity, Inflammation and Cancer. *Curr. Opin. Immunol.* **2007,** *19*, 209–216.
22. Gordon, S. Alternative Activation of Macrophages. *Nat. Rev. Immunol.* **2003,** *3*, 23–35.
23. Mills, C. D.; Kincaid, K.; Alt, J. M.; Heilman, M. J.; Hill, A. M. M-1/M-2 Macrophages and theTh1/Th2 Paradigm. *J. Immunol.* **2000,** *164*, 6166–6173.
24. Mantovani, A.; Sica, A.; Locati, M. New Vistas on Macrophage Differentiation and Activation. *Eur. J. Immunol.* **2007,** *37*, 14–16.
25. Mantovani, A.; Schioppa, T.; Porta, C.; Allavena, P.; Sica, A. Role of tumor-Associated Macrophages in Tumor Progression and Invasion. *Cancer Metastasis Rev.* **2006,** *25*, 315–322.
26. Lin, E. Y.; Li, J. F.; Gnatovskiy, L.; Deng, Y.; Zhu, L.; Grzesik, D. A.; Qian, H.; Xue, X. N.; Pollard, J. W. Macrophages Regulate the Angiogenic Switch in a Mouse Model of Breast Cancer. *Cancer Res.* **2006,** *66*, 11238–11246.

27. Wyckoff, J. B.; Wang, Y.; Lin, E. Y.; Li, J. F.; Goswami, S.; Stanley, E. R.; Segall, J. E.; Pollard, J. W.; Condeelis, J. Direct Visualization of Macrophage-Assisted Tumor Cell Intravasation in Mammary Tumors. *Cancer Res.* **2007,** *67*, 2649–2656.
28. Kryczek, I.; Zou, L.; Rodriguez, P.; Zhu, G.; Wei, S.; Mottram, P.; Brumlik, M.; Cheng, P.; Curiel, T.; Myers, L. et al. B7-H4 Expression Identifies a Novel Suppressive Macrophage Population Inhuman Ovarian Carcinoma. *J. Exp. Med.* **2006,** *203*, 871–881.
29. Biswas, S. K.; Gangi, L.; Paul, S.; Schioppa, T.; Saccani, A.; Sironi, M.; Bottazzi, B.; Doni, A.; Vincenzo, B.; Pasqualini, F. et al. A Distinct and Unique Transcriptional Program Expressed by Tumor-Associated Macrophages (Defective NF-kappaB and Enhanced IRF-3/STAT1 Activation). *Blood* **2006,** *107*, 2112–2122.
30. Poeck, H.; Besch, R.; Maihoefer, C.; Maihoefer, C.; Renn, M.; Tormo, D. et al. 5′-Triphosphate-siRNA Turning Gene Silencing and Rig-I Activation Against Melanoma. *Nat. Med.* **2008,** *14*, 1256–1263.
31. Anz, D.; Koelzer, V. H.; Moder, S.; Thaler, R.; Schwerd, T.; Lahl, K. et al. Immunostimulatory RNA Blocks Suppression by Regulatory T Cells. *J. Immunol.* **2010,** *184*, 939–946.
32. Pan, M.; Geng, S.; Xiao, S.; Ren, J.; Liu, Y.; Li, X. et al. Apoptosis Induced by Synthetic Retinoic Acid CD437 on Human Melanoma A375 Cells Involves RIG-I Pathway. *Arch. Dermatol. Res.* **2009,** *301*, 15–20.
33. Ting, J. P.; Duncan, J. A.; Lei, Y. How the Non-Inflammasome NLRs Function in the Innate Immune System? *Science* **2010,** *327*, 286–290.
34. Chen, G.; Shaw, M. H.; Kim, Y. G.; Nunez, G. NOD-Like Receptors: Role in Innate Immunity and Inflammatory Disease. *Annu. Rev. Pathol.* **2009,** *4*, 365–398.
35. Ogura, Y.; Bonen, D. K.; Inohara, N.; Nicolae, D. L.; Chen, F. F.; Ramos, R. et al. A Frame Shift Mutation in NOD2 Associated with Susceptibility to Crohn's Disease. *Nature* **2001,** *411*, 603–606.
36. Hugot, J. P.; Chamaillard, M.; Zouali, H.; Lesage, S.; Cezaed, J. P.; Belaiche, J. et al. Association of NOD2 Leucin-Rich Repeat Variants with Susceptibility to Crohn's Disease. *Nature* **2001,** *411*, 599–603.
37. Da Silva Correia, J.; Miranda, Y. Y.; Austin-Brown, N.; Hsu, J.; Mathison, J.; Xiang, R. et al. Nod1-Dependent Control of Tumor Growth. *Proc. Natl. Acad. Sci. USA* **2006,** *103*, 1840–1845.
38. Chen, G. Y.; Shaw, M. H.; Redondo, G.; Nunez, G. The Innate Immune Receptor Nod1 Protects the Intestine from Inflammation-Induced Tumorigenesis. *Cancer Res.* **2008,** *68*, 10060–10067.
39. Martinon, F.; Petrilli, V.; Mayor, A.; Tardivel, A.; Tschopp, J. Gout-Associated Uric Acid Crystals Activate the NALP3 Inflammasome. *Nature* **2006,** *440*, 237–241.
40. Schroder, K.; Zhou, R.; Tschopp, J. The NLRP3 Inflammasome: A Sensor for Metabolic Danger? *Science* **2010,** *327*, 296–300.
41. Schroder, K.; Tshopp, J. The Inflammasome. *Cell* **2010,** *140*, 821–832.
42. Allen, I. C.; TeKippe, E. M.; Woodford, R. M.; Uronis, J. M.; Holl, E. K.; Rogers, A. B. et al. The NLRP3 Inflammasome Functions as a Negative Regulator of Tumorigenesis During Colitis-Associated Cancer. *J. Exp. Med.* **2010,** *207*, 1045–1056.
43. Elliott, M. R.; Chekeni, F. B.; Trampont, P. C.; Lazarowski, E. R.; Kadl, A.; Walk, S. G. et al. Nucleotides Released by Apoptotic Cells Act as a Find-Me Signal to Promote Phagocytic. DOI: 10.1038/nature08296.

44. Zhou, R.; Tardivel, A.; Thorens, B.; Choi, I.; Tschopp, J. Thioredoxin-Interacting Protein Links Oxidative Stress to Inflammasome Activation. *Nat. Immunol.* **2010,** *11*, 136–140.
45. Ghiringhelli, F.; Apetoh, L.; Tesniere, A.; Aymeric, L.; Ma, Y.; Ortiz, C. et al. Activation of theNLRP3 Inflammasome in Dendritic Cells Induces IL-1beta-Dependent Adaptive Immunity Against Tumors. *Nat. Med.* **2009,** *15*, 1170–1178.
46. Faustin, B.; Chen, Y.; Zhai, D.; Negrate, G. L.; Lartigue, L.; Satterthwait, A. et al. Mechanisms ofBcl-2 and Bcl-X(L) Inhibition of NLRP1 Inflammasome: Loop Domain-Dependent Suppression of ATP Binding and Oligomerization. *Proc. Natl. Acad. Sci. USA* **2009,** *106*, 3935–3940.
47. Saitoh, T.; Fujita, N.; Jang, M. H.; Uematsu, S.; Yang, B. G.; Satoh, T. et al. Loss of Autophagy Protein Atg16L1 Enhances Endotoxin-Induced IL-1beta Production. *Nature* **2008,** *456*, 264–268.
48. Bürckstummer, T.; Baumann, C.; Bluml, S.; Dixit, E.; Durnberger, G.; Jahn, H. et al. *Nat. Immunol.* **2009,** *10*, 266–272.
49. Fernandes-Alnemri, T.; Yu, J. W.; Datta, P.; Wu, J.; Alnemri, E. S. AIM2 Activates the Inflammasome and Cell Death in Response to Cytoplasmic DNA. *Nature* **2009,** *458*, 509–513.
50. Hornung, V.; Ablasser, A.; Charrel-Dennis, M.; Bauernfeind, F.; Horvath, G.; Caffrey, D. R. et al. AIM2 Recognizes Cytosolic dsDNA and Forms a Caspase-1-Activating Inflammasome with ASC. *Nature* **2009,** *458*, 514–518.
51. Roberts, T. L.; Idris, A.; Kelly, J. A.; Burnton, C. M.; Hodgson, S.; Hardy, L. L. et al. HIN-200 Proteins Regulate Caspase Activation in Response to Foreign Cytoplasmic DNA. *Science* **2009,** *323*, 1057–1060.
52. DeYoung, K. L.; Ray, M. E.; Su, Y. A.; Anzick, S. L.; Johnstone, R. W.; Trapani, J. A. et al. Cloning a Novel Member of the Human Interferon-Inducible Gene Family Associated with Control of Tumorigenicity in a Model of Human Melanoma. *Oncogene* **1997,** *15*, 453–457.
53. Bergsbaken, T.; Fink, S. L.; Cookson, B. T. Pyroptosis: Host Cell Death and Inflammation. *Nat. Rev. Microbiol.* **2009,** *7*, 99–109.
54. Patsps, G.; Germann, A.; Gebert, J.; Dihlmann, S. Restoration of Absent in Melanoma 2 (AIM2)induces G2/M Cell Cycle Arrest and Promotes Invasion of Colorectal Cancer Cells. *Int. J. Cancer* **2009,** *126*, 1838–1849.
55. Coussens, L. M.; Werb, Z. Inflammation and Cancer. *Nature* **2002,** *420*, 860–867.
56. Grivennikov, S. I.; Greten, F. R.; Karin, M. Immunity, Inflammation, and Cancer. *Cell* **2010,** *140*, 883–899.
57. Hiraoka, K.; Miyamoto, M.; Cho, Y. et al. Concurrent Infiltration by CD8+ T Cells and CD4+ T Cells Are a Favourable Prognostic Factor in Non-Small-Cell Lung Carcinoma. *Br. J. Cancer* **2006;** 94: 275–280.
58. Curiel, T. J.; Coukos, G.; Zou, L. et al. Specific Recruitment of Regulatory T Cells in Ovarian Carcinoma Fosters Immune Privilege and Predicts Reduced Survival. *Nat. Med.* **2004;** 10, 942–949.
59. Gajewski, T. F.; Schreiber, H.; Fu Y-X. Innate and Adaptive Immune Cells in the Tumor Microenvironment. *Nat. Immunol.* **2013,** *14*, 1014–1022.
60. Hervas-Stubbs, S.; Perez-Gracia, J. L.; Rouzaut, A. et al. Direct Effects of Type I Interferons on Cells of the Immune System. *Clin. Cancer Res.* **2011;** 17: 2619–2627.
61. Mellman, I.; Coukos, G.; Dranoff, G. Cancer Immunotherapy Comes of Age. *Nature* **2011,** *480*, 480–489.

62. Tu J-F, Ding Y-H, Ying X-H, et al. Regulatory T cells, Especially ICOS+ FOXP3+ Regulatory T Cells, are Increased in the Hepatocellular Carcinoma Microenvironment and Predict Reduced Survival. *Sci. Rep.* **2016,** *6*, 35056.
63. Kumar, V.; Patel, S.; Tcyganov, E. et al. The Nature of Myeloid-Derived Suppressor Cells in the Tumor Microenvironment. *Trends Immunol.* **2016,** *37*, 208–220.
64. Nagarsheth, N.; Wicha, M. S.; Zou, W. Chemokines in the Cancer Microenvironment and Their Relevance in Cancer Immunotherapy. *Nat. Rev. Immunol.* **2017,** *17*, 559–572.
65. Dangaj, D.; Bruand, M.; Grimm, A. J. et al. Cooperation Between Constitutive and Inducible Chemokines Enables T Cell Engraftment and Immune Attack in Solid Tumors. *Cancer Cell* **2019,** *35*, 885–900.e10.
66. Spranger, S.; Luke, J. J.; Bao, R. et al. Density of Immunogenic Antigens Does Not Explain the presence or Absence of the T-Cell–Inflamed Tumor Microenvironment in Melanoma. *Proc. Natl. Acad. Sci.* **2016,** *113*, E7759–E7768.
67. Nywening, T. M.; Belt, B. A.; Cullinan, D. R. et al. Targeting Both Tumour-Associated CXCR2 + Neutrophils and CCR2 + Macrophages Disrupts Myeloid Recruitment and Improves Chemotherapeutic Responses in Pancreatic Ductal Adenocarcinoma. *Gut* **2018,** *67*, 1112–1123.
68. Binnewies, M.; Roberts, E. W.; Kersten, K. et al. Understanding the Tumor Immune Microenvironment (TIME) for Effective Therapy. *Nat. Med.* **2018;** 24: 541–550.
69. Tsujikawa, T.; Kumar, S.; Borkar, R. N. et al. Quantitative Multiplex Immunohistochemistry Reveals Myeloid-Inflamed Tumor-Immune Complexity Associated with Poor Prognosis. *Cell Rep.* **2017,** *19*, 203–217.
70. DeNardo, D. G.; Brennan, D. J.; Rexhepaj, E. et al. Leukocyte Complexity Predicts Breast Cancer Survival and Functionally Regulates Response to Chemotherapy. *Cancer Discov.* **2011,** *1*, 54–67.
71. Beatty, G. L.; Winograd, R.; Evans, R. A. et al. Exclusion of T Cells from Pancreatic Carcinomas in Mice Are Regulated by Ly6Clow F4/80+ Extratumoral Macrophages. *Gastroenterology* **2015**, *149*, 201–210.
72. Joyce, J. A.; Fearon, D. T. T Cell Exclusion, Immune Privilege, and the Tumor Microenvironment. *Science* **2015**, *348*, 74–80.
73. Molon, B.; Ugel, S.; Del Pozzo, F. et al. Chemokine Nitration Prevents Intra Tumoral Infiltration of Antigen-Specific T Cells. *J. Exp. Med.* **2011**, *208*, 1949–1962.
74. Mariathasan, S.; Turley, S. J.; Nickles, D. et al. TGFβ Attenuates Tumour Response to PD-L1 Blockade by Contributing to Exclusion of T cells. *Nature* **2018**, *554*, 544–548.
75. Tauriello DVF, Palomo-Ponce, S.; Stork, D. et al. TGFβ Drives Immune Evasion in Genetically Reconstituted Colon Cancer Metastasis. *Nature* **2018**, *554*, 538–543.
76. Rooney, M. S.; Shukla, S. A.; Wu, C. J. et al. Molecular and Genetic Properties of Tumors Associated with Local Immune Cytolytic Activity. *Cell* **2015**, *160*, 48–61.
77. Leone, P.; Shin E-C, Perosa, F. et al. MHC Class I Antigen Processing and Presenting Machinery: Organization, Function, and Defects in Tumor Cells. *J. Natl. Cancer Inst.* **2013**, *105*, 1172–1187.
78. Donnell, J. S.; Teng MWL, Smyth, M. J. Cancer Immunoediting and Resistance to T Cell-Based Immunotherapy. *Nat. Rev. Clin. Oncol.* **2019**, *16*, 151–167.
79. Kloor, M.; Becker, C.; Benner, A. et al. Immunoselective Pressure and Human Leukocyte Antigen Class I Antigen Machinery Defects in Microsatellite Unstable Colorectal Cancers. *Cancer Res.* **2005**, *65*, 6418–6424.

80. Mehta, A. M.; Jordanova, E. S.; Kenter, G. G. et al. Association of Antigen Processing Machinery and HLA Class I defects with Clinicopathological Outcome in Cervical Carcinoma. *Cancer Immunol. Immunother*. **2007**, *57*, 197–206.
81. Torres, M. J.; Ruiz-Cabello, F.; Skoudy, A. et al. Loss of an HLA Haplotype in Pancreas Cancer Tissue and Its Corresponding Tumor Derived Cell Line. *Tissue Antigens* **1996**, *47*, 372–381.
82. McGranahan, N.; Rosenthal, R.; Hiley, C. T. et al. Allele-Specific HLA Loss and Immune Escape in Lung Cancer Evolution. *Cell* **2017**, *171*, 1259–1271.e11.
83. The TRACERx consortium, R Rosenthal, Cadieux, E. L. et al. The TRACERx consortium. Neoantigen-Directed Immune Escape in Lung Cancer Evolution. *Nature* **2019**, *567*, 479–485.
84. Spranger, S.; Bao, R.; Gajewski, T. F. Melanoma-Intrinsic β-Catenin Signalling Prevents Anti-Tumour Immunity. *Nature* **2015**, *523*, 231–235.
85. Spranger, S.; Gajewski, T. F. A New Paradigm for Tumor Immune Escape: β-Catenin-Driven Immune Exclusion. *J. Immunother. Cancer* **2015**, *3*, 43.
86. Villablanca, E. J.; Raccosta, L.; Zhou, D. et al. Tumor-Mediated Liver X Receptor-α Activation Inhibits CC Chemokine Receptor-7 Expression on Dendritic Cells and Dampens Antitumor Responses. *Nat. Med.* **2010**, *16*, 98–105.
87. Oyama, T.; Ran, S.; Ishida, T. et al. Vascular Endothelial Growth Factor Affects Dendritic Cell Maturation Through the Inhibition of Nuclear Factor-Kappa B Activation in Hemopoietic Progenitor Cells. *J Immunol.* **1998**, 1950: 1224–1232.
88. Dong, H.; Strome, S. E.; Salomao, D. R. et al. Tumor-Associated B7-H1 Promotes T-Cell Apoptosis: A Potential Mechanism of Immune Evasion. *Nat. Med.* **2002**, *8*, 793–800.
89. Chen, G.; Huang, A. C.; Zhang, W. et al. Exosomal PD-L1 Contributes to Immunosuppression and Is Associated with Anti-PD-1 Response. *Nature* **2018**, *560*, 382–386.
90. Munn, D. H.; Mellor, A. L. IDO in the Tumor Microenvironment: Inflammation, Counter-Regulation, and Tolerance. *Trends Immunol.* **2016**, *37*, 193–207.
91. Uyttenhove, C.; Pilotte, L.; Théate, I. et al. Evidence for a Tumoral Immune Resistance Mechanism Based on Tryptophan Degradation by Indoleamine 2,3-Dioxygenase. *Nat. Med.* **2003**, *9*, 1269–1274.
92. Yang, L.; Pang, Y.; Moses, H. L. TGF-β and Immune Cells: An Important Regulatory Axis in the Tumor Microenvironment and Progression. *Trends Immunol.* **2010**, *31*, 220–227.
93. Tachibana, T.; Onodera, H.; Tsuruyama, T.; Mori, A.; Nagayama, S.; Hiai, H. et al. Increased Intratumor Valpha24-Positive Natural Killer T Cells: A Prognostic Factor for Primary Colorectal Carcinomas. Clin *Cancer Res.* **2005,** *11*, 7322–7327.
94. Hishiki, T.; Mise, N.; Harada, K.; Ihara, F.; Takami, M.; Saito, T. et al. Invariant Natural Killer T Infiltration in Neuroblastoma with Favorable Outcome. *Pediatr Surg Int.* **2018,** *34*, 195–201.
95. Bricard, G.; Cesson, V.; Devevre, E.; Bouzourene, H.; Barbey, C.; Rufer, N. et al. Enrichment of Human CD4+ V(alpha)24/Vbeta11 Invariant NKT Cells in Intrahepatic Malignant Tumors. *J. Immunol.* **2009,** *182*, 5140–5151.
96. Kenna, T.; Golden-Mason, L.; Porcelli, S. A.; Koezuka, Y.; Hegarty, J. E.; Farrelly, C. et al. NKT Cells from Normal and Tumor-Bearing Human Livers Are Phenotypically and Functionally Distinct from Murine NKT Cells. *J. Immunol.* **2003,** *171*, 1775–1179.

97. Li, X. F.; Dai, D.; Song, X. Y.; Liu, J. J.; Zhu, L.; Zhu, X. et al. A Different Representation of Natural T Cells and Natural Killer Cells Between Tumor-Infiltrating and Periphery Lymphocytes in Human Hepatocellular Carcinoma. *Oncol. Lett.* **2017,** *13*, 3291–3298.
98. Fallarini, S.; Paoletti, T.; Orsi Battaglini, N.; Lombardi, G. Invariant NKT Cells Increase Drug-Induced Osteosarcoma Cell Death. *Br. J. Pharmacol.* **2012,** *167*, 1533–1549.
99. Terabe, M.; Khanna, C.; Bose, S.; Melchionda, F.; Mendoza, A.; Mackall, C. L. et al. CD1d-Restricted Natural Killer T Cells Can Down-Regulate Tumor Immunosurveillance Independent of Interleukin-4 Receptor-Signal Transducer and Activator of Transcription 6 or Transforming Growth Factor-Beta. *Cancer Res.* **2006,** *66*, 3869–3875.
100. Krijgsman, D.; Hokland, M.; Kuppen, P. J. K. The Role of Natural Killer T Cells in Cancer—A Phenotypical and Functional Approach. *Front. Immunol.* **2018,** *9*, 367.
101. Ishikawa, A.; Motohashi, S.; Ishikawa, E.; Fuchida, H.; Higashino, K.; Otsuji, M. et al. A Phase I Study of Alpha-Galactosylceramide (KRN7000)-Pulsed Dendritic Cells in Patients with Advanced and Recurrent Non-Small Cell Lung Cancer. *Clin. Cancer Res.* **2005,** *11*, 1910–1917.
102. Motohashi, S.; Nagato, K.; Kunii, N.; Yamamoto, H.; Yamasaki, K.; Okita, K. et al. A Phase I–II Study of Alpha-Galactosylceramide-Pulsed IL-2/GM-CSF-Cultured Peripheral Blood Mononuclear Cells in Patients with Advanced and Recurrent Non-Small Cell Lung Cancer. *J. Immunol.* **2009,***182*, 2492–2501.
103. Watarai, H.; Fujii, S.; Yamada, D.; Rybouchkin, A.; Sakata, S.; Nagata, Y. et al. Murine Induced Pluripotent Stem Cells Can be Derived from and Differentiate into Natural Killer T Cells. *J. Clin. Invest.* **2010,** *120*, 2610–2618.
104. Yamada, D.; Iyoda, T.; Vizcardo, R.; Shimizu, K.; Sato, Y.; Endo, T. A. et al. Efficient Regeneration of Human Valpha24(+) Invariant Natural Killer T Cells and Their Anti-Tumor Activity In Vivo. *Stem Cells* **2016,** *34*, 2852–2860.
105. Tahir, S. M.; Cheng, O.; Shaulov, A.; Koezuka, Y.; Bubley, G. J.; Wilson, S. B. et al. Loss of IFN-Gamma Production by Invariant NK T Cells in Advanced Cancer. *J. Immunol.* **2001,** *167*, 4046–4050.
106. Laurent, X.; Bertin, B.; Renault, N.; Farce, A.; Speca, S.; Milhomme, O. et al. Switching Invariant Natural Killer T (iNKT) Cell Response from Anticancerous to Anti-Inflammatory Effect: Molecular Bases. *J. Med. Chem.* **2014,** *57*, 5489–5508.
107. Girardi, M.; Oppenheim, D. E.; Steele, C. R.; Lewis, J. M.; Glusac, E.; Filler, R. et al. Regulation of Cutaneous Malignancy by Gamma Delta T Cells. *Science* **2001,** *294*, 605–609.
108. Paul, S.; Lal, G. Regulatory and Effector Functions of Gamma-Delta (Gamma delta) T Cells and Their Therapeutic Potential in Adoptive Cellular Therapy for Cancer. *Int. J.ournal of Cancer* **2016,** *139*, 976–985.
109. Brandes, M.; Willimann, K.; Moser, B. Professional Antigen-Presentation Function by Human Gammadelta T Cells. *Science* **2005,** *309*, 264–268.
110. Brandes, M.; Willimann, K.; Bioley, G.; Levy, N.; Eberl, M.; Luo, M. et al. Cross-Presenting Human Gammadelta T Cells Induce Robust CD8+ Alphabeta T Cell Responses. *Proc. Natl. Acad. Sci. USA* **2009,** *106*, 2307–2312.
111. Altvater, B.; Pscherer, S.; Landmeier, S.; Kailayangiri, S.; Savoldo, B.; Juergens, H. et al. Activated Human Gammadelta T Cells Induce Peptide-Specific CD8+ T-Cell Responses to Tumor-Associated Self-Antigens. *Cancer Immunol. Immunother.* **2012,** *61*, 385–396.

112. Lozupone, F.; Pende, D.; Burgio, V. L.; Castelli, C.; Spada, M.; Venditti, M. et al. Effect of Human Natural Killer and Gammadelta T Cells on the Growth of Human Autologous Melanoma Xenografts in SCID Mice. *Cancer Res.* **2004,** *64*, 378–385.
113. Yuasa, T.; Sato, K.; Ashihara, E.; Takeuchi, M.; Maita, S.; Tsuchiya, N. et al. Intravesical administration of gammadelta T cells successfully prevents the growth of bladder cancer in the murine model. Cancer Immunol. Immunother. **2009,** *58*, 493–502.
114. Fournie, J. J.; Sicard, H.; Poupot, M.; Bezombes, C.; Blanc, A.; Romagne, F. et al. What Lessons Can Be Learned from Gamma Delta T Cell-Based Cancer Immunotherapy Trials? *Cell Mol. Immunol.* **2013,** *10*, 35–41.
115. Kato, Y.; Tanaka, Y.; Miyagawa, F.; Yamashita, S.; Minato, N. Targeting of Tumor Cells for Human Gamma Delta T Cells by Nonpeptide Antigens. *J. Immunol.* **2001,** *167*, 5092–5098.
116. Muraro, M.; Mereuta, O. M.; Carraro, F.; Madon, E.; Fagioli, F. Osteosarcoma Cell Line Growth Inhibition by Zoledronate-Stimulated Effector Cells. *Cell Immunol.* **2007,** *249*, 63–72.
117. Li, Z.; Xu, Q.; Peng, H.; Cheng, R.; Sun, Z.; Ye, Z. IFN-gamma enhances HOS and U2OS Cell Lines Susceptibility to Gammadelta T Cell-Mediated Killing Through the Fas/Fas Ligand Pathway. *Int. Immunopharmacol.* **2011,** *11*, 496–503.
118. Li, Z.; Zhang, J.; Tang, J.; Wang, R. Celastrol Increases Osteosarcoma Cell Lysis by Gamma Delta T Cells Through Up-Regulation of Death Receptors. *Oncotarget* **2016,** *7*, 84388–84397.
119. Rosen, L. S.; Gordon, D.; Tchekmedyian, S.; Yanagihara, R.; Hirsh, V.; Krzakowski, M. et al. Zoledronic Acid Versus Placebo in the Treatment of Skeletal Metastases in Patients with Lung Cancer and Other Solid Tumors: A Phase III, Double-Blind, Randomized Trial—The Zoledronic Acid Lung Cancer and Other Solid Tumors Study Group. *J. Clin. Oncol.* **2003,** *21*, 3150–3157.
120. Dass, C. R.; Choong, P. F. Zoledronic Acid Inhibits Osteosarcoma Growth in an Orthotopic Model. *Mol. Cancer Ther.* **2007,** *6*, 3263–3270.
121. Chen, T.; Berenson, J.; Vescio, R.; Swift, R.; Gilchick, A.; Goodin, S. et al. Pharmacokinetics and pharmacodynamics of zoledronic acid in cancer patients with bone metastases. J Clin Pharmacol. **2002,** *42*, 1228–1236.
122. Scotlandi, K.; Manara, M. C.; Hattinger, C. M.; Benini, S.; Perdichizzi, S.; Pasello, M. et al. Prognostic and Therapeutic Relevance of HER2 Expression in Osteosarcoma and Ewingsarcoma. *Eur J Cancer.* **2005,** *41*, 1349–1361.
123. Ebb, D.; Meyers, P.; Grier, H.; Bernstein, M.; Gorlick, R.; Lipshultz, S. E. et al. Phase II Trial of Trastuzumab in Combination with Cytotoxic Chemotherapy for Treatment of Metastatic Osteosarcoma with Human Epidermal Growth Factor Receptor 2 Overexpression: A Report from the Children Oncology Group. *J Clin Oncol.* **2012,** *30*, 2545–2551.
124. Oberg, H. H.; Peipp, M.; Kellner, C.; Sebens, S.; Krause, S.; Petrick, D. et al. Novel Bispecific Antibodies Increase Gamma Delta T-Cell Cytotoxicity Against Pancreatic Cancer Cells. *Cancer Res.* **2014,** *74*, 1349–1360.
125. Wang, Z.; Li, B.; Ren, Y.; Ye, Z. T-cell-based immunotherapy for osteosarcoma: challenges and opportunities. Front Immunol. **2016,** *7*, 353.
126. van der Veken, L. T.; Coccoris, M.; Swart, E.; Falkenburg, J. H.; Schumacher, T. N.; Heemskerk, M. H. Alpha Beta T Cell Receptor Transfer to Gamma Delta T Cells

Generates Functional Effector Cells Without Mixed TCR Dimers In Vivo. *J. Immunol.* **2009,** *182*, 164–170.
127. Saito, T.; Hochstenbach, F.; Marusic-Galesic, S.; Kruisbeek, A. M.; Brenner, M.; Germain, R. N. Surface expression of only gamma delta and/or alpha beta T cell receptor heterodimers by cells with four (alpha, beta, gamma, delta) functional receptor chains. *J. Exp. Med.* **1988,***168*, 1003–1020.
128. Moser, B.; Eberl, M. Gammadelta T-APCs: A Novel Tool for Immunotherapy? *Cell Mol. Life Sci.* **2011,** *68*, 2443–2452.
129. Li, H.; Pauza, C. D. Rapamycin Increases the Yield and Effector Function of Human Gammadelta T Cells Stimulated In Vitro. *Cancer Immunol. Immunother.* **2011,** *60*, 361–370.
130. Dao, V.; Liu, Y.; Pandeswara, S.; Svatek, R. S.; Gelfond, J. A.; Liu, A. et al. Immune-Stimulatory Effects of Rapamycin Are Mediated by Stimulation of Antitumor Gammadelta T Cells. *Cancer Res.* **2016,** *76*, 5970–5982.
131. Cao, G.; Wang, Q.; Li, G.; Meng, Z.; Liu, H.; Tong, J. et al. mTOR Inhibition Potentiates Cytotoxicity of Vgamma4 Gammadelta T Cells via Up-Regulating NKG2D and TNF-alpha. *J. Leukocyte Biol.* **2016,** *100*, 1181–1189.
132. Twyman-Saint Victor, C.; Rech, A. J.; Maity, A.; Rengan, R.; Pauken, K. E.; Stelekati, E. et al. Radiation and Dual Checkpoint Blockade Activate Non-Redundant Immune Mechanisms in Cancer. *Nature* **2015,** *520*, 373–377.
133. Wherry, E. J. T Cell Exhaustion. *Nat. Immunol.* **2011,** *12*, 492–499. doi: 10.1038/ni.2035
134. Jan, C. I.; Tsai, W. C.; Harn, H. J.; Shyu, W. C.; Liu, M. C.; Lu, H. M. et al. Predictors of Response to Autologous Dendritic Cell Therapy in Glioblastoma Multiforme. *Front Immunol.* **2018,** *9*, 727.
135. Ribas, A.; Wolchok, J. D. Cancer Immunotherapy Using Checkpoint Blockade. *Science* **2018,***359*, 1350–1355.
136. Rosenblatt, J.; Glotzbecker, B.; Mills, H.; Vasir, B.; Tzachanis, D.; Levine, J. D. et al. PD-1blockade by CT-011, Anti-PD-1 Antibody, Enhances Ex Vivo T-Cell Responses to Autologous Dendritic Cell/Myeloma Fusion Vaccine. J. Immunother. **2011,** *34*, 409–418.
137. Schendel, D. J., Gansbacher, B., Oberneder, R. et al. Tumor-Specific Lysis of Human Renal Cell Carcinomas by Tumor-Infiltrating Lymphocytes: I. HLA-A2-restricted recognition of autologous and allogeneic Tumor Lines. *J. Immunol.* **1993,** *151* (8), 4209–4220.
138. Schnurr, M., Galambos, P., Scholz, C. et al. Tumor Cell Lysate-Pulsed Human Dendritic Cells Induce a T-Cell Response Against Pancreatic Carcinoma Cells: An In Vitro Model for the Assessment of Tumor Vaccines. *Cancer Res.* **2001,** *61* (17), 6445–6450.
139. Würtzen, P. A.; Pedersen, L. Ø.; Poulsen, H. S.; Claesson, M. H. Specific Killing of p53 Mutated Tumor Cell Lines by a Cross-Reactive Human HLA-A2-Restricted p53-Specific CTL Line. *Int. J. Cancer* **2001,** *93* (6), 855–861.
140. Zirlik, K. M.; Zahrieh, D.; Neuberg, D.; Gribben, J. G. Cytotoxic T Cells Generated Against Heteroclitic Peptides Kill Primary Tumor Cells Independent of the Binding Affinity of the Native Tumor Antigen Peptide. *Blood* **2006,** *108* (12), 3865–3870.
141. Wang, B.; J. M. Kuroiwa, Y.; He, L.-Z.; Charalambous, A.; Keler, T.; Steinman, R. M. The Human Cancer Antigen Mesothelin Is More Efficiently Presented to the Mouse Immunesystem When Targeted to the DEC-205/CD205 Receptor on Dendritic Cells. *Ann. NY Acad. Sci.* **2009,** *1174*, 6–17.

142. Galili, U. Autologous Tumor Vaccines Processed to Express α-gal Epitopes: A Practical Approach to Immunotherapy in Cancer. *Cancer Immunol. Immunother.* **2004,** *53* (11), 935–945.
143. Hermans, I. F.; Ritchie, D. S.; Yang, J.; Roberts, J. M.; Ronchese, F. CD8+ T Cell-Dependent Elimination of Dendritic Cells In Vivo Limits the Induction of Antitumor Immunity. *J. Immunol.* **2000,** *164* (6), 3095–3101.
144. Kilpatrick, D. C. Animal Lectins: A Historical Introduction and Overview. *Biochim. Biophys. Acta* **2002,** *1572* (2–3), 187–197.
145. East, L.; Isacke, C. M. The Mannose Receptor Family. *Biochim. Biophys. Acta* **2002,** *1572*, (2–3), 364–386.
146. Shrimpton, R. E.; Butler, M.; A.-S. Morel, Eren, E.; Hue, S. S.; Ritter, M. A. CD205 (DEC-205): A Recognition Receptor for Apoptotic and Necrotic Self. *Mol. Immunol.* **2009,** *46* (6), 1229–1239.
147. Burgdorf, S.; V. Lukacs-Kornek, and, Kurts, C. The Mannose Receptor Mediates Uptake of Soluble But Not of Cell-Associated Antigen for Cross-Presentation. *J. Immunol.* **2006,** *176* (11), 6770–6776.
148. Ramakrishna, V.; Treml, J. F.; Vitale, L. et al. Mannose Receptor Targeting of Tumor Antigenpmel17 to Human Dendritic Cells Directs Anti-Melanoma T Cell Responses via Multiple HLA Molecules. *J. Immunol.* **2004,** *172* (5), 2845–2852.
149. He, L.-Z.; Crocker, A.; Lee, J. et al. Antigenic Targeting of the Human Mannose Receptor Induces Tumor Immunity. *J. Immunol.* **2007,** *178* (10), 6259–6267.
150. Dasgupta, S.; Navarrete, A.-M.; Bayry, J. et al. A Role for Exposed Mannosylations in Presentation of Human Therapeutic Self-Proteins to CD4+ T Lymphocytes. *Proc. Natl. Acad. Sci. USA* **2007,** *104* (21), 8965–8970.
151. Chieppa, M.; Bianchi, G.; Doni, A. et al. Cross-Linking of the Mannose Receptor on Monocyte-Derived Dendritic Cells Activates an Anti-Inflammatory Immunosuppressive Program. *J. Immunol.* **2003,** *171* (9), 4552–4560.
152. Idoyaga, J.; Cheong, C.; Suda, K. et al. Cutting Edge: Langerin/CD207 Receptor on Dendritic Cells Mediates Efficient Antigen Presentation on MHC I and II Products In Vivo. *J. Immunol.* **2008,** *180* (6), 3647–3650.
153. Singh, S. K.; Stephani, J.; Schaefer, M. et al. Targeting Glycan Modified OVA to Murine DC-SIGN Transgenic Dendritic Cells Enhances MHC Class I and II Presentation. *Mol. Immunol.* **2009,** *47* (2–3), 164–174.
154. Takahara, K.; Omatsu, Y.; Yashima, Y. et al. Identification and Expression of Mouse Langerin (CD207) in Dendritic Cells. *Int. Immunol.* **2002,** *14* (5), 433–444.
155. van Die, I.; van Vliet, S. J.; Nyame, A. K. et al. The Dendritic Cell-Specific C-type lectin DC-SIGN Is a Receptor for Schistosoma Mansoni Egg Antigens and Recognizes the Glycan Antigen Lewis X. *Glycobiology* **2003,** *13* (6), 471–478.
156. Galustian, C.; Park, C. G.; Chai, W. et al. High and Low Affinity carbohydrate Ligands Revealed for Murine SIGN-R1 by Carbohydrate Array and Cell Binding Approaches, and Differing Specific Cities for SIGN-R3 and Langerin. *Int. Immunol.* **2004,** *16* (6), 853–866.
157. Monzavi-Karbassi, B.; Luo, P.; Cunto-Amesty, G. et al. Fucosylated Lactosamines Participate in Adhesion of HIV-1 Envelope Glycoprotein to dendritic cells. *Arch. Virol.* **2004,** *149* (1), 75–91.

158. Ujita, M.; Nagayama, H.; Kanie, S. et al. Carbohydrate Binding Specificity of Recombinant Human Macrophage β-Glucan Receptor Dectin-1. *Biosci. Biotechnol. Biochem.* **2009,** *73* (1), 237–240.
159. Ferwerda, G.; Meyer-Wentrup, F.; Kullberg, B.-J.; Netea, M. G.; Adema, G. J. Dectin-1 Synergizes with TLR2 and TLR4 for Cytokine Production in Human Primary Monocytes and Macrophages. *Cell. Microbiol.* **2008,** *10* (10), 2058–2066.
160. Dillon, S.; Agrawal, S.; Banerjee, K. et al. Yeast Zymosan, a Stimulus for TLR2 and Dectin-1, Induces Regulatory Antigen-Presenting Cells and Immunological Tolerance. *J. Clin. Investig.* **2006,** *116* (4), 916–928.
161. Hernanz-Falcón, P.; Joffre, O.; Williams, D. L.; Reis e Sousa, C. Internalization of Dectin-1 Terminates Induction of Inflammatory Responses. *Eur. J. Immunol.* **2009,** *39* (2), 507–513.
162. Weck, M. M.; Appel, S.; Werth, D. et al. HDectin-1 Is Involved in Uptake and Cross-Presentation of Cellular Antigens. *Blood* **2008,** *111* (8), 4264–4272.
163. Sancho, D.; Mourão-Sá, D.; Joffre, O. P. et al. Tumor Therapy in Mice via Antigen Targeting to a Novel, DC-Restricted C-Type Lectin. *J. Clin. Investig.* **2008,** *118* (6), 2098–2110.
164. Sancho, D.; Joffre, O. P.; Keller, A. M. et al. Identification of a Dendritic Cell Receptor That Couples Sensing of Necrosis to Immunity. *Nature* **2009,** *458* (7240), 899–903.

Index

K

L

M

N

P